MANY VOICES, ONE VISION:
THE EARLY YEARS OF THE WORLD
HERITAGE CONVENTION

Heritage, Culture and Identity

Series Editor: Brian Graham,
School of Environmental Sciences, University of Ulster, UK

Other titles in this series

Partitioned Lives: The Irish Borderlands
Catherine Nash, Bryonie Reid and Brian Graham
ISBN 978 1 4094 6672 7

Ireland's 1916 Rising
Explorations of History-Making, Commemoration
& Heritage in Modern Times
Mark McCarthy
ISBN 978 1 4094 3623 2

Cosmopolitan Europe: A Strasbourg Self-Portrait
John Western
ISBN 978 1 4094 4371 1

Heritage from Below
Edited by Iain J.M. Robertson
ISBN 978 0 7546 7356 9

Fortifications, Post-colonialism and Power
Ruins and Imperial Legacies
João Sarmento
ISBN 978 1 4094 0303 6

Towards World Heritage
International Origins of the Preservation Movement 1870–1930
Edited by Melanie Hall
ISBN 978 1 4094 0772 0

Selling EthniCity
Urban Cultural Politics in the Americas
Edited by Olaf Kaltmeier
ISBN 978 1 4094 1037 9

Many Voices, One Vision: The Early Years of the World Heritage Convention

CHRISTINA CAMERON
Université de Montréal, Canada

MECHTILD RÖSSLER
UNESCO and chercheur associé, EGHO, France

Routledge
Taylor & Francis Group

LONDON AND NEW YORK

First published 2013 by Ashgate Publishing

2 Park Square, Milton Park, Abingdon, Oxon OX14 4RN
711 Third Avenue, New York, NY 10017, USA

Routledge is an imprint of the Taylor & Francis Group, an informa business

First issued in paperback 2016

British Library Cataloguing in Publication Data
Cameron, Christina.
  Many voices, one vision : the early years of the World Heritage Convention. – (Heritage, culture and identity)
  1. World Heritage Convention (1972) 2. World Heritage Committee.
  3. World Heritage areas. 4. Cultural property – Protection – History – 20th century.
  5. Historic preservation – History – 20th century.
  I. Title II. Series III. Rössler, Mechtild.
  363.6'9–dc23

The Library of Congress has cataloged the printed edition as follows:
Cameron, Christina.
  Many voices, one vision : the early years of the World Heritage Convention / by Christina Cameron and Mechtild Rössler.
     pages cm.—(Heritage, culture and identity)
  Includes bibliographical references and index.
  ISBN 978-1-4094-3765-9 (hardback)—ISBN 978-1-4094-3766-6 (ebook)—
ISBN 978-1-4094-8477-6 (epub) 1. World Heritage Convention (1972) 2. Cultural property—Protection (International law)—History—20th century. I. Rössler, Mechtild. II. Title.
  K3791.A41972C36 2013
  363.6'9—dc23

2013000855

ISBN 978-1-4094-3765-9 (hbk)
ISBN 978-1-138-24808-3 (pbk)

Transferred to Digital Printing 2014

This book is dedicated to the selfless men and women who cherish the world's heritage places and foster a culture of conservation

Contents

List of Figures   ix
Foreword   xiii
Acknowledgements   xvii
List of Abbreviations   xix

1   Creation of the World Heritage Convention   1

2   Process for Identifying World Heritage Sites   27

3   Populating the World Heritage List: 1978–2000   45

4   Conserving World Heritage Sites   103

5   The Players   155

6   Assessment of the World Heritage System: 1972–2000   221

Appendix: Vignettes of Interviewees   247
Bibliography   293
Index   299

List of Figures

1.1   Joseph L. Fisher. © Special Collections & Archives, George
       Mason University   6
1.2   Michel Batisse in 1981 (far right) with Francesco Di Castri behind.
       © UNESCO/Michel Claude. http://photobank.unesco.org/library/
       image/380/1C48USF4E7wWW87B0t2JULM6.jpg   9
1.3   Russell E. Train. © Library of Congress, Prints and Photographs
       Division   18
1.4   René Maheu, Director-General of UNESCO 1961–1974.
       © UNESCO. http://www.unesco.org/new/en/unesco/about-us/who-
       we-are/history/directors-general/rene-maheu   22

2.1   Michel Parent in about 1980. © ICOMOS   34
2.2   Reconstructed Market Square in Warsaw, Poland. © UNESCO/
       A. Husarska. http://photobank.unesco.org/library/image/277/
       15425.jpg   40

3.1   Number of inscriptions per year to the World Heritage List
       1978–2000. © UNESCO/Mechtild Rössler and Anand Kanitkar   46
3.2   Balance among cultural, natural and mixed World Heritage Sites in
       the year 2000. © UNESCO/Mechtild Rössler and Anand Kanitkar   46
3.3   Distribution of World Heritage Sites by regions in the year 2000.
       © UNESCO/Mechtild Rössler and Anand Kanitkar   47
3.4   1981 World Heritage Committee in Sydney, Australia from left to
       right: Anne Raidl, Gérard Bolla and Margaret van Vliet. © Bernd
       von Droste   48
3.5   Lake District National Park, United Kingdom. © Michael Turner,
       United Kingdom. http://www.lakedistrict.gov.uk/learning/
       photogallery   64
3.6   Michel Batisse's proposal for rural landscapes presented to the
       1987 World Heritage Committee. © UNESCO. http://whc.unesco.
       org/archive/1987/sc-87-conf005-inf4e.pdf   65
3.7   1990 World Heritage Committee in Banff, Canada from left
       to right: James Chamberlain (USA), Adul Wichiencharoen
       (Thailand), Magdalina Stantcheva (Bulgaria), Azedine Beschaouch
       (Tunisia), Christina Cameron (Canada, Chairperson), Licia Vlad
       Borrelli (Italy), Seydina Sylla (Senegal), Leslie Taylor (Mayor of
       Banff) and Salvador Diaz-Berrio (Mexico). © Christina Cameron   66

3.8   Tongariro National Park, New Zealand. © UNESCO/
        S.A. Tabbasum. http://whc.unesco.org/en/list/421/gallery/   69
3.9   Participants at the 1994 Nara conference on authenticity in Nara,
        Japan. © Bernd von Droste   86
3.10   Raymond Lemaire. © Raymond Lemaire International Centre for
        Conservation   87
3.11   Herb Stovel and Henry Cleere, Riomaggiore, Cinque Terre, Italy,
        19 March 2001. © ICCROM   88

4.1   Ichkeul National Park, Tunisia. © UNESCO/Marc Patry.
        http://whc.unesco.org/en/list/8/gallery   107
4.2   Evolution of the number of state of conservation reports and
        the List of World Heritage in Danger 1978–2000. © UNESCO/
        Mechtild Rössler and Chloe Bigio   116
4.3   Kahuzi-Biega National Park, Democratic Republic of the Congo.
        © UNESCO/Guy Debonnet. http://whc.unesco.org/en/list/137/
        gallery   117
4.4   Whale Sanctuary of El Vizcaino, Mexico. © UNESCO/Jim
        Thorsell. http://photobank.unesco.org/library/image/304/16917.jpg 118
4.5   Twentieth anniversary experts meeting in Paris (October 1992)
        from left to right: Azedine Beschaouch (rapporteur), Bernd von
        Droste (Director, World Heritage Centre), Christina Cameron
        (Chairperson) and Laurent Levi-Strauss (UNESCO).
        © Bernd von Droste   129
4.6   Angkor, Cambodia. © UNESCO/Francesco Bandarin. http://whc.
        unesco.org/en/list/668/gallery   138
4.7   Galapagos Islands, Ecuador. © UNESCO/Marc Patry   138

5.1   Old City of Jerusalem and its Walls in 1980. © Michael Turner,
        Israel   168
5.2   W National Park, Niger. © UNESCO/Matthias Kunert. http://whc.
        unesco.org/en/list/749/gallery   171
5.3   Jim Thorsell on mission in Canaima National Park, Venezuela.
        © Jim Thorsell   180
5.4   Meteora, Greece. © UNESCO/Yvon Fruneau. http://photobank.
        unesco.org/library/image/439/FC74w4rar6NsDfo2807QmO7a.jpg   182
5.5   Piero Gazzola. © ICCROM   186
5.6   Léon Pressouyre on mission in Mostar, Bosnia and Herzegovina in
        the late 1990s. © Katérina Stenou   189
5.7   Henry Cleere on mission in China. © Henry Cleere   192
5.8   1994 World Heritage Committee meeting in Phuket, Thailand from
        left to right: Jukka Jokilehto (ICCROM) and Henry Cleere, Jean-
        Louis Luxen and Carmen Añón Feliu (ICOMOS). © ICCROM   199

5.9   1983 World Heritage Committee meeting in Florence, Italy from
       left to right: Jürgen Hillig (Science Sector, UNESCO), Makaminan
       Makagiansar (Assistant Director-General Culture, UNESCO),
       Jorge Gazaneo (Argentina, acting Chairperson), Bernd von Droste
       (Ecological Sciences, UNESCO) and Anne Raidl
       (Cultural Heritage, UNESCO). © Bernd von Droste   202
5.10   World Heritage Centre staff in the mid-1990s. © Bernd von Droste   206
5.11   1998 launch of *World Heritage in Young Hands* in Paris from
       left to right: Sarah Titchen, Taro Komatsu, Elisabeth Khawaykie,
       Bernd von Droste and Breda Pavlic. © Bernd von Droste   216

6.1   Old City of Dubrovnik, Croatia. © UNESCO/Francesco Bandarin.
       http://whc.unesco.org/en/list/95/gallery   227
6.2   Jabiluka mine site at Kakadu National Park, Australia. © Dominic
       O'Brien   229
6.3   Three Directors-General of UNESCO who oversaw
       implementation of the World Heritage Convention from left to
       right: Koichiro Matsurra, Japan (1999–2010), Amadou-Mahtar
       M'Bow, Senegal (1974–1987) and Federico Mayor, Spain
       (1987–1999). Predecessors on screen, from left to right: Jaime
       Torres Bodet, Mexico (1948–1952), Julian Huxley, United
       Kingdom (1946–1948), René Maheu, France (1961–1974),
       Luther Evans, United States of America (1953–1958) and
       Vittorino Veronese, Italy (1958–1960). © UNESCO/Michel
       Ravassard. http://photobank.unesco.org/library/image/377/
       DFA806xh7S3lYkc57S1detQo.jpg   243

A.1   Carmen Añón Feliu in Madrid, 2009. © Christina Cameron   247
A.2   Azedine Beschaouch in Brasilia, 2010. © Christina Cameron   248
A.3   Gérard Bolla in Paris, 2007. © Christina Cameron   249
A.4   Mounir Bouchenaki in Paris, 2009. © Christina Cameron   251
A.5   Hans Caspary in Mainz, 2011. © Christina Cameron   252
A.6   Lucien Chabason in Paris, 2012. © Christina Cameron   253
A.7   Henry Cleere in London, 2008. © Christina Cameron   254
A.8   Jim Collinson in Windsor, 2010. © Christina Cameron   255
A.9   Bernd von Droste. © Bernd von Droste   256
A.10   Catherine Dumesnil in Paris, 2009. © Christina Cameron   257
A.11   Regina Durighello in Paris, 2009. © Christina Cameron   258
A.12   Hal Eidsvik in Ottawa, 2009. © Christina Cameron   260
A.13   Sir Bernard Feilden. © Archant   261
A.14   Francesco Francioni in Rome, 2010. © Christina Cameron   262
A.15   Guo Zhan in Brasilia, 2010. © Christina Cameron   263
A.16   Natarajan Ishwaran in Paris, 2009. © Christina Cameron   264
A.17   Nobuo Ito in Kyoto, 2012. © Christina Cameron   265

A.18   Jukka Jokilehto in Rome, 2010. © Christina Cameron   266
A.19   François Leblanc in Ottawa, 2009. © Christina Cameron   268
A.20   Francisco Lopez Morales in Brasilia, 2010. © Christina Cameron 269
A.21   Jean-Louis Luxen in Leuven, 2009. © Christina Cameron   270
A.22   Koïchiro Matsuura in Paris, 2009. © Christina Cameron   271
A.23   Federico Mayor in Madrid, 2009. © Christina Cameron   272
A.24   Amadou-Mahtar M'Bow in Paris, 2009. © UNESCO/
       Mechtild Rössler   273
A.25   Jeff McNeely. © Jeff McNeely   274
A.26   Rob Milne in Paris, 2009. © Christina Cameron   275
A.27   Dawson Munjeri in Brasilia, 2010. © Christina Cameron   277
A.28   Adrian Phillips in London, 2008. © Christina Cameron   278
A.29   Harald Plachter in Mainz, 2011. © Christina Cameron   279
A.30   Léon Pressouyre in Paris, 2008. © Christina Cameron   280
A.31   Anne Raidl in Vienna, 2008. © Christina Cameron   281
A.32   John Reynolds in Springfield, 2011. © Christina Cameron   282
A.33   Jane Robertson Vernhes in Paris, 2009. © Christina Cameron   283
A.34   Roland Silva in Victoria, 2011. © Christina Cameron   284
A.35   Herb Stovel in Rome, 2002. © ICCROM   285
A.36   Jim Thorsell in Banff, 2010. © Christina Cameron   287
A.37   Abdelaziz Touri in Paris, 2011. © Christina Cameron   288
A.38   Russell Train in Springfield, 2008. © Christina Cameron   289
A.39   Licia Vlad Borrelli in Rome, 2010. © Christina Cameron   290
A.40   Ray Wanner in Springfield, 2011. © Christina Cameron   291

Foreword

An optimistic vision of shared human values and collective international behaviour inspired the Convention concerning the Protection of the World Cultural and Natural Heritage (World Heritage Convention), adopted by the General Conference of UNESCO on 16 November 1972. With the celebration of the fortieth anniversary of the World Heritage Convention, that initial optimism is coming progressively under question and those ideals are being compromised by geopolitical considerations and rivalries. One of several instruments put forward by UNESCO to encourage a culture of international cooperation and peaceful co-existence, it is rooted in the belief that certain special places are so important that they form part of humanity's birthright. As such, all countries are expected to contribute to the protection and conservation of these exceptional properties in order to make them accessible to present and future generations. Considered a flagship programme of UNESCO, the World Heritage Convention celebrated its fortieth anniversary in 2012. It has influenced heritage activity in virtually every country in the world. With nearly 1,000 properties identified on the World Heritage List and participation from almost all countries, it is time to ask whether the optimistic vision embraced by its founders has been fulfilled.

This book has its roots in a 2005 international symposium held in Paris on the sixtieth anniversary of the creation of UNESCO.[1] Koïchiro Matsuura, then Director-General of UNESCO and Françoise Rivière, then Assistant Director-General for Culture encouraged the academic community to document the history of UNESCO's diverse programmes and in particular to capture the voices of the past. The authors, both active participants in World Heritage at that time, identified the need for a history of the World Heritage system. Aware that the generation who had pioneered the World Heritage Convention was aging and that some had already passed away, the authors launched the World Heritage Oral Archives project at the University of Montreal under the auspices of the Canada Research Chair on Built Heritage, in collaboration with the UNESCO oral archives initiative. It aims at recording precious knowledge that still exists in the memories of participants in the creation and implementation of the 1972 Convention before they fade and disappear.

---

1 UNESCO, Programme for international symposium on 60 years of UNESCO's history, Paris, 2005, ARC/2005/ann/pi/1/rev. 9. Retrieved from http://unesdoc.unesco.org/images/0014/001401/140122e.pdf; UNESCO, *60 ans d'histoire de l'UNESCO. Actes du colloque international*, Paris, 16–18 novembre 2005 (Paris, 2007), pp. 1–611.

This study covers the period leading to the adoption of the World Heritage Convention in 1972 and its implementation until the year 2000. The choice of this time frame is based on the urgency of conducting interviews and the culmination of a major reform agenda adopted by the World Heritage Committee in Cairns, Australia in 2000 that set the World Heritage system on a new path. The authors believe that the early years set the stage for future activity and provide a foil for understanding the subsequent evolution of the Convention.

This account of the creation of the World Heritage Convention and its early implementation develops two different streams, marrying documentary evidence with the voices of the pioneers. There is a considerable body of published literature on the subject of World Heritage, ranging from academic work to coffee table books with beautiful images of World Heritage Sites. The first scholarly research on World Heritage is a thoughtful doctoral thesis by Sarah Titchen completed in 1995.[2] Since that time, research on World Heritage has focused principally on various heritage concepts, conservation and management issues as well as legal aspects. Sources available to support this research include scholarly publications, unpublished doctoral theses, UNESCO records and various archives. One of the challenges is to draw meaningful stories out of the enormous volume of documentation generated by UNESCO and the World Heritage system.

An essential complement to the rich documentary record is the collection of in-depth interviews with those involved in the creation and implementation of the Convention.[3] Forty pioneers have been interviewed so far, selected on the basis of their key contribution to the Convention. They have been chosen to capture different perspectives of the three constituent bodies established in the Convention text: States Parties, technical advisors and UNESCO. The first interview was recorded in 2006. With more than sixty hours of audio recording, the World Heritage oral archives project is well established. It remains open-ended to accommodate other interviews that the authors intend to carry out in the future. The interviews will eventually be made available on-line by the University of Montreal and by the UNESCO archives.

This book looks at the international aspect of the creation and implementation of the Convention. It is not intended to cover the essential supporting work being done by individual States Parties and site managers except insofar as national and local issues came to the attention of the international constituent bodies. For this reason, the authors have not interviewed World Heritage site managers nor consulted personal archives of individuals involved at the time. Both these avenues offer promising directions for further research.

---

2   Sarah Titchen, *On the construction of outstanding universal value: UNESCO's World Heritage Convention (Convention concerning the Protection of the World Cultural and Natural Heritage, 1972) and the identification and assessment of cultural places for inclusion in the World Heritage List* (Canberra, 1995), pp. 1–309.

3   Christina Cameron and Mechtild Rössler, "Voices of the Pioneers: UNESCO's World Heritage Convention 1972–2000," *Journal of Cultural Heritage Management and Sustainable Development*, 1/1 (2011), pp. 42–54.

The book is structured in six chapters. Chapters 1 and 2 present a chronological history of the development of the World Heritage Convention and its implementation up to 1980. Chapters 3, 4 and 5 cover parallel timeframes from different thematic perspectives after which chapter 6 pulls the various strands together into an overall assessment of the period up to 2000.

The authors come from different academic disciplines and have been cast in different roles in the World Heritage system, one as a member of the UNESCO secretariat, the other as the head of a State Party delegation and chairperson of the fourteenth and thirty-second sessions of the World Heritage Committee. As of 2013 they have been involved with World Heritage for a cumulative forty-eight years. They are conscious of the pitfalls of bringing their own bias to this study but have made best efforts to reflect the views of the pioneers and the documentary evidence and not to be influenced unduly by the period beyond 2000.

The World Heritage Convention has been affected significantly by forty years of history. Although the text of the Convention remains unchanged, the way it has been implemented reflects global trends like mass tourism and climate change as well as evolving perceptions of the nature of heritage itself and approaches to conservation. Some are sounding the alarm, claiming that the system is imploding under its own weight. Others, like *The Economist,* believe that the system itself is in danger because "the UN cultural agency is torn between its own principles and its members' wishes; the principles are losing ground."[4] At the thirty-sixth session of the World Heritage Committee held in 2012 in St. Petersburg, the Director-General of UNESCO called for a rejuvenation of the Convention and a return to its scientific base. Perhaps this account of the high-minded ideals that inspired the origins of the World Heritage Convention will stimulate reflection on its meaning in the twenty-first century and mobilize renewed commitment to its lofty vision on the occasion of the fortieth anniversary of the Convention in 2012.

Christina Cameron and Mechtild Rössler
April 2013

4   "UNESCO's world heritage sites a danger list in danger," *The Economist,* 26 August 2010. Retrieved from http://www.economist.com/node/16891951

Acknowledgements

Research for this book has only been possible through the support of many institutions and individuals. The authors acknowledge with humility the collaboration of so many people who have chased down almost forgotten documents and searched through old archives to fill gaps in the story. We wish in particular to thank those pioneers who graciously accepted our invitation to share their stories and documents with us. They are, in alphabetical order, Carmen Añón Feliu, Azedine Beschaouch, Gérard Bolla, Mounir Bouchenaki, Hans Caspary, Lucien Chabason, Henry Cleere, Jim Collinson, Bernd von Droste, Catherine Dumesnil, Regina Durighello, Hal Eidsvik, Sir Bernard Feilden, Francesco Francioni, Guo Zhan, Natarajan Ishwaran, Nobuo Ito, Jukka Jokilehto, François Leblanc, Francisco Lopes Morales, Jean-Louis Luxen, Koïchiro Matsuura, Federico Mayor, Amadou-Mahtar M'Bow, Jeff McNeely, Rob Milne, Dawson Munjeri, Adrian Phillips, Harald Plachter, Léon Pressouyre, Anne Raidl, John Reynolds, Jane Robertson Vernhes, Roland Silva, Herb Stovel, Jim Thorsell, Abdelaziz Touri, Russell Train, Licia Vlad Borrelli and Ray Wanner. Photographs and short biographies of the interviewees are in Appendix 1.

We express our gratitude to the flexibility and enthusiasm of audio technicians who have recorded our interviews at different studios around the world: UNESCO, Paris, France (Farid Zidour, Edwin Murillo-Mercado); Banff (Mark Tierney); Linha Direta, Brasilia, Brasil (Luis Augusto Mendonça); Polaris Recording Studio, Windsor, Canada (Joe Collins and George Hellow); Bova Sound, Ottawa, Canada (Phil and Janet Bova); Audio Studio, K.U. University, Leuven, Belgium (Werner Mathius); Bias Studios, Springfield, Virginia, United States of America (Cory Foley-Marsello and Gloria Dawson); Aquarium Studio, London, United Kingdom; Office for the Conservation of Monuments, Mainz, Germany; Sharp's Audio Visual, Victoria, Canada and T-Born Studio, Kyoto, Japan. Access to these studios was facilitated by David Martel (UNESCO), Ana Lúcia Dias Guimarães (Brasilian Commission for UNESCO), Deborah Miller, Elizabeth Leblanc and Jennifer Duquette (Parks Canada), Blanca Vargas (Fundacion Cultura de Paz, Madrid), Carmen Anon Feliu (Madrid), Parviz Koohafkan and Britta Killermann (FAO, Rome), Gabriele Eschig (Austrian Commission for UNESCO), Hannelore De Keyser (Raymond Lemaire Centre for Conservation, K.U. University, Leuven) and Hans Caspary (Mainz).

The authors also wish to thank the many archivists, librarians and records managers who have collaborated on this project: Jens Boel, Adele Torrance, Petra van den Born and Phan Sang (UNESCO library and archives, Paris); Carole Darmouni (UNESCO Division of Public Information, Paris); Katherine Rewinkel

El-Darwish (IUCN library and documentation centre, Gland); Lucile Smirnov (ICOMOS documentation centre, Paris); Paul Arenson and Maria Mata Caravaca (ICCROM library and archives, Rome); Bastian Bertzky (World Conservation Monitoring Centre, Cambridge); Mark Derez (Raymond Lemaire Archives, Leuven); John Pinkerton (Parks Canada World Heritage archives, Gatineau); Stephen Morris and Jonathan Putnam (United States National Park Service World Heritage archives, Washington); and Joanne Archer (University of Maryland Hornbake library, College Park).

In addition the authors are grateful for the diligence of interns and research assistants who have traced photographs, prepared documentation and transcribed interviews: Chloe Bigio, Snejana Athanova and Anand Kanitkar at UNESCO, and Claudette Chapdelaine, Roha Khalaf, Myriam St.-Denis, Ève Wertheimer and Sarah Youngblutt at the Université de Montréal. In particular, we wish to acknowledge the extensive and thorough bibliography on World Heritage undertaken by doctoral candidate Judith Herrmann at the Université de Montréal.

We appreciate the help of friends and colleagues who have furthered the project by various means, including Henry Cleere, Phyllis Ellin, Tina Feilden, Frank Hodsoll, Claudine Houbart, Nobuko Inaba, Jukka Jokilehto, Joe King, François Leblanc, Hugh Miller, Meryl Oliver, Bénédicte Selfslagh, Peter Stott, Herb Stovel, Mike Turner, Koenraad van Balen, Bernd von Droste and James Warden. We particularly acknowledge the support of Christopher Young and Dixi Lambert who agreed to review the manuscript and provided insightful comments.

Of course, this project would not have been completed without the quiet support of our spouses Hugh and Thomas. Thank you.

Disclaimer

List of Abbreviations

ALECSO	Arab Educational, Cultural and Scientific Organization
DoCoMoMo	International Committee for Documentation and Conservation of Buildings, Sites and Neighbourhoods of the Modern Movement
ECOSOC	United Nations Economic and Social Council
FAO	Food and Agricultural Organization
GEF	Global Environment Facility
IBE	International Bureau of Education
ICCROM	International Centre for the Study of the Preservation and Restoration of Cultural Property
ICME	International Council on Metals and the Environment
ICOM	International Council of Museums
ICOMOS	International Council of Monuments and Sites
IFLA	International Federation of Landscape Architects
IFPC	International Fund for the Promotion of Culture
IIEP	International Institute for Educational Planning
ILO	International Labour Organization
ITUC	Integrated Territorial and Urban Conservation
IUCN	International Union for Conservation of Nature and Natural Resources
IUPN	International Union for the Protection of Nature (later IUCN)
IWGC	Intergovernmental Working Group on Conservation
MAB	Man and the Biosphere Programme
OMMSA	Organization for Museums and Sites of Africa
OUV	Outstanding universal value
OWHC	Organization of World Heritage Cities
PADU	Protected Areas Data Unit
TICCIH	International Committee for the Conservation of the Industrial Heritage
UIA	International Union of Architects
UNDP	United Nations Development Programme
UNEP	United Nations Environment Programme
UNESCO	United Nations Educational, Scientific and Cultural Organization
UNF	United Nations Foundation
UNHCR	United Nations High Commissioner for Refugees
UNICEF	United Nations International Children's Emergency Fund
UNPROFOR	United Nations Protection Force

UNSCCUR	United Nations Conference on the Conservation and Utilization of Resources
USAID	U.S. Agency for International Development
WCMC	World Conservation Monitoring Centre
WCPA	World Commission on Protected Areas
WHO	World Health Organization
WWF	World Wildlife Fund

Chapter 1

Creation of the World Heritage Convention

The creation of the World Heritage Convention is a complex story that involves drafts, counter-drafts, dramatic debates and institutional rivalries. The key objective of establishing an international system of cooperation to protect globally significant heritage places was never in doubt, but the means to accomplish it were subject to institutional positioning and diplomatic manoeuvres. Building on an international discourse that began in the cultural field in the 1920s and continued after the Second World War with an added focus on natural resources protection, the Convention was clearly a product of its time in its reflection of a new global sensitivity to urban development and environmental degradation. Through its collective measures to identify and conserve the world's most significant places, the 1972 World Heritage Convention represents an extraordinary achievement in the annals of international agreements.

With hindsight, the creation of the World Heritage Convention seems inevitable. No single person or group can claim parenthood for the achievement of this international treaty because it is the result of decades of discussion and several separate independent initiatives. One could argue that the emergence of the World Heritage Convention reflects the zeitgeist or spirit of the era. During the 1920s and 1930s under the auspices of the League of Nations, as explained by Sarah Titchen in her unpublished doctoral thesis, concepts of common heritage and international cooperation as well as a distinctive style of international diplomacy emerged. Titchen points to the 1931 Athens Conference organized by the League's International Museums Office as an important marker for these ideas which eventually found their way into the World Heritage Convention.[1] As part of the 1931 conference, the first International Congress of Architects and Technicians of Historic Monuments developed the Athens Charter for the Restoration of Historic Monuments which includes the statement that "the conservation of the artistic and archaeological property of mankind is one that interests the community of the States, which are wardens of civilisation."[2] Titchen concludes that the League of Nations activities promoted "the idea of a common heritage of humankind

---

1   Sarah Titchen, *On the construction of outstanding universal value: UNESCO's World Heritage Convention (Convention concerning the Protection of the World Cultural and Natural Heritage, 1972) and the identification and assessment of cultural places for inclusion in the World Heritage List* (Canberra, 1995), pp. 12–24.

2   *The Athens Charter for the Restoration of Historic Monuments*, art. VII. Retrieved from http://icomos.org/index.php/en/charters-and-texts?id=167:the-athens-charter-for-the-restoration-of-historic-monuments&catid=179:charters-and-standards

deserving of international conservation through international cooperation and collaboration – a style and an idea that were to feature again when the functions of the League were taken over by UNESCO in December 1946. From these origins came the development of the World Heritage Convention."[3]

In the late 1940s, inter-twined initiatives reveal the gathering strength of the environmental movement. In 1948, preparatory work was underway to establish the International Union for the Protection of Nature (IUPN, later IUCN).[4] The 1949 United Nations Conference on the Conservation and Utilization of Resources (UNSCCUR) at Lake Success, New York, was organized by a powerful line-up of international bodies including the Food and Agricultural Organization (FAO), UNESCO, the World Health Organization (WHO) and the International Labour Organization (ILO). In his overview of the subject, McCormick regrets that "most environmental historians unfairly ignore UNSCCUR" and concludes: "Without question, it was a major step in the rise of the global environmental movement."[5]

The move to protect special places can be seen as part of a more general international response to the unparalleled destruction of heritage in two world wars. Bombardments, looting and illicit trafficking of cultural property mobilized UNESCO and other organizations to take measures to prevent such loss in the future. In the 1960s, as memories of war grew distant, new concerns arose. Industrialization and urban development occurred at a dizzying pace, threatening the survival of ecosystems and cultural monuments. International institutions were gaining influence as communications improved and long-distance travel became more accessible. In this period, momentum to create international agreements for the protection of special places in the world occurred simultaneously and unconnectedly. The emergence of parallel international initiatives, one for natural heritage, the other for cultural heritage, each apparently unknown to the other until 1970, reflects the decade's heightened awareness of environmental degradation and cultural loss. They popped up at about the same time in different organizations involved in the protection of culture and the environment.[6]

Independent yet similar, these proposals reflect a growing interaction among conservation professionals from different countries and an exciting swirl of new holistic approaches to environmental protection and territorial planning. They also reflect the isolation of the disciplines of natural and cultural sciences. The creative solutions and innovative approaches that emerged in the decade leading up to the 1972 Convention stand unrivalled in the following 40 years. An analysis of the documents from the 1960s and the recollections of the pioneers confirm that it was this dynamic period of the 1960s that laid the foundation for the subsequent conceptual and operational development of the World Heritage system.

---

3   Titchen, *Construction*, p. 35.

4   See chapter 5 for the establishment of IUPN, later IUCN.

5   John McCormick, *The Global Environmental Movement* (Chichester, 1995), p. 41.

6   For a view from one participant, see Michel Batisse, "The struggle to save our world heritage," *Environment*, 34/10 (1992), pp. 12–32.

The Natural Heritage Initiative

The history of the natural component of the World Heritage Convention involves the International Union for Conservation of Nature and Natural Resources (IUCN), the Science Sector of UNESCO and a group of influential American environmentalists. IUCN, as an international environmental organization founded through support from UNESCO in 1948, brought a scientific focus to its concern for the careful use and conservation of natural resources. UNESCO, as an intergovernmental organization, viewed use and protection of natural resources from a scientific as well as from a more political perspective. To further their goals, both IUCN and UNESCO concentrated on research, information exchange and targeted programmes. It was only in 1965, when the White House Conference on International Cooperation brought forward the idea of a World Heritage Trust, that a proposal for a formal international agreement gained momentum. The United Nations Conference on the Human Environment in Stockholm in 1972 was the catalyst that led IUCN to prepare a draft convention on the conservation of the world's heritage places.

In his history of IUCN, Martin Holdgate refers to the 1960s as a decade of "environmental explosion".[7] It was during this period that IUCN spearheaded an initiative for international protection of natural heritage. The origins of an international effort to protect ecologically important areas can be traced to IUCN's 1958 proposal to the United Nations Economic and Social Council (ECOSOC) to create a list of the world's most important national parks and equivalent reserves. Endorsed by the United Nations General Assembly in 1962, the first *United Nations List of Protected Areas and Equivalent Reserves* was prepared by IUCN and published that same year.[8] It is interesting to note that right from the start the list included several cultural heritage parks and landscapes, including the vast Khmer archaeological fields at Angkor (Cambodia) and the mediaeval open parliament site at Thingvellir (Iceland), both now listed as World Heritage Sites.[9]

The initiative was further developed at IUCN's First World Conference on National Parks held in Seattle, Washington in 1962. The purpose of this large international gathering of conservationists from over sixty countries, including many American delegates, was to improve global understanding and to encourage a national parks movement on a worldwide basis.[10] Among the twenty-eight

---

7   Martin Holdgate, *The Green Web: A Union for World Conservation* (London, 1999), p. 106.

8   Alexander Gillespie, *Protected Areas and International Environmental Law* (Leiden/Boston, 2008), p. 111.

9   IUCN International Commission on National Parks, *United Nations List of National Parks and Equivalent Reserves: Part Two and Addenda to Part One* (Morges, 1962), pp. 5, 11.

10   Alexander B. Adams, *First World Conference on National Park* (Washington, 1962), p. xxxii.

wide-ranging recommendations that covered endangered species, animal habitats, agricultural lands, terrestrial and marine parks as well as interdisciplinary research, management and training matters, two stand out from a World Heritage perspective. Recommendation 12 proposes further work on park planning to include among others "nature reserves, scientific areas, prehistoric, historic and cultural sites."[11] In a surprisingly early recognition of the links between culture and nature, recommendation 4 encourages support for UNESCO's proposal from the Cultural Sector to safeguard the beauty and character of landscapes because of the obvious connection to national parks and equivalent reserves.[12] In his reflections on World Heritage, UNESCO staff member Michel Batisse (who did not attend the Seattle meeting) argues that "this intimate association of natural and cultural sites could only be conceived in the United States where the protection of these two types of sites is the responsibility of the National Park Service."[13]

A World Heritage Trust: 1965

IUCN might well have continued with its ten-year programme approved in Seattle if American President Lyndon Johnson had not decided to celebrate the 20th anniversary of the United Nations by designating 1965 as International Cooperation Year. This declaration led to the White House Conference on International Cooperation during which the idea of a World Heritage Trust first took shape. The conference challenged different economic and social sectors, including the natural resources sector, to explore how international cooperation could be improved.

Historian Peter Stott, in his carefully documented study of the Committee on Natural Resources Conservation and Development, makes a compelling case that it was the Chairperson Joseph Fisher who conceived the idea.[14] Russell Train, then head of the Conservation Foundation and a member of Fisher's committee, recalls in an interview how Fisher told him to work up the idea of a World Heritage Trust:

> Joe was then president of an organisation called Resources for the Future which was a Ford Foundation-funded economic think-tank, I guess you might call it, dealing with resource issues primarily. And as I said, he chaired the committee. And he brought into one of our meetings a proposal for a World Heritage Trust. I don't think it had been fleshed out in any way. It was conceptual, what he had put together and quite brief – it may have been in his own hand-writing, I don't

---

11   Adams, *First World*, p. 380.

12   Adams, *First World*, p. 377.

13   Michel Batisse and Gérard Bolla, *The Invention of World Heritage* (Paris, 2005), p. 17.

14   Peter H. Stott, "The World Heritage Convention and the National Park Service, 1962–1972," *The George Wright Forum*, 28/3 (2011), pp. 281–3.

recall. And he gave me a copy and asked me, what did I think of it? And it was a new idea as far as I was concerned.[15]

Fisher's committee presented several concepts that eventually became part of the World Heritage Convention. The final report speaks of "unique and irreplaceable resources ... of legitimate international concern [that] should be maintained for the study and enjoyment of all peoples of the world and for the benefit of the country in which they lie." In terms of identifying sites, the report calls for "the compilation of a basic list of areas and sites that might be of international concern" and the need "to evaluate the basic list and select those few areas and sites that meet the high standards." In this vision, a World Heritage Trust would include "only those areas and sites that are absolutely superb, unique and irreplaceable." For such sites, international cooperation would be made available through funding and technical advice, education and tourism promotion. Fisher's committee formally recommended the establishment of "a Trust for the World Heritage that would be responsible to the world community for the stimulation of international cooperative efforts to identify, establish, develop, and manage the world's superb natural and scenic areas and historic sites for the present and future benefit of the entire world citizenry."[16] It is remarkable how many of these concepts found their way into the World Heritage Convention.

IUCN and a World Heritage Trust: 1966–1967

Holdgate describes this period as "bursting with environmental activities" in the governmental and intergovernmental sectors, producing laws, setting up departments and commissioning state of environment reports.[17] In 1966, the idea of World Heritage was introduced to the IUCN community at its General Assembly in Lucerne. Galvanizing the assembly was keynote speaker Fisher, the American who had chaired the Committee on Natural Resources Conservation and Development a few months earlier at the 1965 White House conference. Also present were two other influential Americans, Harold J. Coolidge, a founder of IUCN who was elected its president that year, and Train, a member of Fisher's committee. In his speech at Lucerne entitled "New Perspectives on Conservation for the Years Ahead," Fisher made five specific institutional proposals "as examples of the kind of ventures we must dare to make in any conservation

---

15   Canada Research Chair on Built Heritage, Université de Montréal, audio interview of Russell E. Train by Christina Cameron, Springfield, Virginia, 2 December 2008.

16   Report of the Committee on Natural Resources Conservation and Development to the White House Conference on International Cooperation, John F. Kennedy Presidential Library and Museum, Samuel E. Belk personal papers, box 10, National Citizens' Commission, Washington, D.C., 28 November – 1 December 1965, pp. 17–19. The authors are grateful to Peter H. Stott for his guidance in locating this document.

17   Holdgate, *Green*, p. 110.

Figure 1.1   Joseph L. Fisher

programme worthy of the future's challenge and potential." Among the five was
the proposal for a World Heritage Trust imported directly from the 1965 White
House Conference on International Cooperation.[18]

Fisher's proposal formulates concepts and language that eventually find their
way into the text of the World Heritage Convention. He repeats the exact wording
from the White House conference recommendation. Using both natural and
cultural sites, Fisher gives examples of what he considers to be World Heritage,
such as "the Grand Canyon of the Colorado; the Serengeti Plains; Angel Falls;
Mt. Everest; archaeological sites such as Angkor, Petra, or the ruins of Inca,
Mayan, and Aztec cities; historic structures such as the pyramids, the Acropolis,
or Stonehenge." He proposes two lists, "a basic list of areas and sites that might
be of international concern" and a selection from it of "those few areas and sites
that meet the high standards that would be required. It is essential that the criteria
for selection be highly refined and that the Trust include only those areas and sites
that are absolutely superb, unique and irreplaceable." The concept of the two lists

18   IUCN, *Proceedings of the ninth General Assembly at Lucerne 25 June to 2 July
1966* (Morges, 1967), p. 73.

survives in the text of the World Heritage Convention as "an inventory of property forming part of the cultural and natural heritage" and the "World Heritage List" (article 11.1–2). Finally, Fisher calls for international cooperation to raise funds and provide technical assistance to support "proper use of such areas as a means towards economic growth."[19]

UNESCO's science representative Batisse, who attended the 1966 Lucerne meeting, later recalls: "I heard of this proposal for the first time in Joseph Fisher's speech. I found it both appealing and timely that the protection of nature and of culture should be placed on the same level." He goes on to admit that he missed the implications for UNESCO because he thought Fisher meant a private philanthropic foundation, not an intergovernmental mechanism.[20]

The next year, the idea of a World Heritage Trust continued to be promoted by the Americans. As President of the Conservation Foundation, Train gave a speech at the World Wildlife Fund's International Congress on Nature and Man in Amsterdam and "spelled out the concept of the World Heritage that had been put forward by Joe Fisher and then our committee and pushed for its development and adoption by the international community."[21] Arguing that the works of humankind are inextricably linked to the physical environment, Train advocated for "an international cooperative effort that brings together in a unified programme a common concern for both man's natural heritage and his cultural heritage."[22]

UNESCO and the Natural Sciences in the 1960s

At the same time that this American proposal was being pushed through the IUCN network, UNESCO became involved in natural heritage conservation through its mandate for fostering international standards and cooperation in the natural sciences. An early achievement is the 1962 UNESCO resolution to provide assistance to developing countries for the conservation of natural resources, flora and fauna.[23] From the outset, UNESCO's programmes in this area were highly dependent on the scientific capacity of IUCN.[24] Until the middle of the 1960s, their focus was on providing support for research in the basic sciences, earth

---

19 IUCN, *Proceedings of the ninth*, pp. 73–4.

20 Batisse and Bolla, *Invention*, p. 17. It is interesting to note that a private philanthropic foundation for cultural sites called the World Monuments Fund was founded in the same era in 1965.

21 Canada Research Chair, interview Train.

22 Russell E. Train, "World Heritage: A Vision for the Future," *World Heritage 2002: Shared Legacy, Common Responsibility* (Paris, 2003), p. 36; Batisse and Bolla, *Invention*, p. 17; Stott, "The World Heritage Convention," p. 283.

23 UNESCO, Records of the General Conference twelfth session, Paris, 1963, p. 36. Retrieved from http://unesdoc.unesco.org/images/0011/001145/114582e.pdf

24 Batisse and Bolla, *Invention*, pp. 15–16.

sciences and life sciences as well as on the application of science and technology to development.

It is only in 1966 that a new area of scientific interest emerges, namely research focused on natural resources per se. At the UNESCO General Conference of 1966, members approved a broad programme of research and education on various separate scientific disciplines. In addition, they approved a new research and training programme "relating to the natural environment and resources of the land areas and their conservation," specifically noting the need for ecological studies and interdisciplinary work. To further this new agenda, which reflects an emerging global realization that environmental issues needed to be examined in a holistic way, UNESCO received authorization for a 1968 experts meeting.[25] This is a key event that marks the emergence of an environmental perspective at UNESCO. In her recent history, Chloé Maurel considers it to be the first global scientific meeting at UNESCO to consider jointly the problems of the environment and development, a precursor for the later concept of sustainable development.[26]

The 1968 intergovernmental conference of experts on the scientific basis for rational use and conservation of the resources of the biosphere and a parallel UNESCO cultural experts meeting in the same year were responding to similar global conditions: the accelerating rate of environmental deterioration, rapid urbanization, rural exodus and exploding industrialization. Oddly enough, mutually unaware of the other, the natural and cultural heritage initiatives proceeded in parallel within UNESCO until 1970, a reflection of the compartmentalization of the secretariat at that time. Batisse, then Director of the Natural Resources Division in the Science Sector of UNESCO, admits that he "had unfortunately no knowledge of these events until 1970."[27] In comparison to the small 1968 meeting of cultural heritage experts, the biosphere conference was on an entirely different scale. Reflecting the surge of interest in the environment in the 1960s, the biosphere conference assembled well over three hundred participants from sixty-three countries, mainly experts from universities, academies of science and international organizations. Among the delegates were the American environmentalists and proponents of a World Heritage Trust: Coolidge, Train and ecologist Lee Talbot.[28]

The fundamental premise of the 1968 meeting was that man is a key factor in the biosphere. The discussions generated a wealth of new creative ideas on how to improve the human environment through research, education and policy

---

25   UNESCO, Records of the General Conference fourteenth session, Paris, 1967, p. 45. Retrieved from http://unesdoc.unesco.org/images/0011/001140/114048e.pdf

26   Chloé Maurel, *Histoire de l'UNESCO – les trente premières années. 1945–1974* (Paris, 2012), p. 287.

27   Batisse and Bolla, *Invention*, p. 16.

28   UNESCO, Final report of intergovernmental conference of experts on the scientific basis for rational use and conservation of the resources of the biosphere, Paris, 6 January 1969, SC/MD/9, annex 5. Retrieved from http://unesdoc.unesco.org/images/0001/000172/017269EB.pdf

Figure 1.2   Michel Batisse in 1981 (far right) with Francesco Di Castri behind

development. Experts noted the need to establish parks and nature reserves as a means of providing baselines for comparison with managed and artificial ecosystems. In an early elaboration of the concept of sustainable development, the experts linked use to conservation, stating that "conservation, while including preservation, has come generally to mean the wise use of resources." Of particular importance were specific recommendations to create a Man and the Biosphere programme at UNESCO, to preserve natural areas and endangered species through establishing natural protected areas and national parks, and to bring these ideas to the broader United Nations system at the upcoming United Nations Conference on the Human Environment in Stockholm in 1972. At this time, the experts were not envisaging a binding intergovernmental treaty but only a "Universal Declaration on the Protection and Betterment of the Human Environment."[29]

On the advice of this expert meeting, UNESCO's 1968 General Conference authorized the development of a long-term intergovernmental and interdisciplinary programme on rational use and conservation of the natural environment and its resources. It also encouraged cooperation with other sectors of the United Nations system and international non-governmental organizations.[30] As a result, UNESCO's sixteenth General Conference officially adopted on 23 October 1970

---

29   UNESCO, Final report biosphere, SC/MD/9, paras. 18, 64, 113, 120, 128.

30   UNESCO, Records of the General Conference fifteenth session, Paris, 1969, pp. 39–40. Retrieved from http://unesdoc.unesco.org/images/0011/001140/114047e.pdf

the establishment of the Man and the Biosphere (MAB) programme whose goals were to study the structure and functioning of ecological regions, to maintain genetic diversity and to encourage research and education in the natural sciences.[31] It provided for an International Coordinating Council and a UNESCO secretariat to set up and monitor a global network of protected biosphere reserves covering the diverse natural regions of the world. Significant was the holistic approach to ecosystems that included flora and fauna as well as human use of natural regions. The powers of the programme were quite limited. Although the MAB programme could encourage action through moral suasion and appeals to national pride, it did not have the weight of an international treaty.

While UNESCO was committed to reviewing the MAB programme after the proposed Stockholm summit, it had no intention of developing its own international convention for natural heritage. It is worth noting that UNESCO's Batisse served as Secretary General for the 1968 experts meeting and subsequently contributed to the IUCN draft convention text.

IUCN and a proposed convention: 1970–1972

The catalyst that motivated IUCN to work on an international convention was the 1968 decision of the United Nations General Assembly to accept Sweden's invitation to host the Stockholm Conference on the Human Environment. This was a ground-breaking event in the growth of international environmentalism.[32] IUCN saw the Stockholm summit as an opportunity to get endorsements for new international environmental laws. It initiated the preparation of four new conventions targeted at protecting certain islands for scientific research, controlling trade in wild species and plants, conserving wetlands and conserving world heritage.[33]

American ecologist and IUCN staff member at that time, Lee Talbot, who had also worked on the American proposal, made a formal submission to IUCN's executive in September 1970 that, according to Stott, became the basis for the IUCN draft convention.[34] The text was further developed by IUCN Deputy Director-General Frank Nicholls who headed a task force that included Train, Batisse and a representative from FAO. The draft Convention on the Conservation of the World Heritage leaned towards natural areas but nonetheless included cultural sites as well, using the *United Nations List of Protected Areas and Equivalent Reserves* as a starting point. In the document, there was no identification of a

31   UNESCO, Records of the General Conference sixteenth session, Paris, 1971, pp. 35–8. Retrieved from http://unesdoc.unesco.org/images/0011/001140/114046e.pdf

32   For context on the rise of a global environmental movement and the key significance of the Stockholm conference, see John McCormick, *Reclaiming Paradise: the Global Environmental Movement* (Bloomington, 1991), pp. 88–106.

33   Holdgate, *Green*, p. 113.

34   Stott, "The World Heritage Convention," p. 284.

secretariat although it was assumed to be IUCN.[35] There is some debate over the date when Batisse first realized that the IUCN initiative might overlap with the cultural heritage convention underway in the Cultural Sector of UNESCO. He himself writes that he found out by accident in May 1970 and reported back to his colleagues in the Cultural Sector who he claims were unconcerned.[36] Talbot says in his recent interview with Stott that it was only in April 1971 that Batisse "suddenly registered what we were doing," noting that Batisse remarked that UNESCO "already had such a convention we've been working on, because our General Assembly several years ago directed us to do so."[37] In any event, IUCN finalized its draft convention text in 1971 for submission to the international working group responsible for the Stockholm summit process.

The Cultural Heritage Initiative

UNESCO's initial mandate for culture gradually expanded to include the safeguarding of cultural heritage within its sphere of interest. With regard to heritage places, modest activities gradually built foundational elements for a cooperative system to protect, preserve and restore monuments and sites. In 1948, the organization decided to explore the creation of an international committee of experts and the establishment of an international fund to subsidize preservation activities for "monuments and sites of historical value".[38] During the 1950s, its work grew to include information-sharing on preservation and presentation techniques, mechanisms to protect monuments from the dangers of armed conflict and specific measures to protect landscapes. The idea of creating an international committee of experts "to serve as an advisory body for UNESCO on the conservation, protection and restoration of monuments, artistic and historical sites and archaeological excavations"[39] materialized in 1951 as the International Committee on Monuments, Artistic and Historical Sites and Archaeological Excavations.

With the committee's collaboration, UNESCO's early achievements during this decade included the Convention for the Protection of Cultural Property in the Event of Armed Conflict (1954),[40] the creation of the International Centre for the Study of the Preservation and Restoration of Cultural Property (ICCROM)

35   Holdgate, *Green*, p. 114; Stott, "The World Heritage Convention," p. 284.

36   Batisse and Bolla, *Invention*, pp. 20–21.

37   Stott, "The World Heritage Convention," p. 284.

38   UNESCO, Records of the General Conference third session, Paris, 1949, p. 28. Retrieved from http://unesdoc.unesco.org/images/0011/001145/114593e.pdf

39   UNESCO, Records of the General Conference fifth session, Paris 1950, p. 44. Retrieved from http://unesdoc.unesco.org/images/0011/001145/114589e.pdf

40   UNESCO, Records of the General Conference eighth session, Paris, 1955, p. 33. Retrieved from http://unesdoc.unesco.org/images/0011/001145/114586e.pdf

(1956),[41] and a Recommendation on International Principles applicable to Archaeological Excavations (1956).[42] Of special note is the first international campaign, championed by UNESCO Director-General René Maheu, to save the Nubian monuments in Abu Simbel and Philae in Egypt (1960–1968), putting into practice the principle of shared international responsibility for conserving the outstanding heritage of humanity.[43] Maurel claims that the prestige and visibility of this amazing project consolidated UNESCO's leadership in the field of cultural heritage.[44] Foreshadowing the development of a more holistic approach to cultural conservation in the decade to follow, the 1960 General Conference of UNESCO also decided that a proposed countryside recommendation "shall cover the safeguarding of the beauty and character of sites as well as of landscape."[45]

UNESCO's Cultural Heritage Initiatives in the 1960s

As a result of the massive destruction of cultural property through the two world wars, then through rapid urbanization and industrial growth particularly in developed countries, momentum grew in the 1960s for new UNESCO initiatives to preserve cultural heritage sites and to foster international cooperation. These additional tools, while still piece-meal and ad hoc responses to specific situations, were rooted in ideas and values aimed at creating a climate of peace and friendship. While the notion of a common heritage belonging to all humanity persisted, one can observe a philosophical shift from the previous decade. The idea of conservation as an end in itself is replaced by a new concept, that of valuing cultural monuments and sites for their social and economic role in daily life. Seeing heritage in its context, especially its urban context, brought forth the challenge of conserving individual static monuments in an evolving environment. This is the decade that introduces the notion of urban and rural setting, and proposes the integration of cultural heritage into land planning at a territory-wide level, a curious parallel to the development of an ecosystems-wide approach for natural heritage protection. As an extension of this thinking, the 1960s also witness consideration of natural areas as part of culture, not as real wilderness but in the social sense of setting for community life.

A series of new UNESCO instruments and initiatives demonstrates the gradual development of international action. The first is a Recommendation

---

41   UNESCO, Records of the General Conference ninth session, Paris, 1957, p. 24. Retrieved from http://unesdoc.unesco.org/images/0011/001145/114585e.pdf

42   UNESCO, Records of the General Conference ninth session, Paris, 1957, p. 40. Retrieved from http://unesdoc.unesco.org/images/0011/001145/114585e.pdf

43   UNESCO, Records of the General Conference eleventh session, Paris, 1961, p. 51. Retrieved from http://unesdoc.unesco.org/images/0011/001145/114583e.pdf

44   Maurel, *Histoire,* pp. 286–7.

45   UNESCO, Records of the General Conference eleventh session, Paris, 1961, p. 51. Retrieved from http://unesdoc.unesco.org/images/0011/001145/114583e.pdf

concerning the Safeguarding of the Beauty and Character of Landscapes and Sites (1962) setting out general principles that affirm the cultural and aesthetic values of landscapes and sites, and recognizes their "powerful physical, moral and spiritual regenerating influence."[46] In addition to adding nature protection to a cultural instrument, the significant innovation of this Recommendation lies in its application to entire territories, not selected landscapes or sites, and its attention to the lands around historic buildings, anticipating a later doctrinal interest in context and setting. It is limited by its non-binding nature and its dependence on national legal measures and rules.

Two years later in 1964, the Second International Congress of Architects and Specialists of Historic Monuments adopted an international code for conservation practice, the International Charter for the Conservation and Restoration of Monuments and Sites (known as the Venice Charter). The close connection to UNESCO is evident in the presence of the Director-General René Maheu at the opening session and the on-going participation of Hiroshi Daifuku, then head of UNESCO's Cultural Sector.[47] In terms of ideas, the Venice Charter invokes the concept of a common heritage, it widens the scope of interest beyond specific monuments to their wider urban and rural setting, and it recognizes the valid contribution of all periods. But while promoting the desirability of adapting to new societal needs, the Venice Charter gave precedence to aesthetic and historical values. It is worth noting that this meeting also adopted a resolution put forward by UNESCO for the creation of the International Council of Monuments and Sites (ICOMOS) which came into being in 1965.

In this exciting atmosphere marked by innovation and international professional exchange, new ideas emerged on the relationship between tourism and cultural heritage, the need to deal with threats posed by new building projects and a vision for a comprehensive international system for heritage protection. Following a UNESCO study to ascertain how far the preservation of monuments contributed to tourism and hence to economic development, the 1966 General Conference adopted a curious resolution declaring tourism to be of outstanding cultural interest and noted the link between preservation of cultural property and effective development programmes.[48] In response to concerns about the increasing threat to cultural heritage caused by uncontrolled growth, urban development and engineering works, another international instrument was created, a Recommendation concerning the Preservation of Cultural Property endangered

---

46 UNESCO, Records of the General Conference twelfth session, Paris, 1963, pp. 139–42. Retrieved from http://unesdoc.unesco.org/images/0011/001145/114582e.pdf

47 Russell V. Keune, "An interview with Hiroshi Daifuku," *CRM: The Journal of Heritage Stewardship*, 8/1 and 2 (2011), pp. 31–45. Retrieved from http://crmjournal. cr.nps.gov/03_spotlight_sub.cfm?issue=Volume%208%20Numbers%201%20and%20 2%20Winter%2FSummer%202011&page=4&seq=1

48 UNESCO, Records of the General Conference fourteenth session, Paris, 1967, pp. 62–4. Retrieved from http://unesdoc.unesco.org/images/0011/001140/114048e.pdf

by Public or Private Works (1968).[49] This standard illustrates how far ideas about cultural heritage had evolved in this decade, for it defines immoveable cultural heritage in terms of historic districts and traditional structures, not as selected isolated monuments; as well, it emphasizes the importance of urban and rural settings on a territory-wide basis. Intended for implementation at a national level, the Recommendation does not convey the notion of international heritage value for cultural sites.

Origins of the UNESCO Draft Convention

While these concepts and initiatives are important contributors to World Heritage, UNESCO's draft convention can be traced to 1966 when the General Conference authorized the Director-General "to study the possibility of arranging an appropriate system of international protection, at the request of the States concerned, for a few of the monuments that form an integral part of the cultural heritage of mankind."[50] Significantly, this proposal involves an international system, in other words a systematic, not ad hoc, approach, and a concept of heritage that extends beyond national borders to the international sphere. What is also noteworthy is the notion of selectivity; only a few monuments were targeted for this international system.

What followed the 1966 resolution sheds light on the text of the World Heritage Convention. Hiroshi Daifuku, who headed UNESCO's Section for the Development of the Cultural Heritage, began commissioning papers from international experts for a manual on the restoration of monuments.[51] Early in 1968, UNESCO hosted a meeting of invited experts to develop a proposal for an international system. The unwieldy title is self-explanatory: "Meeting of experts to co-ordinate, with a view to their international adoption, principles and scientific, technical and legal criteria applicable to the protection of cultural property, monuments and sites." While purporting to be a global project, the work was dominated by European participation. More than half of the thirteen experts came from European countries;[52] the commissioned technical papers were authored by Robert Brichet (France), Professor Guglielmo De Angelis d'Ossat (Italy) and Professor Jan Zachwatowicz (Poland); with the exception of the

---

49   UNESCO, Records of the General Conference fifteenth session, Paris, 1969, pp. 139–45. Retrieved from http://unesdoc.unesco.org/images/0011/001140/114047e.pdf

50   UNESCO, Records of the General Conference fourteenth session, Paris, 1967, p. 61. Retrieved from http://unesdoc.unesco.org/images/0011/001140/114048e.pdf

51   Letter from Hiroshi Daifuku to Ernest A. Connally, Paris, 4 July 1967, University of Maryland, Hornbake Library, Papers of Ernest A. Connally, box 5. Experts and their topics included Sanpaolesi on theory and practice, Gazzola on history and training, Foramitti on measured drawing and photogrammetry, Sekino on preservation of wooden structures in Asia, Connally on American conservation practice and Daifuku on legislation.

52   Austria, France, Ghana, India, Italy, Japan, Poland, Spain, USSR, United Arab Republic, United Kingdom, United States of America, Yugoslavia. For a detailed description of the 1968 and 1969 experts meetings, see Titchen, *Construction*, pp. 53–8.

League of Arab States, all the non-governmental organizations were represented by Europeans.[53] Experts compared heritage protection systems from twelve countries: Austria, Spain, United States of America, France, Ghana, India, Italy, Japan, United Arab Republic, United Kingdom, Soviet Union and Yugoslavia. In addition, they analysed underlying scientific concepts and operational principles as well as legal provisions required to support both national and international protection systems.

The final report contains concepts, approaches and wording that eventually appear in the 1972 World Heritage Convention, including the taxonomic structure of "monuments, groups of buildings and sites" as well as the notion of cultural landscapes that embodies the combined work of man and nature. Natural sites were also included, not for their importance as ecosystems but for their cultural, aesthetic and picturesque values. The most important recommendation from these experts was their conclusion, oddly similar to the natural heritage proposal, that a comprehensive international system would require two components: a national protection system to cover all cultural properties and an international protection system for selected monuments and sites of universal interest that were in need of technical or financial assistance. This key proposal explains the existence of the little-known 1972 Recommendation concerning the Protection, at the National Level, of the Cultural and Natural Heritage which was adopted at same time as the World Heritage Convention but largely ignored to this day.[54] The experts from the 1968 meeting did not envisage a general list of sites of universal interest nor did they necessarily identify UNESCO as the administrative institution. Their report refers to the need for an "international protection body, or possibly UNESCO" to oversee implementation of the international protection system, to be made up of experts and technicians from States and permanent staff.[55]

The essential framework and core concepts of an international system were fleshed out at this time. The shared excitement and enthusiasm of the group is captured in the report: "It surely is time for a new approach, replacing the traditional by something more modern and dynamic … contemporary opinion is favourable. Never has there been so much enlightened interest in cultural

---

53 UNESCO, Final report of meeting of experts to co-ordinate, with a view to their international adoption, principles and scientific, technical and legal criteria applicable to the protection of cultural property, monuments and sites, Paris, 31 December 1968, SCH/CS/27/8. Retrieved from http://whc.unesco.org/archive/1968/shc-cs-27-8e.pdf. The commissioned technical papers came from a particular European approach to conservation which would not necessarily be shared in other European countries, particularly in the north and the west.

54 Concern for heritage places in general is also reflected in article 5 of the World Heritage Convention.

55 UNESCO, Final report of meeting of experts to co-ordinate, with a view to their international adoption, principles and scientific, technical and legal criteria applicable to the protection of cultural property, monuments and sites, Paris, 31 December 1968, SCH/CS/27/8, para 49. Retrieved from http://whc.unesco.org/archive/1968/shc-cs-27-8e.pdf

property."[56] The next year, another experts meeting, involving many of the 1968 participants, further explored the concepts and practical measures for establishing the proposed system. Participation was again predominantly European.[57] The 1969 meeting considered two new working papers that UNESCO had commissioned: one by Raymond Lemaire (Belgium) and François Sorlin (France) on the basic premises related to protection of monuments, groups and sites of universal value and interest; the other by Robert Brichet (France) and Mario Matteucci (Italy) on practical steps to establish an appropriate international system. The experts recommended the preparation of a UNESCO convention. They reiterated the need for two instruments: a recommendation for national systems and a convention for an international system.[58] The final report was deemed of such importance that it was subsequently distributed through UNESCO's publishing house.

This section on the extraordinary achievements of the 1960s would be incomplete without paying tribute to the intellectual leadership of those who built many of the foundations of twentieth-century conservation theory and practice in the cultural heritage field. While there were many contributors, several key players had a pervasive influence. Italian architect Piero Gazzola, superintendent of historical monuments in Verona, drafted the Venice Charter, was founding President of ICOMOS and chaired the epic 1968 UNESCO experts meeting. Belgian architect Raymond Lemaire, professor in Leuven, contributed to the draft Venice Charter, was founding Secretary-General for ICOMOS, attended UNESCO's experts meetings and authored one of its working papers. French architect François Sorlin, inspector-general of historical monuments, took part in the drafting of the Venice Charter and attended the UNESCO experts meetings, preparing one of the working papers.[59] Two other key contributors who were part of the drafting group for the Venice Charter and later wrote technical papers for the UNESCO experts meetings were Polish professor of architecture, Jan Zachwatowicz and Italian jurist Mario Matteucci.

After studying the work of the experts meetings, UNESCO's 1970 General Conference considered and endorsed the document on the "desirability of

---

56   UNESCO, Final report of meeting of experts to co-ordinate, with a view to their international adoption, principles and scientific, technical and legal criteria applicable to the protection of cultural property, monuments and sites, Paris, 31 December 1968, SCH/ CS/27/8, paras. 32–3. Retrieved from http://whc.unesco.org/archive/1968/shc-cs-27-8e.pdf

57   Austria, Belgium, Czechoslovakia, France, Ghana, India, Italy, the Netherlands, Peru, Poland, Spain, Tunisia, the United Kingdom and the United States of America.

58   UNESCO, Final report of meeting of experts to establish an international system for the protection of monuments, groups of buildings and sites of universal interest, Paris, 10 November 1969, SHC/MD/4. Retrieved from http://whc.unesco.org/archive/1969/shc-md-4e.pdf

59   Raymond Lemaire and François Sorlin, The appropriate system for the international protection of monuments, groups of buildings and sites of universal value and interest: basic premises of the question, Paris, 30 June 1969, SCH/conf.43/4, pp. 1–10. Retrieved from http://whc.unesco.org/archive/1969/shc-conf-43-4-e.pdf

adopting an international instrument for the protection of monuments and sites of universal value." The meeting in fact approved the preparation of "two new complementary instruments on the protection of monuments and sites of universal interest."[60] It is remarkable that this initiative was developing in isolation from parallel efforts by IUCN and the United States of America. Little notice appears to have been taken of references to the American idea for a World Heritage Trust, explicitly mentioned in documents for the 1969 expert meeting, including Brichet and Matteuci's paper in which the authors refer to Russell Train and the 1965 American proposal for a World Heritage Trust involving IUCN.[61] National government representatives at the General Conference do not appear to have connected the two initiatives either. UNESCO's Batisse says that he only found out about the IUCN proposed convention in 1970 and claims that he immediately informed his Cultural Sector colleagues. Noting the compartmentalization of the UNESCO secretariat, he recalls that the Cultural Sector was not concerned with a competing convention nor was it aware of the powerful movement awakened by the Stockholm conference.[62] UNESCO therefore continued to draft its two cultural heritage instruments, unaware that the entire organization would soon be swept up in the byzantine negotiations that brought the various strands together into the 1972 World Heritage Convention.

The American Draft Convention

A third convention text prepared by the United States of America took shape through serendipity. According to Train, following the election of Richard Nixon in 1968, the President was looking for an environmental hook to respond to polls indicating that significant numbers of Americans had concerns about the degradation of the environment. Train recounts that the President had "created a system of task forces on various subjects to advise him as the incoming President on policy matters" and that a task force on the environment was something of an afterthought. Train's friend, Henry Loomis, deputy head of this operation, asked him to lead it. According to Train, "our key recommendation was that there should be a focal point for environmental policy-making in the administration, in the Executive Branch of the government. And that eventually translated into what became the Council on Environmental Quality."[63] In

---

60   UNESCO, Records of the General Conference sixteenth session, Paris, 1971, pp. 54–5. Retrieved from http://unesdoc.unesco.org/images/0011/001140/114046e.pdf These two legal instruments eventually became the 1972 Convention and Recommendation.

61   Robert Brichet and Mario Matteuci, Practical steps to facilitate the possible establishment of an appropriate international system, Paris, 13 June 1969, SCH/conf.43/5, para. XIX. Retrieved from http://unesdoc.unesco.org/images/0021/002151/215153fo.pdf

62   Batisse and Bolla, *Invention*, pp. 20–21.

63   Canada Research Chair, interview Train.

Figure 1.3   Russell E. Train

1970, Nixon set up the Environmental Protection Agency and the President's Council on Environmental Quality. Train was appointed to chair the council.

Train later recounts the curious situation that resulted in Nixon's 1971 endorsement of World Heritage:

> I became the first Chairman of the council and that then led to the fact that we took the lead in putting together for the President his annual environmental message to the Congress. And Nixon – nobody still really believes this when I tell them this – Nixon had a remarkable environmental record. He seemed to have no particular interest in the subject but he was extremely supportive ... the reasons probably were complex, reflecting Nixon's personality ... He also was quite aware that the younger generation was extremely interested in the environment and he made this one of his priority areas. And it has been described by an author named Richard Reeves, in a book called *Nixon alone in the White House,* in which Nixon sat with a yellow pad in a rocking chair in a back room and listed the things he thought were high priority, some ten areas of interest such as defence, that he wanted to be very much involved with. And then down at the bottom he added a couple of

other things and he said, "I think these things are high priority but I want nothing to do with them." One was the environment.[64]

Train and the other advocates for a World Heritage Trust skilfully linked their initiative to the centennial of the creation of Yellowstone National Park. The President's 1971 message to Congress celebrates the centennial of Yellowstone, the world's first national park and the beginning of the national parks movement worldwide. Just as Yellowstone was important to all Americans, the President wrote:

> It would be fitting by 1972 for the nations of the world to agree to the principle that there are certain areas of such unique worldwide value that they should be treated as part of the heritage of all mankind and accorded special recognition as part of a World Heritage Trust. Such an arrangement would impose no limitations on the sovereignty of those nations which choose to participate, but would extend special international recognition to the areas which qualify and would make available technical and other assistance where appropriate to assist in their protection and management. I believe that such an initiative can add a new dimension to international cooperation.[65]

The President's environmental message directed government officials to "develop initiatives for presentation in appropriate international forums to further the objective of a World Heritage Trust."[66] The upcoming United Nations Summit on the Human Environment in Stockholm fit the bill. Aware of the two other convention texts from IUCN and UNESCO, the Americans nonetheless attempted to produce a third integrative text. By August 1971, the United States had in hand a draft Convention on the Establishment of a World Heritage Trust for submission to the Stockholm preparatory meeting to be held in New York in September 1971. The American text picked up the notion of an overall inventory of sites and a selective list of the most significant ones, an idea that could be found in both IUCN's and UNESCO's drafts. It introduced the idea of creating a board of experts elected by the contracting parties to set conservation standards, select and remove sites from the inventory and allocate technical assistance through a proposed fund for needy areas and sites. What was unusual about the United States draft was the blank space in the document for the identification of the secretariat of the World Heritage Trust. Did the Americans not wish to choose between the competing

---

64   Canada Research Chair, interview Train; Richard Reeves, *President Nixon alone in the White House* (New York, 2001), pp. 172–3.

65   Richard Nixon, Special message to the Congress proposing the 1971 environmental program, 8 February 1971, art. IV. Retrieved at http://www.presidency.ucsb.edu/ws/?pid=3294

66   A.L. Doud, Memorandum of Law, Circular 175: Request for Authority to negotiate a multilateral treaty creating a World Heritage Trust, 9 February 1971, United States National Park Service, World Heritage archives, file 1973–1975, p. 1.

interests of IUCN and UNESCO? An internal memo may have been tongue-in-cheek or may in fact have indicated a preference for keeping the secretariat at IUCN. On the matter of a secretariat, the memo explains that IUCN "which is advocating a proposal similar to the Trust may undertake the duties which would generate these expenses. Since in its own proposal, it omitted such an article, it may feel that its resources are adequate."[67] In other words, IUCN could use its own resources to administer the general organization of the treaty.

Bringing the Threads Together: The Final Stages

The catalyst for bringing the three overlapping conventions together was a preparatory meeting for the Stockholm summit, the first large gathering of world leaders to consider environmental concerns.[68] In September 1971 in New York, the Intergovernmental Working Group on Conservation (IWGC) for the Stockholm conference had initially expected only IUCN's proposal. Instead it was faced with three draft texts: a Convention on Conservation of the World Heritage (IUCN), a Convention concerning the International Protection of Monuments, Groups of Buildings and Sites of Universal Value (UNESCO) and a Convention on the Establishment of a World Heritage Trust which the American submitted during the meeting. All three included both cultural and natural sites albeit with differing emphasis. IUCN's text focused mainly on national parks and ecosystems while including some cultural parks and landscapes; UNESCO's draft favoured monuments, groups of buildings and sites but also covered landscapes for their cultural, aesthetic and picturesque values; the American proposal gave equal weight to natural areas and cultural or historic sites. Batisse characterized the situation as absurd.

The IUCN and UNESCO texts had parallel goals. When Batisse presented the UNESCO draft to environmentalists, he acknowledged its cultural bias and argued that it was open for amendment at an intergovernmental meeting of experts scheduled at UNESCO the following April.[69] In spite of UNESCO's proposal, the IWGC took the position that it could only examine the IUCN draft since the UNESCO one was not part of the Stockholm process. It adopted two principles for the re-drafting of the IUCN text: the definition of World Heritage should focus on the conservation of natural areas "without excluding cultural sites;" and that preparation should proceed as planned for completion at the Stockholm conference.[70]

---

67 Doud, Memorandum, p. 3.
68 Holdgate, *Green*, p. 114.
69 Batisse and Bolla, *Invention*, pp. 25–7.
70 Titchen, *Construction*, pp. 64–5.

Diplomatic Manoeuvres

Following the New York meeting, international diplomacy kicked in. At stake was the threat of overlapping international agreements as well as control of the administrative structure. In all three drafts, the identification of a secretariat was not spelled out. Would it be housed at a non-governmental body like IUCN or within the intergovernmental United Nations system? Those involved in the IUCN draft were reluctant to see control shift to an intergovernmental bureaucracy. On the other hand, UNESCO argued that the subject fell within its mandate for education, science and culture.[71] At this stage, a possible solution was to strip the cultural elements from the IUCN draft which could then be approved at Stockholm; in the same vein, the UNESCO draft could strip out references to natural heritage and continue its course for the General Conference meeting in fall 1972.

As proponents of a single merged convention, the Americans remained dissatisfied with this solution. At UNESCO, Director-General Maheu directed his officials to pursue a single convention, convinced that the UNESCO draft could be enlarged to include natural heritage as well. By November 1971, Gérard Bolla had left his position as chief of staff in Maheu's office to take over the Cultural Sector. Batisse, "who understood full well what was happening on the other side of the Atlantic," advised Bolla to discuss this approach with the Americans. In a later interview, Bolla describes his diplomatic trip to the United States to seek their support:

> I had only been named for fifteen days and I asked for approval to go on mission to Washington ... I met with interested American groups ... and I said, listen, it is entirely possible to revise the convention in a really fundamental way. An expert committee is going to meet in April 1972. We have already invited them. We are going to say to governments, "Listen, you have to send experts in nature as well as in culture because we are going to have the problem of including natural properties in a really serious way."[72]

Bolla met with officials from the State Department, the Department of the Interior and the United States National Commission for UNESCO. He particularly

---

71   Canada Research Chair on Built Heritage, Université de Montréal, audio interview of Gérard Bolla by Christina Cameron and Mechtild Rössler, Paris, 26 October 2007.

72   Canada Research Chair, interview Bolla. "J'avais été nommé depuis quinze jours, que j'ai demandé un ordre de mission pour Washington et j'y suis allé et cela avait été organisé très bien: une réunion où il y avait tous les milieux intéressés américains ... Et moi, je leur ai dit : Écoutez, c'est tout à fait possible de réviser la Convention, le projet d'une façon vraiment fondamentale. Un comité d'experts gouvernementaux va siéger en avril 72. Nous les avons déjà convoqués. Nous allons dire aux gouvernements: Écoutez, il faut des experts de la nature comme de la culture parce qu'on va avoir le problème d'inclure d'une façon sérieuse les biens naturels."

Figure 1.4   René Maheu, Director-General of UNESCO 1961–1974

appreciated the role played by a young State Department lawyer, Carl Salans in facilitating the meetings. Before Bolla left Washington, he received formal notice of American support for restructuring the UNESCO draft.[73]

The final stage in determining the way forward was the establishment of communications between the leaders of the two United Nations agencies, namely Maheu from UNESCO and Canadian Maurice Strong, Secretary-General of the United Nations Conference on the Human Environment. In response to Strong's request for views on the IUCN draft, Maheu flexed muscle and claimed the domain as the responsibility of UNESCO. Pointing out that duplication within the United Nations was not desirable, he called on Strong to do his utmost to spare countries from "facing two rival propositions on the same subject." He did acknowledge that the UNESCO draft placed too much importance on the monumental aspect of cultural heritage and offered an opportunity to amend the draft at the upcoming meeting of experts.[74]

---

73   Batisse and Bolla, *Invention,* p. 71.
74   Batisse and Bolla, *Invention*, pp. 72–3.

Maheu's appeal was successful. Batisse describes a subsequent meeting of the IWGC in March 1972:

> A delicate negotiation was conducted to reconcile the wishes of all parties concerned and, in particular, those of environmentalists, who were urging that the conservation of nature at long last be given the same attention and weight as cultural preservation. Eventually, it was agreed by the preparatory committee that a single text should be worked out for adoption by the general conference of UNESCO, which was to meet that same year.[75]

The next month, UNESCO hosted a committee of government experts to prepare the two international instruments: a draft convention and a draft recommendation. This is the key meeting where the World Heritage Convention was effectively created. It assembled government experts from sixty-seven countries as well as observers and international organizations including ICOMOS, ICCROM, IUCN, ICOM, IFLA and others. Once again European nations were the most represented, with other participants evenly distributed among other regions of the world.[76] Unfortunately, some documents, including the manuscript notes from Bolla and from France's Michel Parent, who chaired the drafting committee at this meeting, were destroyed by a 1984 fire in the UNESCO archives.[77] It is clear however that there was a procedural hassle over working documents. The Americans had submitted their draft which gave equal attention to cultural and natural heritage, arguing that it should therefore serve as the basis for the discussion. But UNESCO insisted that, from a procedural point of view, the working document had to be its own draft on the protection of monuments and sites of universal interest. The American delegation was thus forced to propose many amendments in order to

---

75   Michel Batisse, "The struggle to save our world heritage," *Environment*, 34/10 (1992), p. 15.

76   UNESCO, List of participants at the special committee of government experts to prepare a draft convention and a draft recommendation to member states concerning the protection of monuments, groups of buildings and sites, Paris, 4–22 April 1972, SHC.72/conf.37/22. Retrieved from http://unesdoc.unesco.org/images/0021/002151/215100mb. pdf Experts came from the following countries: Afghanistan, Algeria, Argentina, Australia, Austria, Belgium, Brazil, Bulgaria, Cameroon, Canada, Central African Republic, Chad, Chile, Colombia, Costa Rica, Denmark, Dominican Republic, Ecuador, Egypt, Federal Republic of Germany, Finland, France, Ghana, Guatemala, Holy See, Hungary, India, Indonesia, Iran, Iraq, Israel, Italy, Japan, Jordan, Khmer Republic, Lebanon, Mexico, Monaco, Netherlands, Nicaragua, Nigeria, Pakistan, Peoples Republic of the Congo, Peru, Philippines, Poland, Romania, Rwanda, San Marino, Senegal, Sweden, Switzerland, Tunisia, Turkey, Soviet Republic of Belarus, Soviet Republic of Ukraine, Spain, Sudan, Tanzania, Togo, Union of Soviet Socialist Republics, United Kingdom, United States of America, Uruguay, Venezuela, Vietnam and Yugoslavia. Several countries sent natural heritage experts.

77   Canada Research Chair, interview Bolla.

insert its ideas into the UNESCO draft. Batisse did acknowledge the importance of the American contribution, remarking that the experts worked with the comparative table of the two proposals[78] and that "this comparison turned out to be positive because certain points could be studied more deeply."[79]

Bolla describes the three-week meeting as tense and arduous. Over 60 member states of UNESCO participated in 27 plenary sessions.[80] There were differences among governments with regard to the threshold of heritage value, the concepts of authenticity and integrity, and the funding formula. Much of the first week was spent in corridor discussions about the idea itself. Could cultural properties and natural heritage be combined into a single convention? Several draft amendments proposing parallel provisions for nature and for culture were rejected in favour of a combined approach. Cultural experts greatly outnumbered natural experts. According to Bolla, it was the majority cultural group that needed convincing. Both Bolla and Batisse credit Parent with bringing together "the partisans and adversaries of a marriage between culture and nature."[81] Eventually, two working groups were created to tackle the definitions of cultural and natural heritage. As Bolla tells it, once they got started, it only took about 48 hours to complete the task. The IUCN articles on natural heritage were incorporated into the UNESCO draft.[82] At the end of the three weeks, an astonishingly short period of time, the experts had produced the essence of a final text. Bolla credits the Americans, especially Salans, for having simplified and clarified the wording. It is this negotiated text that the Stockholm summit endorsed and recommended to governments for approval within the UNESCO system.[83]

Adopting the World Heritage Convention

At the 1972 General Conference of UNESCO, the proposal was first discussed at Commission V for General Programme Matters under the chairmanship of

---

78   UNESCO, Comparative table of the provisions of the revised draft convention concerning the protection of monuments, groups of buildings and sites of universal value, submitted by the Director-General of UNESCO, and the provisions of the World Heritage Trust draft convention concerning the preservation and protection of natural areas and cultural sites of universal value, submitted by the United States of America, SHC-72/conf.37/inf. 3. Retrieved from http://whc.unesco.org/archive/1972/shc-72-conf37-inf3e.pdf

79   Batisse and Bolla, *Invention*, p. 31.

80   UNESCO, Draft report of the special committee of government experts to prepare a draft convention and a draft recommendation to member states concerning the protection of monuments, groups of buildings and sites, Paris, 4–22 April 1972, Paris, 21 April 1972, SHC.72/conf.37/19, para. 2. Retrieved from http://whc.unesco.org/archive/1972/shc-72-conf37-19e.pdf

81   Batisse and Bolla, *Invention*, p. 75.

82   Holdgate, *Green*, p. 114.

83   Batisse and Bolla, *Invention*, p. 32. At the Stockholm summit, Batisse led the UNESCO contingent and Train headed the American delegation.

Jean Thomas from France. Although the key objectives were not in dispute, Bolla describes the debate as "violent" to the point that "a procedural battle and a profound divergence on the nature of the States' commitments risked making the whole thing collapse."[84]

The main sticking point was the funding formula, specifically the question of whether contributions to the proposed World Heritage Fund should be on a compulsory or a voluntary basis.[85] The vast majority, mainly developing countries, wanted compulsory contributions of 1 per cent of a member state's annual contribution to the overall UNESCO budget. In support of voluntary contributions were mostly developed countries, including the United States of America, the United Kingdom, Canada, the Federal Republic of Germany and the Soviet bloc. While many believe that the United States opposed an obligatory contribution because it would commit them to paying too much, internal government correspondence reveals the opposite. In a 1974 letter, R.G. Sturgill of the United States National Park Service's Office of International Activities writes to a colleague that the Americans fought hard for the voluntary, as opposed to the involuntary, formula for contributions to the fund, based on a judgement that the involuntary formula would not produce sufficient funds to make the Convention a success, since money would come only from governments in the amount of 1 per cent of their total contribution to UNESCO.[86] Failure to reach an agreement on the funding formula led the Director-General of UNESCO to propose postponing a vote on that specific article to the 1974 General Conference, and by implication delaying approval of the entire Convention. A compromise was eventually reached at Commission V.

When the negotiated text was tabled for endorsement at the General Conference, the anticipated easy passage was disrupted by a proposal from the United Kingdom to amend the contribution formula. Its proposed amendment destabilized the compromise text and resulted in an impasse. During the acrimonious debate that ensued, the American delegation suddenly changed its position "like a thunder bolt in a clear sky" according to Bolla.[87] What followed next is described by Canadian delegate Peter Bennett:

> Whereupon to everyone's amazement and without any prior warning the U.S.A. said it would accept the compromise text ... and requested an immediate vote on the Convention. This took place and the Convention was adopted by 79 votes to 1 with 16 abstentions (Canada, the U.K., the Soviet bloc and several western

---

84   Batisse and Bolla, *Invention*, pp. 80–81.

85   Federico Lenzerini, "Articles 15–16 World Heritage Fund," *The 1972 World Heritage Convention: a Commentary*, eds. Francesco Francioni and Federico Lenzerini (Oxford, 2008), pp. 271–7.

86   R.G. Sturgill, letter to E.A. Connally, 26 April 1974, US National Park Service, World Heritage archives, file 1973–1975.

87   Batisse and Bolla, *Invention*, p. 85.

European States). This American volte-face won them no friends and did not
enhance their reputation at the Conference. It was worth noting that the next
morning the U.S. delegate could be seen apologizing humbly to all and sundry
for his delegation's precipitate change of mind. [88]

Irritation with the American delegation is confirmed by Bolla, who noted that
among others the German delegate was not happy that the Americans had "cut
the grass from under his feet."[89] With American support declared, the General
Conference immediately adopted the Convention concerning the Protection of the
Cultural and Natural Heritage in the evening of 16 November 1972 along with the
Recommendation concerning the Protection, at National Level, of the Cultural and
Natural Heritage.

The idea of World Heritage emerged in the 1960s, an era that witnessed an
effervescence of public policy in social, cultural and environmental fields. This
creative period set a benchmark that has not been replicated since that time. The
1972 World Heritage Convention marks one of the last global agreements to
put forward the concept of universal values and international obligations in the
heritage field. It fills an important gap in the international system of agreements
and programmes that protect places of cultural and environmental importance.
The United States of America made the first ratification in 1973 and Switzerland
became the twentieth State Party on 19 September 1975. The Convention then
came into force three months later on 19 December 1975, setting the stage for its
first General Assembly of States Parties in 1976.

---

88   P.H. Bennett, Report on 17th UNESCO General Conference agenda items 22 and
23 on conservation of the cultural and natural heritage at world and national levels,
December 1972, Parks Canada, World Heritage archives, file 1972–1978, p. 3.

89   Canada Research Chair, interview Bolla. "Finalement, le texte de compromis a été
soumis à la Commission: le premier à parler contre, ça a été le délégué de la République
fédérale d'Allemagne en disant qu'il ne voulait pas d'obligatoire et, tout à coup, le délégué
américain – dont je ne me souviens plus le nom, qui n'était pas un spécialiste – s'est
levé, bon, et a dit qu'il acceptait le texte de compromis. Alors, je dois dire que le délégué
permanent allemand n'était pas très content qu'on lui ait coupé l'herbe sous les pieds. "

Chapter 2
Process for Identifying World Heritage Sites

The genius of the World Heritage Convention is its combination of natural and cultural heritage under one global instrument, an unusual approach in the 1970s when most governments dealt with culture and nature in separate departments. This chapter documents the early implementation of the Convention up to about 1980, a period when the focus was almost entirely on the creation of the World Heritage List.

The Convention itself covered much more than just the establishment of the World Heritage List. Of its 29 operational clauses, four articles set up the statutory bodies charged with implementation: the General Assembly of States Parties, the World Heritage Committee and its Bureau, the advisory bodies and the UNESCO secretariat.[1] To attain its international cooperation objectives, the Convention devotes almost half of its operational clauses to either collecting or spending money.[2] Only eleven articles deal specifically with heritage: articles 1 and 2 provide definitions of World Heritage of outstanding universal value; articles 3 and 4 assign responsibility to States Parties for identifying and managing World Heritage; article 5 makes States Parties responsible for the care of all heritage in their territory; articles 6 and 7 identify the duty to cooperate internationally; articles 11 and 12 call for the making of inventories, a list of World Heritage properties and a list of World Heritage in Danger; article 27 encourages education and raising awareness; and article 29 sets out reporting requirements.

As soon as the Convention came into force, UNESCO was faced with the challenge of designing an effective programme. Bernd von Droste, then a junior ecologist at UNESCO who volunteered to help with implementation, describes the situation in 1976 when the first World Heritage Committee of 15 members was elected:

> The criteria had to be fixed, the Operational Guidelines had to be set up as well as the rules of procedures of the Committee. And since this was a lot of work and UNESCO [had] made no financial provisions whatsoever ... people were asked on a voluntary base to serve. And actually for the Cultural Sector there was no voluntary basis really needed to be identified since they had a legal standards

---

1   UNESCO, Convention concerning the protection of the world cultural and natural heritage, Paris, 16 November 1972, articles 8–10, 14. Retrieved from http://whc.unesco.org/archive/convention-en.pdf The development of each of these organs is covered in chapter 5.

2   UNESCO, Convention concerning the protection of the world cultural and natural heritage, Paris, 16 November 1972, articles 13, 15–26, 28. Retrieved from http://whc.unesco.org/archive/convention-en.pdf

section which was headed by a program specialist, Anne Raidl, at this time. So, Anne Raidl, ex officio, got on that duty to assist in the launching of the World Heritage Convention after it came into force. For the Science Sector which had to deal with the natural part, there was no obvious choice and no one wanted to do it, particularly since the World Heritage Convention was seen as something very esoteric and non-scientific. Particularly this was true also from the point of view of the Man and Biosphere programme and the point of view of the Division of Ecological Sciences.[3]

From the outset, the creators of the Convention envisaged a selective international list of sites possessing outstanding universal value. The process for choosing exceptional sites was prepared under the leadership of UNESCO and the technical advisory bodies identified in the Convention text. The process includes the development of selection criteria and other requirements for inscribing properties on the World Heritage List.

Vision for a World Heritage List

It is worth recalling the original vision for a World Heritage List. In keeping with the notion of a restricted number of sites, IUCN in its initial proposal recommended a short list based on a selection from the 1962 *United Nations List of Protected Areas and Equivalent Reserves.* The 1965 American initiative for a World Heritage Trust called for "those few areas and sites that meet the high standards ... only those areas and sites that are absolutely superb, unique and irreplaceable."[4] UNESCO's 1966 General Conference foresaw "a few of the monuments that form an integral part of the cultural heritage of mankind" within a comprehensive international system with two components: a national protection system to cover all cultural properties and an international protection system for selected monuments and sites of universal interest that were in need of technical or financial assistance.[5] The 1971 American draft convention text reiterated the idea of a restrictive list of the most significant places selected from a comprehensive global inventory of sites.[6]

When the World Heritage Convention came into effect in December 1975, the States Parties faced the challenge of creating guidelines and procedures to

---

3   Canada Research Chair on Built Heritage, Université de Montréal, audio interview of Bernd von Droste by Christina Cameron and Mechtild Rössler, Paris, 5 April 2007.

4   John F. Kennedy Presidential Library and Museum, Samuel E. Belk personal papers, box 10, National Citizens' Commission, Report of the Committee on Natural Resources Conservation and Development to the White House Conference on International Cooperation, Washington, D.C., 28 November – 1 December 1965, pp. 17–19.

5   UNESCO, Records of the General Conference fourteenth session, Paris, 1967, p. 61. Retrieved from http://unesdoc.unesco.org/images/0011/001140/114048e.pdf

6   Chapter 1 provides a detailed explanation of the three proposed conventions.

implement it. To initiate the process, UNESCO organized two preparatory meetings in Morges (May 1976) and Paris (March 1977) involving representatives from the three advisory bodies named in the Convention: the International Union for Conservation of Nature and Natural Resources (IUCN), the International Council for Monuments and Sites (ICOMOS), and the International Centre for the Study of the Preservation and the Restoration of Cultural Property (ICCROM). At these meetings, representatives of the advisory bodies were exclusively from Europe and North America.[7] In addition, the March 1977 meeting in Paris also included "a small number of experts from different regions of the world."[8] In terms of a vision for a World Heritage List, the Convention simply states that it would be composed of sites having outstanding universal value. It was left to the preparatory group to articulate general principles for establishing such a list. The vision, remarkably consistent during this early period, was of a list that would have no formal limit but would be highly selective with a good balance of cultural and natural sites.

In their technical papers, the advisory bodies in effect recommended approaches for transforming this vision into operational reality. While ICOMOS foresaw "a limited number of properties," it also encouraged the Committee to interpret the word "universal" in a nuanced way, not restricting it to the best known properties but also considering less well known properties with aesthetic, educational and scientific value.[9] From a slightly different perspective, IUCN justified a "severe limitation of areas" on the grounds that limited funding and assistance should be focused on the highest priority sites. According to IUCN:

> Areas to be considered under the Convention will be restricted to those relatively few which are truly of international significance ... The World Heritage, however, is not intended to provide protection and recognition to all proposed areas, but only to provide protection and recognition for those relatively few which can be identified as unquestionably having international significance.

---

7   UNESCO, Final report of informal consultation of intergovernmental and non-governmental organizations in the implementation of the Convention concerning the protection of the world cultural and natural heritage, Morges, 19–20 May 1976, CC-76/WS/25, annex I. Retrieved from http://unesdoc.unesco.org/images/0002/000213/021374eb.pdf

8   UNESCO, Issues arising in connection with the implementation of the World Heritage Convention, Paris, 9 June 1977, CC-77/conf.001/4, para. 2. Retrieved from http://whc.unesco.org/archive/1977/cc-77-conf001-4e.pdf; in spite of an extensive archival search, the identity of the experts has not been found.

9   UNESCO, Final report of informal consultation of intergovernmental and non-governmental organizations in the implementation of the Convention concerning the protection of the world cultural and natural heritage, Morges, 19–20 May 1976, CC-76/WS/25, annex III. Retrieved from http://unesdoc.unesco.org/images/0002/000213/021374eb.pdf The ICOMOS paper is reproduced in Jukka Jokilehto et al., *The World Heritage List: What is OUV? Defining the Outstanding Universal Value of Cultural World Heritage Properties* (Berlin, 2008), pp. 58–61.

> This is not to deny that all appropriate areas may warrant proper protection and recognition, but it does strictly limit the areas which should be considered under the Convention.[10]

IUCN cited several reasons for suggesting a selective list, including the need to provide rapid response to "threatened areas of highest world priority," the relatively small budget available through the Convention, the likelihood of success with "a small but well balanced programme" and the expectation that international funds would be available only "for areas of highest international significance."[11] Furthermore, IUCN pointed to other existing mechanisms that provided an integrated international system to conserve sites and species of national and international significance including the Pan-American, African and Wetlands Conventions, the *United Nations List of Protected Areas and Equivalent Reserves* and the Biosphere programme. The preparatory group's final report to the Committee expressed the hope that "only those properties which were, without doubt, of true international significance, would be included."[12]

The preparatory group's 1977 document proposed ideas and text which found their way into the Committee's decisions and then to its Operational Guidelines, a formal document that promulgated World Heritage concepts and conservation approaches to a broader public. The Operational Guidelines was a key document for transmitting Committee direction. It is arguable that in the pre-2000 period, given the state of electronic communications, that very few people at the national or site level would have seen other Committee documents and reports. The Operational Guidelines therefore made available the rules for the implementation of the Convention. In that sense, the Operational Guidelines had a greater impact on heritage practice than any other Committee documents.

The 1977 working document influenced the initial version of the Operational Guidelines particularly with regard to the size of the List, its exclusive character and the definition of "universal." The working document emphasized the need for "a gradual process as the proposed criteria are tested and become more clearly defined; this does not imply, however, that any formal limit should be imposed either on the total number of properties included on the List or on the

---

10   UNESCO, Final report of informal consultation of intergovernmental and nongovernmental organizations in the implementation of the Convention concerning the protection of the world cultural and natural heritage, Morges, 19–20 May 1976, CC-76/WS/25, annex IV. Retrieved from http://unesdoc.unesco.org/images/0002/000213/021374eb.pdf

11   UNESCO, Final report of informal consultation of intergovernmental and nongovernmental organizations in the implementation of the Convention concerning the protection of the world cultural and natural heritage, Morges, 19–20 May 1976, CC-76/WS/25, annex IV. Retrieved from http://unesdoc.unesco.org/images/0002/000213/021374eb.pdf

12   UNESCO, Final report of informal consultation of intergovernmental and nongovernmental organizations in the implementation of the Convention concerning the protection of the world cultural and natural heritage, Morges, 19–20 May 1976, CC-76/WS/25, para. 5. Retrieved from http://unesdoc.unesco.org/images/0002/000213/021374eb.pdf

number of properties any individual State can submit."[13] It further stated that the Convention "is not intended to provide for the protection of all properties and areas of great interest, importance, or value, but only for a select list of the most outstanding of these from a world viewpoint."[14] With regard to different views of the term "universal," the document advised that it meant recognition by a "large or significant segment of humanity."[15] The first session of the World Heritage Committee approved these recommendations almost verbatim for inclusion in its Operational Guidelines. "Several members felt strongly that the World Heritage List should be exclusive and that, because of its impact, the List – in which balance would be sought geographically and between cultural and natural properties – should be drawn up with extreme care."[16]

With regard to the size of the List, the text in the first version of the Operational Guidelines comes directly from the working group document:

> Cultural and natural properties shall be included in the World Heritage List according to a gradual process and no formal limit shall be imposed either on the total number of properties included in the List or on the number of properties any individual State can submit at successive stages for inclusion therein.[17]

For the selective character of the list, the text draws heavily on IUCN's technical paper:

> The Convention provides a vehicle for the protection of those cultural or natural properties or areas deemed to be of outstanding universal value. It is not intended to provide for the protection of all properties and areas of great interest,

---

13   UNESCO, Issues arising in connection with the implementation of the World Heritage Convention, Paris, 9 June 1977, CC-77/conf.001/4, para. 15. Retrieved from http://whc.unesco.org/archive/1977/cc-77-conf001-4e.pdf

14   UNESCO, Issues arising in connection with the implementation of the World Heritage Convention, Paris, 9 June 1977, CC-77/conf.001/4, para. 16. Retrieved from http://whc.unesco.org/archive/1977/cc-77-conf001-4e.pdf

15   UNESCO, Issues arising in connection with the implementation of the World Heritage Convention, Paris, 9 June 1977, CC-77/conf.001/4, para. 17. Retrieved from http://whc.unesco.org/archive/1977/cc-77-conf001-4e.pdf

16   UNESCO, Final report of the first session of the intergovernmental committee for the protection of the world cultural and natural heritage in Paris, 27 June–1 July 1977, Paris, 20 October 1977, CC-77/conf.001/9, para. 18. Retrieved from http://whc.unesco.org/archive/1977/cc-77-conf001-9e.pdf

17   UNESCO, Operational Guidelines for the implementation of the World Heritage Convention, Final report of the first session of the intergovernmental committee for the protection of the world cultural and natural heritage in Paris, 27 June–1 July 1977, Paris, 20 October 1977, CC-77/conf.001/8 rev., para. 5. Retrieved from http://whc.unesco.org/archive/1977/cc-77-conf001-8reve.pdf

importance, or value, but only for a select list of the most outstanding of these from an international viewpoint.[18]

Reflecting the complexity of the term "universal", the paragraph on this matter was the most modified between the preparatory group and the World Heritage Committee's decision but nonetheless drew inspiration from ICOMOS's technical paper:

> The definition of "universal" in the phrase "outstanding universal value" requires comment. Some properties may not be recognized by all people, everywhere, to be of great importance and significance. Opinions may vary from one culture or period to another and the term "universal" must therefore be interpreted as referring to a property which is highly representative of the culture of which it forms part.[19]

These texts remained unchanged for years and some portions still survive in contemporary versions of the Operational Guidelines. They underscore a remarkable consensus in the early days for an open-ended World Heritage List comprised of a highly selective group of cultural and natural sites.

Construction of Outstanding Universal Value

At the heart of the World Heritage system is the identification of eligible properties. When the Convention came into effect, the States Parties had no operational tools to identify sites having outstanding universal value. How would sites be selected? The Convention is clear that the threshold for inclusion is outstanding universal value. But it must be observed that the term "outstanding universal value" is used thirteen times in the English version of the Convention but never defined.

Words are important. In this regard, it is interesting to note that there is a subtle shift in wording in the Convention text from "world" in the title "Convention for the Protection of the World Cultural and Natural Heritage" to "outstanding universal value" in the "Intergovernmental Committee for the Protection of the Cultural and Natural Heritage of Outstanding Universal Value, called the 'World

---

18   UNESCO, Operational Guidelines for the implementation of the World Heritage Convention, Final report of the first session of the intergovernmental committee for the protection of the world cultural and natural heritage in Paris, 27 June–1 July 1977, Paris, 20 October 1977, CC-77/conf. 001/8 rev., para. 5. Retrieved from http://whc.unesco.org/archive/1977/cc-77-conf001-8reve.pdf

19   UNESCO, Operational Guidelines for the implementation of the World Heritage Convention, Final report of the first session of the intergovernmental committee for the protection of the world cultural and natural heritage in Paris, 27 June–1 July 1977, Paris, 20 October 1977, CC-77/conf. 001/8 rev., para. 6. Retrieved from http://whc.unesco.org/archive/1977/cc-77-conf001-8reve.pdf

Heritage Committee'," (article 8.1) and a "Fund for the Protection of the World Cultural and Natural Heritage of Outstanding Universal Value, called 'the World Heritage Fund'"(article 15.1). While the term is not specifically defined in the text, articles 1 and 2 indicate that outstanding universal value is to be considered for cultural heritage from the point of view of history, art, science, ethnography or anthropology, and for natural heritage from the point of view of science, conservation or natural beauty.[20] The determination of outstanding universal value is left to the judgement of the World Heritage Committee using "such criteria as it shall have established" (article 11.2).

Establishing the Selection Criteria: 1976–1980

The two preparatory meetings organized by UNESCO in Morges and Paris laid the groundwork for formulating criteria to select sites with outstanding universal value. In their technical papers, the advisory bodies were instrumental in framing the concepts and wording for the criteria.[21] In an effort to tap a diversity of views, the conclusions of the first meeting, including proposed criteria, were circulated for comment "to well over 100 international experts throughout the world." In addition, ICOMOS invited its 55 national committees to comment.[22] Tabled as a working document at the first session of the World Heritage Committee, the results were influential in shaping the criteria. Following plenary debate, two working groups were created during the Committee session to further refine the work. The cultural heritage group was chaired by Michel Parent of France while the natural heritage group was led by David Hales from the United States of America.[23] The

---

20   There are subtle distinctions in the application of these terms, indicating an intention to have different approaches. For cultural heritage, the qualifiers for monuments and groups of buildings are history, art and science; for sites, the list is different, covering historical, aesthetic, ethnographical or anthropological aspects. For natural heritage, each of the groupings has a different point of view. Natural features have aesthetic or scientific qualifiers; geological and physiographical formations are evaluated through science or conservation; and natural sites are judged on the basis of science, conservation or natural beauty.

21   UNESCO, Informal consultation of intergovernmental and non-governmental organizations in the implementation of the Convention concerning the protection of the world cultural and natural heritage, Morges, 19–20 May 1976, CC-76/WS/25, annexes II, III and IV. Retrieved from http://unesdoc.unesco.org/images/0002/000213/021374eb.pdf The papers from ICOMOS and ICCROM are available in Jokilehto, *World Heritage List*, pp. 56–61.

22   UNESCO, Issues arising in connection with the implementation of the World Heritage Convention, Paris, 9 June 1977, CC-77/conf.001/4, para. 3. Retrieved from http://whc.unesco.org/archive/1977/cc-77-conf001-4e.pdf

23   UNESCO, Final report of the first session of the intergovernmental committee for the protection of the world cultural and natural heritage in Paris, 27 June–1 July 1977, Paris, 20 October 1977, CC-77/conf.001/9, paras. 31, 35 Retrieved from http://whc.unesco.

Figure 2.1   Michel Parent in about 1980

criteria, once approved, were published in the first version of the Operational Guidelines. The Committee made it clear that the criteria would probably need further refinement once they were applied to actual nominations.

As a result of this experience, the criteria went through a number of iterations between 1977 and 1980, an early indicator that the assessment of sites from diverse cultures and regions was a complex matter. Nonetheless, it is interesting to note that the fundamental concepts proposed by ICOMOS and IUCN in their 1976 technical papers survived through the various revisions. An important contribution was the 1979 analysis of the first 84 cultural nominations undertaken by Parent, then head of France's delegation and Rapporteur for the Committee.[24]

---

org/archive/1977/cc-77-conf001-9e.pdf; Peter H. Stott, "The World Heritage Convention and the National Park Service: The First Two Decades, 1972–1992," *The George Wright Forum*, 29/1 (2012), pp. 151–3.

24   Michel Parent, Comparative study of nominations and criteria for world cultural heritage, principles and criteria for inclusion of properties on the World Heritage List, Paris, 11 October 1979, CC-79/conf.003/11 annex. Retrieved from http://whc.unesco. org/archive/1979/cc-79-conf003-11e.pdf This paper was eventually modified slightly and published in Michel Parent, "La problématique du Patrimoine Mondial Culturel," *Momentum*, special issue (1984), pp. 33–49.

Cultural heritage criteria

In addition to editorial improvements, the significant changes in cultural criteria between 1976 and 1980 can be characterized as shifts from specific elements to general principles and from an architectural or geographical perspective to an anthropological point of view. Criterion (i) maintained the artistic dimension proposed by ICOMOS in 1976 but moved from achievements of architects and builders to expressions of creative genius. The 1980 version requires properties to "represent a unique artistic achievement, a masterpiece of the creative genius."[25]

Criterion (ii) sustained ICOMOS's idea of influence on developments over time and space while adding the urban dimension and replacing geographical with cultural regions, an exceedingly important distinction. The 1980 version states that properties should "have exerted great influence, over a span of time or within a cultural area of the world, on developments in architectural, monumental arts or town-planning and landscaping."[26]

Criterion (iii) evolved from a focus on antiquity, with its European overtones, to a more anthropological view of civilizations, reading in 1980 as properties that "bear a unique or at least an exceptional testimony to a civilisation which has disappeared."[27]

Criterion (iv) retains the initial concept of typologies from ICOMOS and ICCROM but shifts from the specificity of architectural styles, construction methods and settlement forms to a more general and inclusive formulation. In addition it removes the notion of threat and adds, as a result of Parent's concerns for comparative analysis, the requirement for historical context. The 1980 version of criterion (iv) requires a property to "be an outstanding example of a type of structure which illustrates a significant stage in history."[28]

The original ICOMOS idea for criterion (v) which emphasized traditional styles of architecture, construction methods and human settlements was heavily revised. While retaining ICOMOS's element of threat, criterion (v) shifts from

25 UNESCO, Operational Guidelines for the implementation of the World Heritage Convention, Report of the rapporteur on the fourth session of the World Heritage Committee in Paris, 1–5 September 1980, Paris, 29 September 1980, CC-80/conf.016/WHC/2 rev., para. 18. Retrieved from http://whc.unesco.org/archive/1980/opguide80.pdf

26 UNESCO, Operational Guidelines for the implementation of the World Heritage Convention, Report of the rapporteur on the fourth session of the World Heritage Committee in Paris, 1–5 September 1980, Paris, 29 September 1980, CC-80/conf.016/WHC/2 rev., para. 18. Retrieved from http://whc.unesco.org/archive/1980/opguide80.pdf

27 UNESCO, Operational Guidelines for the implementation of the World Heritage Convention, Report of the rapporteur on the fourth session of the World Heritage Committee in Paris, 1–5 September 1980, Paris, 29 September 1980, CC-80/conf.016/WHC/2 rev., para. 18. Retrieved from http://whc.unesco.org/archive/1980/opguide80.pdf

28 UNESCO, Operational Guidelines for the implementation of the World Heritage Convention, Report of the rapporteur on the fourth session of the World Heritage Committee in Paris, 1–5 September 1980, Paris, 29 September 1980, CC-80/conf.016/WHC/2 rev., para. 18. Retrieved from http://whc.unesco.org/archive/1980/opguide80.pdf

an architectural perspective to an anthropological one with a focus on traditional human settlements from different cultures. The 1980 version requires properties to "be an outstanding example of a traditional human settlement which is representative of a culture and which has become vulnerable under the impact of irreversible change."[29]

Criterion (vi) retains ICOMOS's notion of associative value but removes the eligibility of individuals from its scope, following Parent's observations on sites associated with "scholars, artists, writers or statesmen." After considering proposals for Edison's laboratory in the United States, Parent recommended that the list should not "become a sort of competitive Honours Board for the famous men of different countries" but should concentrate instead on the great works they had created.[30] In addition, he raised concerns about "extreme cases of areas which may have no tangible cultural property on them but which have been the scene of an important historical event." He referred to the particular cases of the slavery site at the Island of Gorée (Senegal) and the internment camp at Auschwitz (Poland). Although he conceded that the concept of listing "an 'idea' which haunts a historic place" was consistent with the Convention, he argued that priority should be given first to "concrete" cultural property.[31] In its discussions, the Committee recognized the potential political dimension of associative values and directed that this criterion should be used sparingly:

> Particular attention should be given to cases which fall under criterion (vi) so that the net result would not be a reduction in the value of the List, due to the large potential number of nominations as well as to political difficulties. Nominations concerning, in particular, historical events or famous people could be strongly influenced by nationalism or other particularisms in contradiction with the objectives of the World Heritage Convention.[32]

---

29   UNESCO, Operational Guidelines for the implementation of the World Heritage Convention, Report of the rapporteur on the fourth session of the World Heritage Committee in Paris, 1–5 September 1980, Paris, 29 September 1980, CC-80/conf.016/WHC/2 rev., para. 18. Retrieved from http://whc.unesco.org/archive/1980/opguide80.pdf

30   Michel Parent, Comparative study of nominations and criteria for world cultural heritage, principles and criteria for inclusion of properties on the World Heritage List, Paris, 11 October 1979, CC-79/conf.003/11 annex, p. 22. Retrieved from http://whc.unesco.org/archive/1979/cc-79-conf003-11e.pdf

31   Michel Parent, Comparative study of nominations and criteria for world cultural heritage, principles and criteria for inclusion of properties on the World Heritage List, Paris, 11 October 1979, CC-79/conf.003/11 annex, pp. 21, 24. Retrieved from http://whc.unesco.org/archive/1979/cc-79-conf003-11e.pdf

32   UNESCO, Report of the rapporteur on the third session of the World Heritage Committee in Cairo and Luxor, 22–26 October 1979, Paris, 30 November 1979, CC-79/conf.003/13, para. 35. Retrieved from http://whc.unesco.org/archive/1979/cc-79-conf003-13e.pdf

The 1980 version of criterion (vi) replaced "historical" with "universal" significance and added "directly or tangibly" as a further restriction that reinforced the tangible aspect of properties, thereby requiring sites to "be directly or tangibly associated with events or with ideas or beliefs of outstanding universal significance (the Committee considered that this criterion should justify inclusion in the List only in exceptional circumstances or in conjunction with other criteria)."[33]

Natural heritage criteria

At the 1976 preparatory meeting, IUCN presented selection criteria for natural sites.[34] This initial proposal survived largely intact in the period from 1976 to 1980 with the exception of efforts to add a cultural dimension, presumably to capture the Convention's fundamental principle of interaction between nature and people. This move by the Committee to add a cultural aspect probably confirmed some of IUCN's worst fears about the feasibility of receiving equal treatment for natural heritage under the new UNESCO Convention. Only some of these cultural additions were retained and were a source of later difficulties.

For criterion (i), IUCN's original proposal focused on evidence of the evolution of the earth through time. Included among the examples were sites that "would serve to demonstrate where natural and cultural heritage come together to illustrate the emergence of pre-man with the context of the plants, animals, climate and other factors influencing evolution."[35] The idea of connecting the environment with human evolution was retained in slightly different form at the first session of the Committee in 1977 but disappeared in the simplified 1980 version of criterion (i) which called on sites to "be outstanding examples representing the major stages of the earth's evolutionary history."[36] Subsequently, fossil hominid sites were considered under cultural criteria.[37]

---

33 UNESCO, Operational Guidelines for the implementation of the World Heritage Convention, Report of the rapporteur on the fourth session of the World Heritage Committee in Paris, 1–5 September 1980, Paris, 29 September 1980, CC-80/conf.016/WHC/2 rev., para. 18. Retrieved from http://whc.unesco.org/archive/1980/opguide80.pdf

34 UNESCO, Informal consultation of intergovernmental and non-governmental organizations in the implementation of the Convention concerning the protection of the world cultural and natural heritage, Morges, 19–20 May 1976, CC-76/WS/25, annex IV, pp. 2–3. Retrieved from http://unesdoc.unesco.org/images/0002/000213/021374eb.pdf

35 UNESCO, Informal consultation of intergovernmental and non-governmental organizations in the implementation of the Convention concerning the protection of the world cultural and natural heritage, Morges, 19–20 May 1976, CC-76/WS/25, annex IV, p. 2. Retrieved from http://unesdoc.unesco.org/images/0002/000213/021374eb.pdf

36 UNESCO, Operational Guidelines for the implementation of the World Heritage Convention, Report of the rapporteur on the fourth session of the World Heritage Committee in Paris, 1–5 September 1980, Paris, 29 September 1980, CC-80/conf.016/WHC/2 rev., para. 21. Retrieved from http://whc.unesco.org/archive/1980/opguide80.pdf

37 See for example South Africa's fossil hominid sites of Sterkfontein, Swartkrans, Kromdraai, and environs, inscribed under cultural criteria in 2005.

IUCN's proposal for criterion (ii) to recognize ongoing natural processes made no reference to culture. During the 1976 preparatory meetings, "cultural evolution" was added, using the example of terraced agricultural landscapes to illustrate interaction between human activities and the land.[38] This addition was retained in the 1980 version of criterion (ii) which required sites to "be outstanding examples representing significant ongoing geological processes, biological evolution and man's interaction with his natural environment," with emphasis on ongoing processes in the development of communities of plants and animals, landforms and marine and fresh water bodies. The example of agricultural terraces was not retained.[39] This criterion was later found to be inconsistent with the definition of natural heritage in the Convention text.[40]

Criterion (iii) underwent the most significant change, beginning with the IUCN proposal which covered only outstanding natural features but had a cultural dimension related to beauty added to it. This change, made in the course of the preparatory meetings, was a new, subjective idea expressed as "areas of exceptional natural beauty."[41] Approved by the 1977 Committee, this revised criterion (iii) is worded in 1980 as a requirement for sites to "contain superlative natural phenomena, formations or features or areas of exceptional natural beauty." The explanatory text refers to outstanding ecosystems and natural features as well as sweeping vistas and exceptional combinations of natural and cultural elements.[42] This criterion was also later found to be inconsistent with the definition of natural heritage in the Convention text.

Criterion (iv), virtually unchanged from IUCN's original proposal, focused on habitats for rare threatened species. The 1980 version called for properties to "contain the most important and significant natural habitats where threatened species of animals or plants of outstanding universal value from the point of view of science or conservation still survive."[43]

---

38   UNESCO, Issues arising in connection with the implementation of the World Heritage Convention, Paris, 9 June 1977, CC-77/conf.001/4, para. 22. Retrieved from http://whc.unesco.org/archive/1977/cc-77-conf001-4e.pdf

39   UNESCO, Operational Guidelines for the implementation of the World Heritage Convention, Report of the rapporteur on the fourth session of the World Heritage Committee in Paris, 1–5 September 1980, Paris, 29 September 1980, CC-80/conf.016/WHC/2 rev., para. 21. Retrieved from http://whc.unesco.org/archive/1980/opguide80.pdf

40   Chapter 3 explains the inconsistencies between article 2 of the World Heritage Convention and natural criteria (ii) and (iii).

41   UNESCO, Issues arising in connection with the implementation of the World Heritage Convention, Paris, 9 June 1977, CC-77/conf.001/4, para. 22. Retrieved from http://unesdoc.unesco.org/images/0003/000309/030934eb.pdf

42   UNESCO, Operational Guidelines for the implementation of the World Heritage Convention, Report of the rapporteur on the fourth session of the World Heritage Committee in Paris, 1–5 September 1980, Paris, 29 September 1980, CC-80/conf.016/WHC/2 rev., para. 21. Retrieved from http://whc.unesco.org/archive/1980/opguide80.pdf

43   UNESCO, Operational Guidelines for the implementation of the World Heritage Convention, Report of the rapporteur on the fourth session of the World Heritage Committee

Other Requirements for Inscription: 1976–1980

In addition to possessing outstanding universal value as determined through the application of six cultural and four natural criteria, sites were also to be evaluated on other grounds, including authenticity or integrity and protective measures. Cultural sites had to meet the conditions of authenticity and natural sites had to possess integrity. Management and conservation issues were largely ignored in this early phase.

Authenticity

The concept of authenticity in the World Heritage system originates with ICOMOS, although curiously the phrase used in its 1976 paper is "unity and integrity of quality". Immediately re-named as authenticity during the meeting, this concept was to be measured, according to ICOMOS, by the attributes of "setting, function, design, materials, workmanship and condition."[44] Through discussions in the preparatory group, the attributes of design, materials, workmanship and setting were retained but the attributes of function and condition were lost. In addition, the notion of valid contributions of all periods was added. The 1977 session of the Committee approved the following definition of authenticity:

> The property should meet the test of authenticity in design, materials, workmanship and setting. Authenticity does not limit consideration to original form and structure but includes all subsequent modifications and additions, over the course of time, which in themselves possess artistic or historical values.[45]

Almost immediately the meaning of authenticity was challenged by the incoming flood of nominations. In particular, Poland's proposal to list the reconstructed city centre of Warsaw forced further reflection on the meaning of authenticity.

---

in Paris, 1–5 September 1980, Paris, 29 September 1980, CC-80/conf.016/WHC/2 rev., para. 21. Retrieved from http://whc.unesco.org/archive/1980/opguide80.pdf

44   UNESCO, Informal consultation of intergovernmental and non-governmental organizations in the implementation of the Convention concerning the protection of the world cultural and natural heritage, Morges, 19–20 May 1976, CC-76/WS/25, annex III, p. 3. Retrieved from http://unesdoc.unesco.org/images/0002/000213/021374eb.pdf; Stovel contends that the initial concept of authenticity was influenced by the United States National Park Service definition in Herb Stovel, "Effective use of authenticity and integrity as World Heritage qualifying conditions," *City & Time*, 2/3, p. 23. Retrieved from http://www.ct.ceci-br.org

45   UNESCO, Operational Guidelines for the implementation of the World Heritage Convention, Final report of the first session of the intergovernmental committee for the protection of the world cultural and natural heritage in Paris, 27 June–1 July 1977, Paris, 20 October 1977, CC-77/conf.001/8 rev., para. 9. Retrieved from http://whc.unesco.org/archive/1977/cc-77-conf001-8reve.pdf

Figure 2.2   Reconstructed market square in Warsaw, Poland

Foreshadowing later debates,[46] the question focused on whether authenticity had to be measured through tangible attributes or whether it could be judged through intangible attributes alone. Warsaw had been destroyed during the Second World War and was painstakingly reconstructed as an act of national pride. Although the work was carried out with technical skill and scientific research, the city remained undeniably a reconstruction. In its advice to the Committee's Bureau in June 1978, ICOMOS questioned whether the historic centre of Warsaw met the test for authenticity and advised that "further expert opinion is required."[47] The Bureau deferred the nomination "for further expert evaluation with regard to the criterion of authenticity."[48] When the case was next considered in October 1979, opinion was divided in the Bureau. The discussion focused on whether the site met the requirement for authenticity and the fact that the secretariat had not

---

46   Christina Cameron, "From Warsaw to Mostar: the World Heritage Committee and Authenticity," *Bulletin of the Association for Preservation Technology*, 39/2–3 (2008), pp. 19–24.

47   ICOMOS, letter from Secretary General Ernest Allen Connally to Firouz Bagerzadeh, Chairperson of World Heritage Committee, Paris, 7 June 1978. Retrieved from http://whc.unesco.org/archive/advisory_body_evaluation/030.pdf

48   UNESCO, Report of the rapporteur on first meeting of Bureau of Intergovernmental Committee for the protection of the world cultural and natural heritage, Paris, 10 June 1978, CC-78/conf.010/3, para. 22. Retrieved from http://whc.unesco.org/archive/1978/cc-78-conf010-3e.pdf

received information on the proportion of original buildings restored, as distinct from reconstructed. At the same time, reference was made to the upcoming Parent report on general issues that had arisen from the first batch of nominations.[49] With regard to authenticity, Parent raised a question in his report:

> The Committee having laid down that <u>authenticity</u> is a <u>sine qua non</u> at first sight the WH List should not include a town or part of a town which has been entirely destroyed or reconstructed, whatever the quality of the reconstruction … the question is whether the latter could nevertheless be placed on the List because of the exceptional historical circumstances surrounding its resurrection.[50]

Noting that the ICOMOS Venice Charter prohibits reconstructions, Parent speculated on the differences between restoration and reconstruction. In the specific case of the "dilemma of Warsaw", Parent posed the question: "Can a systematic 20th century reconstruction be justified for inclusion on grounds, not of Art but of History?"[51] While promising to answer his question in the conclusion, he in fact never does so.

The 1979 Committee debated the general issues raised in Parent's report but not specifically the Warsaw nomination. Instead it established working groups to amend the criteria.[52] ICOMOS then made a surprising reversal. In its recommendation to the May 1980 Bureau, it changed its previous stance and recommended Warsaw for inscription, advising that authenticity might not be applied in its strict sense but stemmed from Poland's unique achievement in the years 1945 to 1966 in reconstructing a sequence of history running from the thirteenth to the twentieth centuries.[53] The Bureau endorsed ICOMOS's position by recommending the inscription of Warsaw "as a symbol of the exceptionally successful and identical reconstruction of a cultural property which is associated with events of considerable

---

49   UNESCO, Consideration of nominations to the World Heritage List, third session of Bureau of the World Heritage Committee, Paris, 1 October 1979, para 4, CC-79/conf.015/2. Retrieved from http://whc.unesco.org/archive/1979/cc-79-conf015-2e.pdf

50   Michel Parent, Comparative study of nominations and criteria for world cultural heritage, principles and criteria for inclusion of properties on the World Heritage List, Paris, 11 October 1979, CC-79/conf.003/11 annex, p. 19. Retrieved from http://whc.unesco.org/archive/1979/cc-79-conf003-11e.pdf

51   Michel Parent, Comparative study of nominations and criteria for world cultural heritage, principles and criteria for inclusion of properties on the World Heritage List, Paris, 11 October 1979, CC-79/conf.003/11 annex, p. 20. Retrieved from http://whc.unesco.org/archive/1979/cc-79-conf003-11e.pdf

52   UNESCO, Report of the rapporteur on the third session of the World Heritage Committee in Cairo and Luxor, 22–26 October 1979, Paris, 30 November 1979, CC-79/conf.003/13, para. 18. Retrieved from http://whc.unesco.org/archive/1979/cc-79-conf003-13e.pdf

53   ICOMOS World Heritage List no. 30, Paris, May 1980. Retrieved from http://whc.unesco.org/archive/advisory_body_evaluation/030.pdf

historical significance" but adding that "there can be no question of inscribing in the future other cultural properties that have been reconstructed."[54] The 1980 Committee listed Warsaw without further comment but amended the definition of authenticity to include a restriction on future reconstructions, an indication of the general unease regarding reconstructed properties.[55] Sites were then required to "meet the test of authenticity in design, materials, workmanship or setting (the Committee stressed that reconstruction is only acceptable if it is carried out on the basis of complete and detailed documentation of the original and to no extent on conjecture)."[56] From a technical point of view, the prescription on conjecture was probably not attainable.

Integrity

With regard to integrity, the 1976 IUCN technical paper proposed that the essence of integrity centred on the need for sites to have sufficient size to contain all or most of the key elements related to significance and continuity. The paper explicitly set out separate paragraphs and examples to explain the meaning of integrity as it applied to each criterion.[57] For example, areas representing major stages of the earth's evolutionary history were to "contain all or most of the key interrelated and interdependent elements in their natural relationships on the site" and areas representing endangered species should "have sufficient size and contain the necessary habitat requirements for the survival of the species."[58] The IUCN wording remained virtually intact for criterion (i) on evolutionary sites, criterion (ii) on ecosystems and criterion (iv) on habitats. A slight amendment was made to criterion (iii) on natural formations and places of natural beauty by the additional example of coral reefs. The 1980 Operational Guidelines also added a paragraph

54   UNESCO, Report of the rapporteur of the fourth session of the Bureau of the World Heritage Committee, 19–22 May 1980, Paris, 28 May 1980, CC-80/conf.o17/4, para. 4. Retrieved from http://whc.unesco.org/archive/1980/cc-80-conf017-4e.pdf

55   UNESCO, Report of the rapporteur on the fourth session of the World Heritage Committee in Paris, 1–5 September 1980, Paris, 29 September 1980, CC-80/conf.016/WHC/2 rev., para. 19. Retrieved from http://whc.unesco.org/archive/1980/cc-80-conf016-10e.pdf

56   UNESCO, Operational Guidelines for the implementation of the World Heritage Convention, Report of the rapporteur on the fourth session of the World Heritage Committee in Paris, 1–5 September 1980, Paris, 29 September 1980, CC-80/conf.016/WHC/2 rev., para. 18. Retrieved from http://whc.unesco.org/archive/1980/opguide80.pdf

57   UNESCO, Informal consultation of intergovernmental and non-governmental organizations in the implementation of the Convention concerning the protection of the world cultural and natural heritage, Morges, 19–20 May 1976, CC-76/WS/25, annex IV, pp. 2–3. Retrieved from http://unesdoc.unesco.org/images/0002/000213/021374eb.pdf

58   UNESCO, Informal consultation of intergovernmental and non-governmental organizations in the implementation of the Convention concerning the protection of the world cultural and natural heritage, Morges, 19–20 May 1976, CC-76/WS/25, annex IV, pp. 3–4. Retrieved from http://unesdoc.unesco.org/images/0002/000213/021374eb.pdf

to address integrity related to migratory species, an idea later subsumed into the paragraph on habitats.[59]

Management and Conservation

It is curious that the Committee's early Operational Guidelines made little mention of management and conservation issues. IUCN's 1976 technical paper raised some compelling management considerations as part of the inscription process. Taking its cue from the Convention (article 5), IUCN proposed that nominations ought to receive more favourable consideration in cases where institutional commitment, demonstrated capacity, trained personnel and financial resources were in place to manage natural areas for their world heritage values. IUCN further recommended a strict provision that anticipated many subsequent situations at World Heritage Sites:

> If the area, or in certain cases, adjacent lands, are programmed for uses which may conflict with nature conservation (lumbering, grazing, mining, hydroelectric projects) or may create negative influences (polluting industries, airline flyways, highways) it may be considered unreasonable to invest scarce World Heritage financial resources and to sacrifice time which will be lost in securing another area to fulfill the objectives.[60]

These suggestions were not retained by the working group. The ICCROM report to the same meeting focused on deterioration of cultural sites as opportunities for research to improve "the science of conservation in general."[61] The ICOMOS technical report made no mention of management or conservation requirements. The report of the preparatory group only looked at conservation issues from an educational perspective, picking up ICCROM's point by stating that "consideration should also be given to the state of preservation of the proposed property and to the opportunities afforded for scientific investigations and training in problems of preservation."[62]

---

59   UNESCO, Operational Guidelines for the implementation of the World Heritage Convention, Report of the rapporteur on the fourth session of the World Heritage Committee in Paris, 1–5 September 1980, Paris, 29 September 1980, CC-80/conf.016/WHC/2 rev., para. 22. Retrieved from http://whc.unesco.org/archive/1980/opguide80.pdf

60   UNESCO, Informal consultation of intergovernmental and non-governmental organizations in the implementation of the Convention concerning the protection of the world cultural and natural heritage, Morges, 19–20 May 1976, CC-76/WS/25, annex IV, pp. 4–5. Retrieved from http://unesdoc.unesco.org/images/0002/000213/021374eb.pdf

61   UNESCO, Informal consultation of intergovernmental and non-governmental organizations in the implementation of the Convention concerning the protection of the world cultural and natural heritage, Morges, 19–20 May 1976, CC-76/WS/25, annex II, pp. 2-3. Retrieved from http://unesdoc.unesco.org/images/0002/000213/021374eb.pdf

62   UNESCO, Issues arising in connection with the implementation of the World Heritage Convention, Paris, 9 June 1977, CC-77/conf.001/4, para. 21. Retrieved from

From 1976 to 1980, a demanding framework was put in place to guide the development of the World Heritage List, guided by a shared vision for a selective international list of sites. Under the leadership of UNESCO and with the advice of IUCN, ICOMOS and ICCROM, the World Heritage Committee formulated, tested and adjusted selection criteria to determine outstanding universal value as well as basic requirements for authenticity and integrity. Management and conservation were given scant attention and legal protective measures were not raised at all. It would appear that all parties were focused on the exciting prospect of populating the World Heritage List.

http://whc.unesco.org/archive/1977/cc-77-conf001-4e.pdf

Chapter 3

Populating the World Heritage List: 1978–2000

In the period up to 2000, the World Heritage List grew from a modest beginning of 12 sites in 1978 to hundreds of inscriptions as global participation increased and enthusiasm soared for obtaining the cherished brand of World Heritage Site. As inscriptions escalated, troubling questions arose about the size of the World Heritage List, an appropriate balance between cultural and natural sites as well as their equitable regional and cultural distribution. In response to these issues, the World Heritage Committee explored strategies for achieving a more credible, balanced and representative World Heritage List and took key policy decisions during the 1980s and 1990s related to landscapes and authenticity. These various initiatives were only marginally successful. By the end of the century, pressure continued to grow for a manageable and equitable system.

In the first two decades, the World Heritage List grew rapidly and unevenly. The Committee regularly debated the state of the list and its perceived problems. Statistics reveal only part of the picture but some factual evidence serves to frame any consideration of a fair and credible representation on the World Heritage List. Three data sets for the period 1978 to 2000 are particularly pertinent: the numerical increase in the number of properties on the list, distribution of sites by regions, and balance among cultural, natural and mixed sites.[1]

Populating the World Heritage List began cautiously and grew at a measured pace until the middle of the 1990s. In 1978, the first year of inscriptions, the Committee listed only 12 sites as it applied its draft criteria for the first time. The next year, 50 nominations flooded in and the Committee inscribed 45 of them. After that, inscriptions maintained a steady pace of about 25 to 30 a year, although nominations were rising more rapidly. For example, the 1987 session looked at 63 nominations and inscribed 41 of them on the World Heritage List. By 1996, the numbers surged beyond the capacity of the system to evaluate them, with 46 inscriptions in 1997 and 48 inscriptions in 1999. This escalation reached a crescendo at the 2000 session when 89 nominations were considered and 60 sites were inscribed. This unsustainable volume highlighted the urgent need for reform.

While number of sites is not the only way to measure balance among cultural, mixed and natural properties, statistics confirm beyond any doubt that cultural sites consistently dominated the World Heritage List. At least 70 per cent of the properties inscribed in any year were cultural, escalating to 80 and even 90 per

---

1   Up-to-date statistics on the World Heritage List are maintained by the World Heritage Centre of UNESCO. They are available at http://whc.unesco.org/en/list/stat

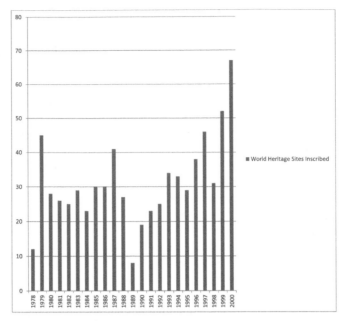

(Source: UNESCO, M. Rössler/A. Kanitkar)

Figure 3.1   Number of inscriptions per year to the World Heritage List 1978–2000

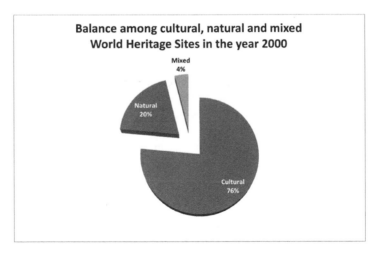

(Source: UNESCO, M. Rössler/A. Kanitkar)

Figure 3.2   Balance among cultural, natural and mixed World Heritage Sites in the year 2000

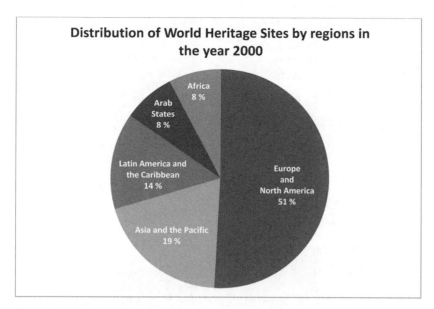

Distribution of World Heritage Sites by regions in the year 2000

(Source: UNESCO, M. Rössler/A. Kanitkar)

Figure 3.3   Distribution of World Heritage Sites by regions in the year 2000

cent in the 1990s. In 2000, the World Heritage List was composed of 76 per cent (529) cultural sites, 20 per cent (136) natural sites and 4 per cent (25) mixed sites.

With regard to regional distribution of World Heritage Sites, Europe and North America consistently dominated the list. At the start, over half of the inscribed sites came from this region. Although in the early 1980s, inscriptions from Europe and North America diminished to the 30 per cent range, they returned to over 50 per cent until the mid-1990s when they reached an astounding 60 to 70 per cent of all inscriptions. By the year 2000, when the World Heritage List had reached a total of 690 sites, Europe and North America dominated all other regions with 51 per cent (351 sites) of the entire list. Two other regions, Asia Pacific and to a lesser extent Latin America and the Caribbean, were successful in listing at least one property almost every year. By the year 2000, these two regions represented 19 per cent (135 sites) and 14 per cent (98 sites) respectively of the World Heritage List. The least well represented areas were the Arab and African regions, both of which started strongly in the first few years of implementation and then faltered. By the year 2000, the Arab and African regions each had only about an 8 per cent share (53 sites) of the list.

Figure 3.4   1981 World Heritage Committee in Sydney, Australia from left to right: Anne Raidl, Gérard Bolla and Margaret van Vliet

The Size of the World Heritage List

The creators of the World Heritage Convention envisaged a highly selective list of sites that could meet the demanding threshold of outstanding universal value.[2] With this objective in mind, the secretariat considered whether to jump-start the process with inscriptions or to begin more slowly by constituting inventories of proposed sites from States Parties as the basis for listing sites, as required by the World Heritage Convention (article 11.1). Anne Raidl, a lawyer employed in the Division of Cultural Heritage of UNESCO and active in World Heritage from 1976 until 1991, reports that the secretariat was divided:

> At that time, I found right from the first phase that it was all going to get out of hand, that there would be far too many inscriptions and that it would be better to follow the letter of the Convention text which foresaw in fact the submission of inventories from States Parties before the inscriptions. I was the only one to affirm this. Mr Bolla and Mr Batisse wanted to proceed forward much more quickly.[3]

---

2   Chapter 2 presents detailed evidence that all versions of draft conventions called for a highly selective list.

3   Canada Research Chair on Built Heritage, Université de Montréal, audio interview of Anne Raidl by Christina Cameron and Mechtild Rössler, Vienna, 28 February 2008.

While acknowledging the difficulty of determining when there would be enough State Party ratifications and completed inventories to proceed, Raidl believes that it was necessary from the beginning to limit the number of inscriptions.

Bernd von Droste, a UNESCO staff member from the Division of Ecological Sciences who was involved with the Convention from the outset, confirms Raidl's recollection that this discussion never went beyond the secretariat. He recalls that before and during the first Committee session in 1977 the secretariat suggested a limit of 100 natural and 100 cultural sites. Explaining this proposal on the grounds that the initial concept of outstanding universal value existed in a simpler form, von Droste says he was concerned at the time about "endless sub-categorization" which could justify any number of entries. He links the idea of restricting the list to the capacity of the support system, noting that the international community has limited resources:

> Conservation … is only operational if there is a boundary set by the numbers. The Convention doesn't say that there are not other properties of outstanding universal value, if they are not on the list. Yes, they are and they must be. But … the obligation of the Committee is to watch also that it remains practicable, credible, achievable, fundable and serviceable.[4]

Jane Robertson Vernhes, who began working at UNESCO in 1979, recalls her impression:

> judging from what Michel Batisse said in particular, that there was definitely an idea of some kind of a limit, that it was very much a prestigious list, and that the idea of really the stars of the world was there and then there was some kind of finite number, I think, attached to that, simply to keep this very high level.[5]

Rob Milne, whose involvement with the United States delegation goes back to the first Committee session in 1977, says that "early on no one anticipated approaching 1,000 properties on the list. I think that people were stretching their imaginations trying to think of the 100 foremost properties."[6]

"À ce moment-là, j'ai trouvé dès la première phase que tout ceci va déborder qu'il va y avoir beaucoup trop d'inscriptions et qu'il aurait mieux valu suivre à la lettre le texte de la Convention qui prévoyait en fait d'abord la soumission d'un inventaire de la part des États membres avant les inscriptions. Et j'ai été seule à affirmer ça. Monsieur Bolla et monsieur Batisse voulaient aller beaucoup plus rapidement en avant."

4 Canada Research Chair on Built Heritage, Université de Montréal, audio interview of Bernd von Droste by Christina Cameron and Mechtild Rössler, Paris, 5 April 2007.

5 Canada Research Chair on Built Heritage, Université de Montréal, audio interview of Jane Robertson Vernhes by Christina Cameron and Mechtild Rössler, Paris, 24 November 2009.

6 Canada Research Chair on Built Heritage, Université de Montréal, audio interview of Rob Milne by Christina Cameron and Mechtild Rössler, Paris, 2 March 2009.

Although the Committee itself did not set a numerical limit for the World Heritage List, concerns were raised early and often about the high number of nominations pouring in to the system. When the first sites were listed in 1978, the Committee debated whether the number of nominations per country and year should be limited.[7] It consistently wavered between the need for restrictions and unwillingness to impose a limit. In 1979, when a batch of 50 nominations arrived, the Committee acknowledged the problem but stated that "no limit on the number of nominations should be imposed."[8]

One explanation for the frenzied listing activity is offered by Azedine Beschaouch, a Tunisian archaeologist involved with World Heritage from 1978 to the present. He recalls the pressure in the early period to move quickly in order to demonstrate the importance of the Convention. He recounts how Belgian conservation architect Raymond Lemaire told him that "it's normal to build up the list ... one cannot promote a Convention with only 20 or 30 sites. You have to get to 100 or more." According to Beschaouch, this explains the Committee's generous interpretation of requirements for early inscriptions and its gradual tightening up of procedures as the list grew.[9]

The advisory bodies developed their own positions. IUCN undertook an evaluation within its own organization and proposed an informal limit for natural heritage sites. Hal Eidsvik, a staff member at IUCN from 1977 to 1980, describes the process for determining this recommendation:

At the time I was at IUCN, there had been no fixed numbers at that point in time. But when I left IUCN in 1980, we prepared something that the State Parties really didn't want and they really didn't look for, which was an inventory of potential World Heritage Sites, an indicative list if you would ... It was done in 1981 because it was a year after I left, but I'd done the preliminary work on it. Now, in preparing that list, I had been involved a year before in preparing what we called a *United Nations List of National Parks and Protected Areas* ... We sat down in IUCN and within the Commission on National Parks and we looked at that. Now what should the World Heritage Convention encompass as World Heritage Sites? And we look at the list and you say "well, it's not going to be 100

7 UNESCO, Report of the rapporteur on the second session of the World Heritage Committee in Washington, 5–8 September 1978, Paris, 9 October 1978, CC-78/conf.010/10 rev., para. 42. Retrieved from http://whc.unesco.org/archive/1978/cc-78-conf010-10reve.pdf

8 UNESCO, Report of the rapporteur on the third session of the World Heritage Committee in Cairo and Luxor, 22–26 October 1979, Paris, 30 November 1979, CC-79/conf.003/13, para. 34. Retrieved from http://whc.unesco.org/archive/1979/cc-79-conf003-13e.pdf

9 Canada Research Chair on Built Heritage, Université de Montréal, audio interview of Azedine Beschaouch by Christina Cameron and Mechtild Rössler, Brasilia, 28 July 2010. "Il m'a dit: c'est que maintenant, c'est normal on veut constituer une liste. On veut constituer les premiers...et on ne peut pas promouvoir une Convention avec seulement vingt ou trente biens. Il faut arriver à cent, etc."

per cent, it's not going to be 50 per cent, should it be 10 per cent, or should it be 1 per cent?" And the decision we arrived at at that time was that it was going to be somewhere in the area of 10 per cent. And that was just a judgment call; there was no scientific basis to it.[10]

The report from the 1981 Committee session confirms Eidsvik's recollection. At the Committee meeting, IUCN stated that "according to the criteria currently adopted, approximately 5 to 10 per cent of the 2000 natural areas which are listed on the *United Nations List of National Parks and Protected Areas* would meet the criteria for inscription on the World Heritage List." In other words, IUCN expected a total of between 100 to 200 natural heritage sites.[11]

Jeff McNeely, IUCN staff member responsible for World Heritage from 1980 to 1983, coordinated the study which was published as *The World's Greatest Natural Areas* (1982). Organized by bio-geographical realms, this first international inventory of superlative natural sites listed 219 properties which were intended, according to McNeely's foreword, "to illustrate the sorts of areas that might be considered of World Heritage quality; it is our hope that this list – far from exhaustive or complete – will stimulate additional ideas about outstanding sites and, more immediately, encourage nominations of the listed sites."[12] In his interview, McNeely reflected on the process, saying that IUCN "took the best sites early" but acknowledged that expansion was inevitable. "I don't think that you can defend any particular number; but I think that we should try to maintain the outstanding universal value aspect of the list."[13] Adrian Phillips, who chaired IUCN's World Commission on Protected Areas (WCPA) from 1994 to 2000, reflects on IUCN's initial assessment:

> That figure, whether it is 250 or 300, is not an arbitrary figure … it's based upon the kind of analysis which IUCN has done on marine sites, on forest sites, on polar sites, and so forth, which gives us a handle on how many sites are really of outstanding significance and would merit inclusion.[14]

Cultural properties presented an even greater challenge. Von Droste perceived initial confusion in the cultural sector, claiming that "the cultural people were

---

10   Canada Research Chair on Built Heritage, Université de Montréal, audio interview of Hal Eidsvik by Christina Cameron, Ottawa, 3 July 2009.

11   UNESCO, Report of the rapporteur on the fifth session of the World Heritage Committee in Sydney, 26–30 October 1981, Paris, 5 January 1982, CC-81/conf/003/6, para. 29. Retrieved from http://whc.unesco.org/archive/1981/cc-81-conf003-6e.pdf

12   IUCN Commission on National Parks and Protected Areas, *The World's Greatest Natural Areas: an indicative inventory of natural sites of World Heritage quality* (Gland, 1982), p. 4.

13   Canada Research Chair on Built Heritage, Université de Montréal, audio interview of Jeff McNeely by Mechtild Rössler, Gland, 17 September 2010.

14   Canada Research Chair on Built Heritage, Université de Montréal, audio interview of Adrian Phillips by Christina Cameron, London, 24 January 2008.

completely lost with the World Heritage Convention, completely. They spoke for hours and fighting with themselves, saying, should we take the best of the best, should we take the best of its category, should we take the most threatened one. They never thought about numbers."[15] His statement on numbers is contradicted by François Leblanc who worked at ICOMOS from 1979 to 1982. Leblanc recalls that in the early years ICOMOS speculated on an overall listing of between 3,000 and 6,000 cultural sites "especially if one considered that one day the list would open up to properties other than those which were considered from European countries. As soon as one opened to other regions of the world ... one would probably discover a good number of things that were not yet identified."[16] Léon Pressouyre, who presented the ICOMOS recommendations in these years, agrees. "I never heard anyone seriously uphold the idea of a fixed number. No one ever said, one must stop at 500, one must stop at 1,000, one must stop at 2,000. The idea was rather to test the waters as one went along."[17] Following a 1982 debate on the listing process, the Committee stood firm in its position that the concept of selectivity should only be based on the threshold of outstanding universal value. "Several delegates argued that the form of words used should not carry any suggestion of restriction on the range and variety of properties which might be inscribed in the List."[18]

By 1985 the system was obviously overloaded. The Assistant Director-General of UNESCO challenged the Committee to deal with "the growing number of nominations," indicating that it was a key issue that was not disappearing but escalating.[19] The Committee's Bureau had boldly recommended guidelines to reduce the number of nominations to be examined each year:

> a limitation by the Bureau of the overall number of properties examined each
> year (a maximum figure of 20 or 25 was put forward during the discussion);

---

15   Canada Research Chair, interview von Droste, 2007.

16   Canada Research Chair on Built Heritage, Université de Montréal, audio interview of François Leblanc by Christina Cameron, Ottawa, 7 April 2009. "Surtout si on considérait qu'un jour la Liste s'ouvrirait à des biens autres que ceux qui étaient considérés particulièrement par les pays européens. Dès qu'on s'ouvrirait sur d'autres régions du monde ... on découvrirait probablement un bon nombre de choses qu'on n'arrive pas à identifier."

17   Canada Research Chair on Built Heritage, Université de Montréal, audio interview of Léon Pressouyre by Christina Cameron and Mechtild Rössler, Paris, 18 November 2008. "Je n'ai jamais entendu personne soutenir sérieusement l'idée d'un 'numérus clausus', c'est-à-dire que personne n'a jamais dit, il faut s'arrêter à 500, il faut s'arrêter à 1000, il faut s'arrêter à 2000. Et l'idée était plutôt d'éprouver le mouvement en marchant. "

18   UNESCO, Report of the rapporteur on the fifth session of the World Heritage Committee in Sydney, 26–30 October 1981, Paris, 5 January 1982, CC-81/conf/003/6, para. 22. Retrieved from http://whc.unesco.org/archive/1981/cc-81-conf003-6e.pdf

19   UNESCO, Report of the rapporteur on the ninth session of the World Heritage Committee in Paris, 2–6 December 1985, Paris, December 1985, SC-85/conf.008/9, para. 4. Retrieved from http://whc.unesco.org/archive/1985/sc-85-conf008-9e.pdf

a limitation of the number of properties that each State Party would be authorized to present (2 properties for example);

a temporary and a voluntarily agreed halt in presenting new nominations by countries already having a high number of properties inscribed on the list.[20]

But the Committee rejected these measures outright, preferring to appeal to countries to restrict nominations voluntarily.[21] Two years later, a significant increase in cultural nominations again raised concerns about the coherence of the World Heritage List and the capacity of the system to support it.

Jim Collinson, Chairperson of the 1986 and 1987 sessions of the World Heritage Committee, speaks about his concerns at listing so many sites:

> My concern at the time was … that we not tie up too much time in the nomination process because there were other matters to be discussed including a review of existing sites and how they were doing and how they were being managed and whether there were any difficulties, whether there was anything that either the Committee could do directly by way of providing technical help or a minimum amount of money or whether it was something that the Member State might be asked to look into. And we didn't have as much time as I thought might have been warranted to cover those subjects. Nor did we have a lot of time to talk about the criteria and particularly the operating procedures. That led into a review of the operating procedures that got started at that time.[22]

He recalls that time constraints meant that some nominations "may not have gotten the attention they deserved" at the 1987 session which received over 60 nominations, 46 of them cultural sites. "The dilemma was, do you defer something for a year and leave the Member State dangling or do you deal with it as best you can and hope you come out with the right conclusion?"[23] The next year, UNESCO warned the Committee of the need "to better manage an increasing number of nominations so as to meet the fundamental objectives of the Convention, i.e. the protection of World Heritage properties, the monitoring of their state of

---

20   UNESCO, Report of the rapporteur on the ninth session of the Bureau of the World Heritage Committee, Paris, 3–5 June 1985, Paris, 12 August 1985, SC-85/conf.007/9, para. 16. Retrieved from http://whc.unesco.org/archive/1985/sc-85-conf007-9e.pdf

21   UNESCO, Report of the rapporteur on the ninth session of the World Heritage Committee, Paris, 2–6 December 1985, Paris, December 1985, SC-85/conf.008/9, para. 15. Retrieved from http://whc.unesco.org/archive/1985/sc-85-conf008-9e.pdf

22   Canada Research Chair on Built Heritage, Université de Montréal, audio interview of Jim Collinson by Christina Cameron, Windsor, 12 July 2010.

23   Canada Research Chair, interview Collinson.

conservation and the mobilization of resources to ensure this."[24] The Committee
stayed the course during its strategic review in 1992, reaffirming its view that there
should be no quantitative limit to the list but that a more thorough evaluation of
the nominations should be undertaken.[25] In addition to a more rigorous evaluation
process, another factor that affected the workload was the increasing demand on
States Parties for more elaborate nomination dossiers. Both these requirements
became more onerous in the period leading up to 2000, affecting the progress of
nominations and the ability of countries to produce satisfactory submissions.

In the mid-1990s, the situation grew worse according to Léon Pressouyre, who
by then was a member of the French delegation:

> Concern with the growth of the World Heritage List, in my opinion, is a recent
> phenomenon that became apparent starting in 1995 or 2000. That is when people
> began to reflect on where it was all going. Will it continue for long like this? Do
> we have the right methodology? Are we making stupid mistakes? But I did not
> feel this restriction in the beginning.[26]

1997 goes into the annals of the Convention's history as the year that forced
a fundamental change in handling nominations. UNESCO's Assistant Director-
General came to the 1997 session in Naples to speak bluntly to the Committee.
Warning that the reputation of the Convention was at stake, he asked for ideas on
how to improve universality while at the same time preventing "a rapid rise in the
total number of sites inscribed on the List."[27] The awkwardness of his admonition
was that the host country had been the State Party that had most contributed to
the crisis, nominating 12 sites that particular year.[28] Indeed Italy had originally

---

24   UNESCO, Report of the rapporteur on the twelfth session of the World Heritage
Committee in Brasilia, 5–9 December 1988, Paris, 23 December 1988, SC-88/conf.001/13,
para. 5. Retrieved from http://whc.unesco.org/archive/1988/sc-88-conf001-13e.pdf

25   UNESCO, Report of the rapporteur on the sixteenth session of the World Heritage
Committee in Santa Fe, 7–14 December 1992, 14 December 1992, WHC-92/conf.002/12,
annex II, paras. 16–17. Retrieved from http://whc.unesco.org/archive/1992/whc-92-
conf002-12e.pdf

26   Canada Research Chair, interview Pressouyre. "L'effarement devant la croissance
de la Liste du patrimoine mondial, à mon avis, est un phénomène récent qui ne s'est dessiné
qu'à partir des années 1995 ou 2000, voilà. C'est là que les gens ont commencé à réfléchir
en disant, où va-t-on? Est-ce qu'on va encore continuer longtemps comme ça? Est-ce qu'on
a la bonne méthode? Est-ce qu'on n'a pas fait des sottises? Et voilà. Mais je n'ai pas senti
au départ cette restriction."

27   UNESCO, Report of the rapporteur on the twenty-first session of the World
Heritage Committee in Naples, 1–6 December 1997, Paris, 27 February 1998, WHC-
97/conf.208/17, para. I.9. Retrieved from http://whc.unesco.org/archive/1997/whc-97-
conf208-17e.pdf

28   UNESCO, Examination of nominations at the twenty-first session of the Bureau
of the World Heritage Committee in Paris, 23–28 June 1997, Paris, 3 May 1997, WHC-97/

proposed 17 nominations and was persuaded by the Director of the World Heritage Centre to reduce the number.[29] Licia Vlad Borrelli, cultural heritage expert with the Italian delegation for over two decades beginning in 1982, notes that Italy presented "an enormous list" to the 1997 session of the Committee in Naples and says she disagreed at the time with this approach: "To inscribe a site requires thorough research; one would not have had time to do the research." She concludes that it would have been preferable to present the sites a few at a time.[30]

In their interviews, World Heritage pioneers often noted that the very size of the list marks the overwhelming success of the World Heritage Convention and its ability to engage countries and communities in heritage issues. Given the concerns raised over the years, they were asked to speculate on its eventual size. Bernard Feilden, renowned conservation specialist and Director-General of ICCROM from 1977 to 1981, believed that the World Heritage List "should be capped at about 1,000. Any list which grows beyond a fairly small number seriously risks de-valuation. When there were only about 300 or so worldwide, it was easy to convince people that designation was something really special and to be valued and guarded."[31] Henry Cleere, ICOMOS coordinator for World Heritage in the 1990s, remarks: "My gut feeling was that it's going to run out around about 1500 which seems to me to be a manageable number. Otherwise you are really going to move into not just the B list, but the C list and down to the Z list."[32] Von Droste is even stricter: "I would think IUCN is on the right track with maximum 200 sites and … culture perhaps 500 sites. I would somehow in this order of magnitude think that this is the maximum of the maximum … but we are very close to being completely overtaken or having over fed the system."[33] Jukka Jokilehto, architect and scholar with years of experience at ICCROM and later at ICOMOS, believes that there is a "natural limit" to the size of the list "which is difficult to foresee if it is 1,500 or 2,000 or 2,500, but I would say that I don't think that it can be limitless."[34]

Koïchiro Matsuura, Chairperson of the 1998 session of the World Heritage Committee and Director-General of UNESCO from 1999 to 2009, believes that the Committee will eventually have to face the question of a ceiling on the number of sites: "It is still too early. Nevertheless now possibly when the number of sites

---

conf.204/3B. Retrieved from http://whc.unesco.org/archive/1997/whc-97-conf204-3be.pdf

29 Personal communication, Alessandro Balsamo to Mechtild Rössler, 20 August 2012.

30 Canada Research Chair on Built Heritage, Université de Montréal, audio interview of Licia Vlad Borrelli by Christina Cameron and Mechtild Rössler, Rome, 6 May 2010. "Puisque pour inscrire un site il faut faire toute une recherche. On n'aurait pas eu le temps de faire la recherche."

31 Canada Research Chair on Built Heritage, Université de Montréal, written response to interview questions by Bernard Feilden, Bawburgh, Norwich, December 2007.

32 Canada Research Chair on Built Heritage, Université de Montréal, audio interview of Henry Cleere by Christina Cameron, London, 24 January 2008.

33 Canada Research Chair, interview von Droste, 2007.

34 Canada Research Chair on Built Heritage, Université de Montréal, audio interview of Jukka Jokilehto by Christina Cameron and Mechtild Rössler, Rome, 5 May 2010.

goes beyond 1,000 the Committee may ask experts to discuss it in a more serious manner ... I would like to introduce a note of caution. Be careful about increasing the number of World Heritage Sites." He further warns against "creating new sites which do not deserve the name of World Heritage Site. Then that would create a problem that would damage the credibility of the World Heritage List."[35]

Several other pioneers highlight the connection between the size of the list and its credibility. According to Collinson:

> I don't think you can simply keep adding sites and areas to the World Heritage List ... At some point it ceases to have a credibility of recognizing, to use the words of the natural criteria, 'superlative, best of the best, representative'. At some point you run out of that and you are starting to duplicate and override and I think that would be a tragedy.[36]

Francioni notes that the solution lies in strict implementation of the guidelines:

> The only way to approach this would be to seriously address the question of a rigorous implementation of the universal outstanding significance of the site. And I think the Convention provides all the tools for that, legal and political and scientific. The question is if we want to use them ... I mean, resisting the political pressure to present items that are already duplicated in the [list].[37]

Historian and archaeologist Guo Zhan, whose long career as a public servant in China included work with World Heritage after the country's ratification of the Convention in 1985, focuses on the quality of the sites. He judges that it is easier to limit the number of natural sites "because the natural world, we know it very well. We know which phenomena are very important in the history of our earth, of the natural biology." He finds the cultural challenge more difficult, "because of the diversity, because of the new perspective of the recognition of the OUV of the heritage but even then there still should be some limitation for that."[38]

Some reflect on the need to remain open to new types of heritage. Regina Durighello, whose involvement with ICOMOS evaluations of World Heritage nominations began in 1990, notes that "if the objective is to promote the protection of properties and the recognition of their outstanding universal value and the transmission of these values to future generations, I don't think that we should

---

35 Canada Research Chair on Built Heritage, Université de Montréal, audio interview of Koïchiro Matsuura by Christina Cameron and Mechtild Rössler, Paris, 24 November 2009.

36 Canada Research Chair, interview Collinson.

37 Canada Research Chair on Built Heritage, Université de Montréal, audio interview of Francesco Francioni by Christina Cameron and Mechtild Rössler, Rome, 5 May 2010.

38 Canada Research Chair on Built Heritage, Université de Montréal, audio interview of Guo Zhan by Christina Cameron, Brasilia, 3 August 2010.

think in terms of numbers but in terms of the quality and interest of sites."[39] Jean-Louis Luxen, ICOMOS Secretary General from 1993 to 2002, also focuses on the quality, not the quantity of inscriptions: "I believe that there are still sites in the world that have outstanding universal value and perhaps there are still sites that we cannot imagine today as qualifying for inscription and that one day they will be."[40]

Others relate the size of the list to the system's capacity for protection and monitoring. Herb Stovel, architect, professor and former staff member at ICOMOS and ICCROM, reflects on this link:

> I think we have to be open to the list continuing to grow, continuing to diversify. I think the numbers question … will sort itself out, because I think at some point the Convention will stop serving its larger purpose of conservation when the numbers get too big. If there are suddenly 2000 sites and UNESCO has no capacity to manage a monitoring system which looks at 2000 sites in any kind of meaningful way, I think the underlying purposes, the conservation purposes, are going to be in such bad repute. The Convention itself will become less meaningful and people will begin to pull away from it or it may fall apart or it may implode.[41]

Occasionally interviewees suggest ways to manage and even reduce the list. Natarajan Ishwaran, who began work in 1986 with UNESCO's Division of Ecological Sciences, notes that there are probably sites which, if nominated now, "would not make it" and asks: "So how would you address that question and how would you use diplomatic means to raise the question, discuss potential solutions and look for conciliatory diplomatically acceptable approaches? That's something that is very interesting."[42] Thorsell refers to an innovative proposal that he made to the 1992 World Parks Congress in Venezuela that World Heritage listing "should be valid for only 25 years after which the site undergoes another

39   Canada Research Chair on Built Heritage, Université de Montréal, audio interview of Regina Durighello by Christina Cameron and Mechtild Rössler, Paris, 25 November 2009. "Si l'objectif est de promouvoir la protection des biens et la reconnaissance de la valeur universelle exceptionnelle et la transmission de ces valeurs aux générations, je ne pense pas qu'il faille raisonner en termes de nombre mais en termes de qualité et d'intérêt des sites qui sont inscrits, des biens qui sont inscrits sur la Liste."

40   Canada Research Chair on Built Heritage, Université de Montréal, audio interview of Jean-Louis Luxen by Christina Cameron, Leuven, 26 March 2009. "Je crois que y a encore des sites dans le monde qui ont une valeur universelle exceptionnelle et peut-être que y encore des sites que nous n'imaginons pas aujourd'hui pouvoir être inscrits et qu'ils le seront un jour."

41   Canada Research Chair on Built Heritage, Université de Montréal, audio interview of Herb Stovel by Christina Cameron, Ottawa, 3 February 2011.

42   Canada Research Chair on Built Heritage, Université de Montréal, audio interview of Natarajan Ishwaran by Christina Cameron and Mechtild Rössler, Paris, 24 November 2009.

evaluation to see if it is still upholding the values, to see if there are any mistakes made in the beginning, to see if the integrity issues have gone away or if they've lost their integrity."[43]

By the late 1990s, the situation reached crisis proportions. UNESCO warned the 1998 session of the Committee that the system was about to collapse: "For 1999, there are 89 new nominations, breaking all records of the past. This poses a very serious problem, testing the capacities of ICOMOS and IUCN, as well as the Secretariat, Bureau and the Committee in giving each case the attention it merits."[44] The Committee finally took action. As part of the Cairns reform agenda of 2000, it endorsed a hotly-debated proposal to limit the number of nominations to a maximum of 30 new nominations per session and a limit of one nomination per country.[45]

Balance between Cultural and Natural Heritage Sites

Equitable representation of cultural and natural heritage sites is a fundamental premise of the World Heritage Convention. The 1978 Committee recognized this idea when it adopted the World Heritage symbol designed by Michel Olyff of Belgium. According to Australian Ralph Slatyer, a delegate at the meeting, "the square can be considered to represent the constructed, man-made cultural heritage linked to and surrounded by the globe of the world's natural heritage."[46] Yet when the 1978 Committee inscribed the first 12 sites on the World Heritage List, two-thirds were cultural properties, setting a pattern that persists to this day.

---

43   Canada Research Chair on Built Heritage, Université de Montréal, audio interview of Jim Thorsell by Christina Cameron, Banff, 11 August 2010; Hemanta Mishra and N. Ishwaran, "Summary and conclusions of the workshop on the World Heritage Convention held during the IV World Congress on national parks and protected areas Caracas, Venezuela, February, 1992," *World Heritage: twenty years later. Based on papers presented at the World Heritage and other workshops held during the IVth World Congress on National Parks and Protected Areas, Caracas, Venezuela, February 1992* (Gland and Cambridge, 1992), p. 16.

44   UNESCO, Report of the rapporteur on the twenty-second session of the World Heritage Committee in Kyoto, 30 November to 5 December 1998, Paris, 29 January 1999, WHC-98/conf.203/18, para. IV.3. Retrieved from http://whc.unesco.org/archive/1998/whc-98-conf203-18e.pdf

45   UNESCO, Report of the rapporteur on the twenty-fourth session of the World Heritage Committee in Cairns, 27 November–2 December 2000, Paris, 16 February 2001, WHC-2000/conf.204/21, para. VI.3.3. Retrieved from http://whc.unesco.org/archive/2000/whc-00-conf204-21e.pdf The annual limit has crept back up to 45 nominations with a maximum of two nominations per country.

46   Ralph O. Slatyer, "The Origin and Evolution of the World Heritage Convention," *Ambio*, 12/3–4 (1983), p. 139.

At the 1980 session, when the United States delegation raised concerns about this imbalance, the Committee adopted a number of measures to improve the situation. Some were aimed at attracting more nominations of natural sites, including the targeted use of preparatory assistance and implementation of the Convention's requirement for inventories or tentative lists of potential sites. Others focused on the Committee itself, in particular the rotation of the chairmanship every two years between cultural and natural heritage experts as well as compliance with the Convention's requirement that Committee representatives be "persons qualified in the field of cultural or natural heritage" (article 9.3). In support of this goal, IUCN announced its plan for a worldwide inventory of natural sites to guide countries in the preparation of their tentative lists.[47]

At its 1981 session, the Committee dodged the issue by adopting a platitudinous statement that the "World Heritage List should be as representative as possible of all cultural and natural properties which meet the Convention's requirement for outstanding universal value."[48] Two years later, the low number persisted and States Parties were encouraged to accelerate their nomination of natural heritage sites.[49]

Landscapes

One way of increasing the number of natural sites was to include more landscapes. At the 1984 session of the Committee in Buenos Aires, many countries welcomed IUCN's global inventory entitled *The World's Greatest Natural Areas*.[50] But it did not satisfy all countries because it limited itself to pristine natural areas untouched by human interaction. Densely populated countries complained that IUCN failed to list natural heritage sites where human beings had modified the environment to create ecologically balanced and culturally interesting landscapes. Could such properties be considered under the World Heritage Convention? This IUCN bias towards wilderness areas was mentioned by several pioneers. IUCN's

---

47　UNESCO, Report of the rapporteur on the fourth session of the World Heritage Committee in Paris, 1–5 September 1980, Paris, 29 September 1980, CC-80/conf.016/WHC/2 rev., paras. 6, 21, 47–9. Retrieved from http://whc.unesco.org/archive/1980/opguide80.pdf

48　UNESCO, Report of the rapporteur on the fifth session of the World Heritage Committee in Sydney, 26–30 October 1981, Paris, 5 January 1982, CC-81/conf/003/6, paras. 22, 29. Retrieved from http://whc.unesco.org/archive/1981/cc-81-conf003-6e.pdf

49　UNESCO, Report of the rapporteur on the seventh session of the World Heritage Committee in Florence 5–9 December 1983, Paris, January 1984, SC/83/conf.009/8, para. 25. Retrieved from http://whc.unesco.org/archive/1983/sc-83-conf009-8e.pdf

50　IUCN Commission on National Parks and Protected Areas, *The World's Greatest Natural Areas: an indicative inventory of natural sites of World Heritage quality* (Gland, 1982), pp. 1–69; UNESCO, Report of the rapporteur on the eighth session of the World Heritage Committee in Buenos Aires, 29 October–2 November 1984, Buenos Aires, 2 November 1984, SC/84/CONF,004/9, para. 20. Retrieved from http://whc.unesco.org/archive/1984/sc-84-conf004-9e.pdf

McNeely gives insight into the organization's approach to human activities within nominated areas. In speaking of the nomination of the Great Barrier Reef (Australia), he recalls that he found the boundaries too large "because there were activities that were taking place ... that really were not consistent with the ideals of World Heritage."[51]

Milne remembers an early incident that shows how nature was perceived differently in some regions of the world. He recalls the opening speech by the wife of Egypt's President Anwar Sadat at the 1979 session in Cairo:

> She addressed the Committee with the beauties of natural areas and the importance of natural heritage as she saw in her backyard garden with roses. And in a way that emphasized the humanity and the sensitivity and recognition of cultural values, aesthetic values, and in a way marginalized the significance and global importance of natural areas, natural ecosystems, of functioning in a natural way. None of us really at the time anticipated that would be reflected down through the years but in a way [it] cast a future of the Convention that has been played out very much along those lines.[52]

Pressouyre believes that IUCN's view to the effect that human beings are not deemed to be part of nature has roots in eighteenth-century philosophy, reinforced in the nineteenth century. Referring to his experience in evaluating nominations of sites with both natural and cultural values, he expresses his frustration at this interpretation of nature. In the case of the Giant's Causeway and Causeway Coast (United Kingdom) IUCN recognized the exceptional geological formation of black basalt columns sticking out of the sea but did not see the relevance of the intangible values in the site's association with Irish mythology. In the case of Meteora (Greece), IUCN judged that there was nothing special about the granite pillars supporting the monastery buildings. Pressouyre fundamentally disagrees on this point: "If you say there is no link between the cultural and natural values, you are making an enormous mistake because it is self-evident. It is set in the landscape. That is its essence."[53]

A member of the French delegation, Lucien Chabason introduced the 1984 Committee to the notion of rural landscapes, a key concept that has had a major influence on the implementation of the Convention. In an interview, Chabason recalls his thinking, pointing out that the vast landscapes depicted in American movies were usually in national parks:

---

51   Canada Research Chair, interview McNeely.

52   Canada Research Chair, interview Milne.

53   Canada Research Chair, interview Pressouyre. "Si vous dites qu'il n'y a pas un lien entre la valeur culturelle et la valeur naturelle, vous faites une énorme bêtise, vous faites une énorme bêtise parce que c'est évident. C'est inscrit dans le paysage, enfin et dans sa raison d'être."

There are questions that we as Europeans, and in particular old rural countries like France, Italy and the United Kingdom, did not find in this Convention. We are going to be in an uncomfortable situation because we do not have the same concept, in particular the American idea of wilderness. We are not going to find it in my country. We are going to find something else that must be put in the Convention.[54]

Chabason explains why he proposed rural landscapes:

Historically, since Neolithic times in Europe at any rate, man has greatly transformed the land to cultivate it, to make it habitable. In transforming the land he ... has modified the ecosystem and at the same time he has not created historical monuments. He has created a new land that often presents outstanding characteristics, for example the rice terraces of Java or the Philippines, which respond to the spirit of the Convention.[55]

In the view of some delegations, properties like rice terraces and vineyards were unlike the natural sites in IUCN's study because they were "exceptional combinations of natural and cultural elements," a phrase that came from natural criterion (iii). Robertson Vernhes, who drafted the minutes of the Buenos Aires meeting, recalls Chabason's intervention: "It was the idea of the human landscape ... I didn't think they even used the word cultural landscape ... I think the French were very much thinking of their own vineyards, having seen the terraces of the Philippines or Bali." Reflecting on the meaning, she adds: "This idea of the people in the environment, you know, that it wasn't savage nature, wilderness nature, and it wasn't the churches. It was something in between which was of World Heritage value."[56]

Vlad Borrelli, Chairperson of the 1983 session and a member of the Italian delegation at the 1984 session, recalls the problem of explaining the notion of humanized landscapes found in countries with a long history. Remarking on the originality of the Convention "in putting together cultural properties and the

---

54   Canada Research Chair on Built Heritage, Université de Montréal, audio interview of Lucien Chabason by Christina Cameron and Petra van den Dorn, Paris, 2 October 2012. "Il y a des questions que nous, les Européens, et en particulier les vieux pays ruraux, comme la France, l'Italie, et le Royaume Uni, nous n'avons pas retrouvé dans cette convention. Nous allons être dans une situation inconfortable parce que nous n'avons pas le même concept, en particulier le concept américain de 'wilderness'. Nous n'allons pas le trouvé dans mon pays. Nous allons trouver autre chose qu'il faut mettre dans la convention."

55   Canada Research Chair, interview Chabason. "Historiquement depuis le néolithique, en Europe en tout cas, l'homme a beaucoup transformé la terre pour la rendre cultivable, pour la rendre habitable, et en transformant la terre il ... a modifié l'écosystème. Et au même temps il n'a pas créé des monuments historiques, il a créé une nouvelle terre qui présente souvent des caractéristiques exceptionnels, par exemple les rizières de Java ou des Philippines, qui répondent à l'esprit de la convention."

56   Canada Research Chair, interview Robertson Vernhes.

environment," she says that when the Committee asked for natural site nominations, "at that moment it opened up the question of humanized natural properties because it was not easy to convince people who presented forests, huge forests, that ... a small site like Capri, for example, could be considered a natural property." [57]

The thoughtful debate at the 1984 session reveals the different reactions of ICOMOS, IUCN and UNESCO to a consideration of what was then called rural landscapes. UNESCO's Michel Batisse stated that in the spirit of the Convention "there should not be a polarization towards either 'culture' or 'nature' although there had perhaps been such a tendency in the past as States Parties had initially nominated the properties which clearly met either the cultural or natural criteria." ICOMOS gave a clear philosophical statement that "the role of the Convention was not to 'fix' such landscapes but rather to conserve their harmony and stability within a dynamic, evolutive framework." On the other hand, IUCN was cautious. While stating that it recognized protected landscapes, exemplified by the national parks of the United Kingdom which "consist essentially of man-modified and man-maintained landscapes," IUCN warned "that care should be taken in the identification of such landscapes to ensure the nomination of only those properties of outstanding universal value."[58] The Committee set up a task force of technical experts, including ICOMOS, IUCN and the International Federation of Landscape Architects (IFLA), to prepare a framework for the identification and nomination of rural landscapes.[59] This request opened an international theoretical discussion that took almost a decade to resolve and in a sense distracted the Committee from the balance issue at a time when two thirds of the listed sites were still cultural.

The task force reported in 1985 about its unexpected discovery of a serious flaw in the Committee's working tools by pointing to inconsistencies between the Convention text and the evaluation criteria. In its definitions of cultural and natural heritage, the Convention clearly leaves a place for natural features associated with cultural properties but refers only to the "aesthetic" value of natural features and the vague notion of "natural beauty" for natural sites. Article 1 identifies two circumstances in which natural attributes can be taken into account in the consideration of cultural properties: first for groups of buildings "because of their

---

57   Canada Research Chair, interview Vlad Borrelli. "Quand on nous a demandé de présenter des sites naturels, à ce moment là, a déclenché la question des biens naturels anthropisés parce que ça n'a pas été facile de faire comprendre à des gens qui présentaient des forêts, des énormes forêts que même un petit endroit comme Capri, par exemple, pouvait entrer entre les biens naturels."

58   UNESCO, Report of the rapporteur on the eighth session of the World Heritage Committee, Buenos Aires, 29 October–2 November 1984, Buenos Aires, 2 November 1984, SC/84/CONF,004/9, paras. 21–4. Retrieved from http://whc.unesco.org/archive/1984/sc-84-conf004-9e.pdf

59   UNESCO, Report of the rapporteur on the eighth session of the World Heritage Committee, Buenos Aires, 29 October-2 November 1984, Buenos Aires, 2 November 1984, SC/84/CONF,004/9, paras. 21–4. Retrieved from http://whc.unesco.org/archive/1984/sc-84-conf004-9e.pdf

place in the landscape" and secondly for sites that illustrate "the combined work of nature and man." Article 2 restricts the scope for natural heritage to scientific or conservation values with the exception of "natural features ... which are of outstanding universal value from an aesthetic or scientific point of view" and "natural sites or precisely delineated natural areas of outstanding universal value from the point of view of science, conservation or natural beauty."

When the Committee created its inscription criteria, it was not consistent with the Convention text and erroneously included in natural criterion (ii) "man's interaction with his natural environment" and in natural criterion (iii) "exceptional combinations of natural and cultural elements."[60] The task force only recommended changes to criterion (iii) to deal with the second inconsistency and suggested additions to the cultural criteria to accommodate rural landscapes.[61] Unsure of the implications of these proposals and cautious about the complexity of the issues, the Committee asked for further study and possible testing if an appropriate nomination came forward.[62] Von Droste accepts responsibility for these errors but points to ambiguity in the Convention text itself:

> It was our fault at the beginning when we formulated in 76 under the leadership of David Hales who was presiding the Commission on Natural Criteria. ... The whole question of beauty, of aesthetics was not entering into the cultural criteria but entering under the natural criteria. In particular the wonderful harmonious interplay between nature and culture came under natural heritage ... If you read the Convention of course you can see that the links between nature and culture are clearly expressed under the cultural part of the Convention, whereas ... the question of beauty and aesthetics only comes under the natural heritage criteria. So there is also some misleading text in the Convention itself but certainly the interpretation of the article 2 ... this translation into the criteria had not been done with full precision and led to endless debates on whether landscapes would come under natural or cultural criteria.[63]

The 1986 Bureau considered the same task force document and remained cautious. Concerned about the definition of rural landscapes, a potential

---

60   UNESCO, Operational Guidelines for the implementation of the World Heritage Convention, WHC/2 rev. January 1984, para. 24. Retrieved, para. 24. Retrieved from http://whc.unesco.org/archive/opguide84.pdf

61   UNESCO, Elaboration of guidelines for the identification and nomination of mixed cultural and natural properties and rural landscapes, Paris, 19 November 1985, SC/85/conf.008/3, paras. 2.2–2.8. Retrieved from http://whc.unesco.org/archive/1985/sc-85-conf008-3e.pdf

62   UNESCO, report of the rapporteur on the ninth session of the World Heritage Committee in Paris, 2–6 December 1985, Paris, December 1985, SC-85/conf.008/9, paras. 25–7. Retrieved from http://whc.unesco.org/archive/1985/sc-85-conf008-9e.pdf

63   Canada Research Chair, interview von Droste, 2007.

Figure 3.5   Lake District National Park, United Kingdom

proliferation in nominations and the problems of managing living landscapes, it felt that it was premature to recommend any alterations to the Committee's guidelines. It also welcomed the proposal from the United Kingdom to present a draft nomination of a rural landscape, the Lake District National Park, as a way to test the applicability of the proposed changes.[64]

For the next three years, the Committee tried to figure out how to deal with rural landscapes using the Lake District as its test case. At the June 1987 Bureau, ICOMOS supported inscription of the Lake District as a cultural property but IUCN was not able "to come to a conclusion as to whether this nomination met the criteria for natural properties since there was a debate within IUCN as to whether this was truly a 'natural' site in the sense of article 2 of the Convention (i.e. nature not modified by man)." The Bureau concluded that the test case had shown the need to reconsider the question of nominations "which contained a synergetic

---

64   UNESCO, Report of the rapporteur on the tenth session of the Bureau of the World Heritage Committee in Paris, 16–19 June 1986, Paris, 15 September 1986, CC-86/conf.001/11, paras. 8–11. Retrieved from http://whc.unesco.org/archive/1986/cc-86-conf001-11e.pdf

combination of cultural and natural elements" and requested a list of questions regarding "the strict application of the definition of cultural and natural heritage as set out in the Convention; the variety and distribution of rural landscapes which might be considered as having outstanding universal value; and the conditions which would have to be met to ensure long-term protection without 'fossilizing' a living rural landscape."[65]

As the Committee continued to flounder, UNESCO initiated work on how outstanding universal value might be interpreted for properties that had natural and cultural values in equal proportion, but not necessarily of exceptional value. Under Batisse's direction, an odd proposal emerged using diagrams of circles and squares to argue that a property that did not meet the threshold under either cultural or natural criteria could nonetheless be considered to possess outstanding universal value. Batisse explained that this category "has natural and cultural elements of considerable value which separately do not meet criteria but their *combination* could qualify the property as a *mixed site.*" He argued that rural landscapes appear to fall into this category "when they possess both cultural and natural attributes which, *by their combination*, offer something exceptional and of universal value."[66]

Case 4:

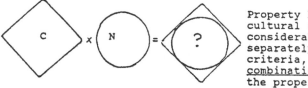

Property has natural and cultural elements of considerable value which separately do not meet criteria, but their combination could qualify the property as a mixed site.

Figure 3.6   Michel Batisse's proposal for rural landscapes presented to the 1987 World Heritage Committee

Robertson Vernhes, who worked on the proposal, describes the unease she felt at the time:

> I didn't like that idea personally because it didn't match my own vision of World Heritage which for me it is of outstanding universal value. Full stop. You don't

65 UNESCO, Report of the rapporteur on the eleventh session of the Bureau of the World Heritage Committee in Paris, 23–26 June 1987, Paris, 8 August 1987, SC-87/conf.004/11, paras. III 6B, 18. Retrieved from http://whc.unesco.org/archive/1987/sc-87-conf004-11e.pdf

66 UNESCO, Note on rural landscapes and the World Heritage Convention, Paris, 12 November 1987, SC-87/conf.005/inf. 4, paras. 5, 8. Retrieved from http://whc.unesco.org/archive/1987/sc-87-conf005-inf4e.pdf

Figure 3.7   **1990 World Heritage Committee in Banff, Canada from left to right: James Chamberlain (USA), Adul Wichiencharoen (Thailand), Magdalina Stantcheva (Bulgaria), Azedine Beschaouch (Tunisia), Christina Cameron (Canada, Chairperson), Licia Vlad Borrelli (Italy), Seydina Sylla (Senegal), Leslie Taylor (Mayor of Banff) and Salvador Diaz-Berrio (Mexico)**

have to be a halfway house put in another halfway house and come together with a whole house ... But this idea, the circles and the triangles and putting the halfway house together, that I was very unhappy with.[67]

The Committee continued to leave its decision open on the Lake District pending clarification on cultural landscapes, a new term that curiously replaced the term "rural landscapes" without explanation. Collinson seems to recall that the reason for this change was that rural was perceived as being too narrow. "There was a concern that rural left out the more remote areas. Rural implies an agrarian kind of society and use in many parts of the world whereas there are some natural areas with population nodes that taken together can still represent an integrated kind of landscape."[68] When in 1988 natural sites hit a new low of about 25 per cent of inscriptions, the Committee expressed regret over the persistent imbalance

---

67   Canada Research Chair, interview Robertson Vernhes.
68   Canada Research Chair, interview Collinson.

of the list. In light of IUCN's lack of enthusiasm for humanized landscapes, the Committee decided on an interim basis to assign to ICOMOS the sole responsibility for evaluating nominations with a combination of natural and cultural elements, using cultural criteria alone and giving IUCN a consultative role.[69] In effect, this decision further distorted the question of balance, since landscapes were then counted as "cultural" sites on the World Heritage List.

In 1990, during yet another consideration of the Lake District nomination, the Committee reached an impasse. Opinion was sharply divided and a passionate debate was eventually settled through a formal vote, a procedure rarely used at the time. In the end, the Lake District was not accepted as a World Heritage Site. The Committee concluded that it did not have sufficiently clear criteria to rule on this type of property and directed the secretariat to develop guidance for the next session.[70] Pressouyre describes the failure to inscribe the Lake District as an "historic error" and proof that the Convention is an archaic instrument which has a generous vision of nature and a narrow perception of culture rooted in architectural monuments. He concludes that "everyone knows how to protect a minaret, a steeple, a house, but no one knows how to protect a way of life, a vast landscape, a fragile ensemble threatened by globalization."[71] The secretariat's new study on cultural landscapes was hotly debated at the 1991 session, resulting in a Committee request for further study among ICOMOS, IUCN and other competent partners.[72]

As a result, an expert group met at La Petite Pierre (France) to prepare the ground for the Committee's 1992 amendments to the inscription criteria and new guidance on definitions and categories of cultural landscapes. It is worth noting that, contrary to the previous period, IUCN participated actively. Indeed, Phillips claims that Bing Lucas, then Chair of IUCN's World Commission on Protected Areas and an enthusiastic advocate for protected landscapes, was "much involved in the drafting of the definition of cultural landscapes and identification of those three categories."[73] After nearly a decade of discussion, the World Heritage framework for cultural landscapes was finally approved on the basis of the group's recommendations. Three types were proposed: designed, organically evolved and

---

69 UNESCO, Report of the rapporteur on the twelfth session of the World Heritage Committee in Brasilia, 5–9 December 1988, Paris, 23 December 1988. SC-88/conf.001/13, paras. 9, 30–31. Retrieved from http://whc.unesco.org/archive/1988/sc-88-conf001-13e.pdf

70 UNESCO, Report of the rapporteur on the fourteenth session of the World Heritage Committee in Banff, 7–12 December 1990, Banff, 12 December 1990, CLT-90/conf.004/13, para. 18. Retrieved from http://whc.unesco.org/archive/1990/cc-90-conf004-13e.pdf

71 Canada Research Chair, interview Pressouyre. "Tout le monde sait protéger un minaret, un clocher, une maison, mais que personne ne sait protéger un mode de vie, un grand paysage, un ensemble fragile du fait de la mondialisation."

72 UNESCO, Report of the rapporteur on the fifteenth session of the World Heritage Committee, Carthage, 9–13 December 1991, Carthage, 12 December 1991, SC-91/conf.002/15, paras. 58–65. Retrieved from http://whc.unesco.org/archive/1991/sc-91-conf002-15e.pdf

73 Canada Research Chair, interview Phillips.

associative cultural landscapes. The last one marks a move towards intangible values, deriving significance "by virtue of the powerful religious, artistic or cultural associations of the natural element rather than material cultural evidence, which may be insignificant or even absent."[74] Four of the existing cultural criteria were slightly modified by adding words to recognize the new category: criteria (iii) added "cultural tradition," (iv) "landscape," (v) "land use" and (vi) "living traditions." Because cultural landscapes embody dynamic processes, their authenticity is hard to assess. For this reason, the test of authenticity for cultural landscapes was amended to focus on "their distinctive character and components."[75] These changes still appear in the Operational Guidelines.

The same year, IUCN convened a special session during the Fourth World Parks Congress in Caracas to prepare proposed revisions to natural criteria to eliminate any confusion. The workshop concluded that "references to man's interaction with nature (criterion ii) and exceptional combinations of natural and cultural elements (criterion iii) are inconsistent with the legal definition of natural heritage in Article 2 of the Convention" and should be removed.[76] After considering these recommendations, the 1992 Committee agreed that the two phrases were inconsistent with the legal definition of natural heritage and removed them. Natural criterion (iii) was further revised with the addition of "aesthetic importance" so that it now required sites to "contain superlative natural phenomena or areas of exceptional natural beauty and aesthetic importance." In line with the changes in criterion (iii), the explanatory text on integrity was re-written to emphasize that sites should be "of outstanding aesthetic value and include areas whose conservation is essential for the long-term maintenance of the beauty of the site."[77]

Further reflection was carried out in 1993 by La Petite Pierre expert group at a meeting in Templin/Schorfheide (Germany). The experts confirmed the applicability of the adjusted criteria and framework for cultural landscapes; as well, they developed a global survey of such sites as a guide to States Parties in identifying

---

74   UNESCO, Report of the expert group on cultural landscapes, La Petite Pierre (France) 24–26 October 1992, Revision of the Operational Guidelines for the implementation of the World Heritage Convention, Paris, 2 November 1992, WHC-92/conf.002/10/add., III 41. Retrieved from http://whc.unesco.org/archive/1992/whc-92-conf002-10adde.pdf

75   UNESCO, Report of the expert group on cultural landscapes, La Petite Pierre (France) 24–26 October 1992, Revision of the Operational Guidelines for the implementation of the World Heritage Convention, Paris,2 November 1992, WHC-92/conf.002/10/add., I-II. Retrieved from http://whc.unesco.org/archive/1992/whc-92-conf002-10adde.pdf

76   Hemanta Mishra and N. Ishwaran, "Summary and conclusions," p. 14; Nora J. Mitchell, "Considering the Authenticity of Cultural Landscapes," *Bulletin of the Association for Preservation Technology*, 39/ 2–3 (2008), pp. 25–32; Mechtild Rössler, "Applying Authenticity to Cultural Landscapes," *Bulletin of the Association for Preservation Technology*, 39/ 2-3 (2008), pp. 47–52.

77   UNESCO, Revision of the Operational Guidelines for the implementation of the World Heritage Convention, Paris, 11 October 1992, WHC-92/conf.002/10, paras. 6-7. Retrieved from http://whc.unesco.org/archive/1992/whc-92-conf002-10e.pdf

Figure 3.8   Tongariro National Park, New Zealand

appropriate candidates.[78] That same year Tongariro (New Zealand), previously listed under natural criteria alone, became the first property to be inscribed as a cultural landscape in recognition of its cultural and spiritual significance for the Maori people. It had taken almost a decade to resolve the theoretical issues.[79]

Looking back, some pioneers reflect on the implications of introducing the cultural landscapes category. Collinson notes the complexity of arriving at a balanced judgement:

> I think the idea is warranted. I just think it's much more difficult to get your mind around. And in all cases, you really have to go back to the criteria and the procedures laid out in the operating procedures to be able to be sure and to be satisfied that you have given it a fair shake.[80]

---

78   UNESCO, Report of the international expert meeting on cultural landscapes of Outstanding Universal Value, Templin, 12–17 October 1993, Paris November 1993, WHC-93/inf. 4, para. 5. Retrieved from http://whc.unesco.org/archive/1993/whc-93-conf002-inf4e.pdf

79   For in-depth discussion of cultural landscapes, see Bernd von Droste et al., *Cultural Landscapes of Universal Value: Components of a Global Strategy* (New York, 1995), pp. 1-464 and two articles by Mechtild Rössler on "World Heritage Cultural Landscapes: A UNESCO Flagship Programme 1992–2006," *Landscape Research*, 31/4 (2006), pp. 333–53 and "Applying Authenticity to Cultural Landscapes," *Bulletin of the Association for Preservation Technology*, 39/ 2–3 (2008), pp. 47–52.

80   Canada Research Chair, interview Collinson.

In like manner, Abdelaziz Touri, a Moroccan archaeologist and State Party representative, supports the concept of cultural landscapes as an enrichment of the list but cautions that the definition needs clarification: "It got us out of the monument, it got us out of the medina … to something larger which is for me an interesting notion. Now it needs some distance. . It is not yet clear what constitutes a landscape … The Committee would benefit from a more precise definition."[81] Pressouyre goes farther, believing that many of the sites that were listed for natural values in the early years would qualify today as cultural landscapes. By way of example he cites Air and Ténéré (Niger), listed only under natural criteria, noting its rich legacy of prehistoric petroglyphs. According to him, "one did not want to mix culture and nature, and in a sense one thought at that time that nature was the consolation prize for countries that lacked monuments."[82] Carmen Añón Feliu, a Spanish expert in landscapes and supporter of La Petite Pierre group, shares Pressouyre's view that many early inscriptions should be revisited: "Geographers say that the history of man is found in the landscape. So how could one separate nature and culture? I do not see it, even in these sites that have been inscribed so far for natural only. This is something I want to study some day."[83]

Although by 1993 cultural landscapes were accepted on the World Heritage List, the imbalance between cultural and natural sites continued apace. In fact, the imbalance was exacerbated by the decision to consider landscapes as cultural sites. Once the dust had settled, the Committee returned to its earlier concerns about imbalance between cultural and natural sites. In 1994, it criticized the low number of natural properties on the list, attributing the imbalance in part to the criteria changes that favoured the cultural dimensions of landscapes to the detriment of their natural aspects.[84] In 1995 a lively debate explored obstacles and proposed solutions to increase the listing of natural sites. The ideas included

---

81   Canada Research Chair on Built Heritage, Université de Montréal, audio interview of Abdelaziz Touri by Christina Cameron and Petra Van Den Born, Paris, 22 June 2011. "C'est une avancée parce qu'on est sorti du monument, on est sorti de la Médina, on est sorti pour avoir quelque chose de plus large, n'est-ce pas, et qui est à mon sens une notion intéressante. Maintenant, il faudrait du recul. … On ne comprend pas ce que c'est encore que le paysage, n'est-ce pas. … le Comité gagnerait à ce que une définition un peu plus précise."

82   Canada Research Chair, interview Pressouyre. "On ne voulait pas mélanger culture et nature et que, en quelque sorte on pensait, à ce moment là, que la nature c'était le prix de consolation des pays qui avaient moins de monuments au sens propre du terme que les autres."

83   Canada Research Chair on Built Heritage, Université de Montréal, audio interview of Carmen Añón Feliu by Christina Cameron, Madrid, 18 June 2009. "Les géographes disent que dans les paysages est l'histoire de l'homme. Alors comment est-ce qu'on peut séparer nature et culture? Moi je ne le vois pas, même dans ces lieux que jusqu'à présent ont été déclarés juste comme naturel et c'est quelque chose que je veux proposer un jour de étudier et de revenir."

84   UNESCO, Report of the rapporteur on the eighteenth session of the World Heritage Committee in Phuket, 12–17 December 1994, Paris, 31 January 1995, WHC-

sending more natural heritage specialists to the Committee, strengthening natural heritage positions within the secretariat and developing a new global inventory of potential natural sites. Some belittled the issue as a form of unproductive competition, advocating for the creation of a single list of common criteria to cover all properties.[85] At the Committee's request, a follow-up expert meeting in 1996 was organized at the Parc national de la Vanoise (France) to work on a global strategy for natural heritage. It went beyond its mandate and recommended an overarching global strategy that would reinforce the unifying concept of World Heritage by embracing both cultural and natural heritage.[86] In the year 2000, after two decades of effort, natural properties represented only 20 per cent of the World Heritage List. At this point, the debate was subsumed by a more general discussion on representativeness in all its dimensions.[87]

Equitable Regional and Cultural Distribution

In addition to an uneasy balance between cultural and natural heritage, the system was confronted with difficulties in achieving equitable regional and cultural representation on the World Heritage List. The initial discussions on representativity began with Committee membership itself and extended later to include the World Heritage List as well. The language in the Convention text that required Committee membership to reflect an "equitable representation of the different regions and cultures of the world" (article 8.2) was later applied to the goal of equitable representation of heritage properties. The former is a political matter; the latter concerns a complex philosophical discourse about the definition of heritage. Nonetheless, there is some evidence that the two issues are related. In 1990, there were 112 States Parties and 21 Committee members. Although the Committee represented only 19 per cent of the members, a surprising 47 per cent of inscriptions came from their countries. Ten years later in 2000, when Committee members represented a mere 13 per cent of the then-162 States Parties, 43 per

---

94/conf.003/16, para. X 8. Retrieved from http://whc.unesco.org/archive/1994/whc-94-conf003-16e.pdf

85   UNESCO, Report of the rapporteur on the nineteenth session of the World Heritage Committee in Berlin, 4–9 December 1995, Paris, 31 January 1996, WHC-95/conf.203/16, paras. X 3, 9–10. Retrieved from http://whc.unesco.org/archive/1995/whc-95-conf203-16e.pdf

86   UNESCO, Report of the rapporteur on the twentieth session of the World Heritage Committee in Merida, 2–7 December 1996, Paris, 10 March 1997, WHC-96/conf.201/21, paras. IX 8–10. Retrieved from http://whc.unesco.org/archive/1996/whc-96-conf201-21e.pdf

87   For an analysis of the ecological representativeness of natural World Heritage Sites see Helen Diane Hazen, *The Role of the World Heritage Convention in Protecting Natural Areas* (St. Paul, 2006), pp. 33–63.

cent of inscriptions were still situated in Committee members' countries.[88] On the basis of these rather uncomfortable statistics, one could conclude that equitable representation on the Committee would lead to equitable cultural and geographic representation on the list.

With regard to Committee membership, dissatisfaction grew as ratification of the Convention by States Parties advanced steadily over the years. Following the 1987 election for Committee membership, some States Parties complained about a "lack of balance in the distribution of seats to different geographical groups" and called for the "universal and cultural representation within the Committee foreseen by the Convention."[89] Two years later, the principle of rotation was affirmed and countries were asked to voluntarily stand back in order to ensure a better balance in Committee membership.[90] Few countries complied. The issue continued to escalate and reached a crescendo in an impassioned debate at the 1991 General Assembly of States Parties, fuelled by statistics on the composition of the World Heritage Committee and the first specific study on equitable representation of different regions and cultures of the world.[91] Yet, in spite of dissatisfaction with the electoral system and the composition of the Committee, States Parties refused to set rigid rules for the allocation of seats. The situation festered for the rest of the decade until it became the subject of one of four task forces preparing the 2000 Cairns reform agenda.[92]

A more complex question concerns the representativity of the World Heritage List. The composition of the Committee may play a role in ensuring equitable representation of sites on the list, but the key determinant is the nomination and site selection process. In 1978, the Committee discussed "the problems of typology,

———————————

88   UNESCO, Report of the rapporteur on the fourteenth session of the World Heritage Committee in Banff, 7–12 December 1990, Banff, 12 December 1990, CLT-90/conf.004/13, paras. I.1, VII.A. Retrieved from http://whc.unesco.org/archive/1990/cc-90-conf004-13e.pdf; UNESCO, Report of the rapporteur of the twenty-fourth session of the World Heritage Committee in Cairns, 27 November–2 December 2000, Paris, 16 February 2001, WHC-2000/conf.204/21, paras. I.1, X.A. Retrieved from http://whc.unesco.org/archive/2000/whc-00-conf204-21e.pdf

89   UNESCO, Summary record of the sixth General Assembly of States Parties in Paris, 30 October 1987, Paris, 31 October 1987, CC-87/conf.013/6, para. 21. Retrieved from http://whc.unesco.org/archive/1987/cc-87-conf013-6e.pdf

90   UNESCO, Summary record of the seventh General Assembly of States Parties in Paris, 9 and 13 November 1989, Paris, 13 November 1989, CC-89/conf.013/6, para. 12. Retrieved from http://whc.unesco.org/archive/1989/cc-89-conf013-6e.pdf

91   UNESCO, Means of ensuring an equitable representation of the different regions and cultures of the world on the World Heritage Committee, Paris, 2 November 1991, CLT-91/conf.013/4. Retrieved from http://whc.unesco.org/archive/1991/clt-91-conf013-4e.pdf

92   UNESCO, Proposals concerning equitable representation in the World Heritage Committee, Paris, 13 November 2000, WHC-2000/conf.204/6. Retrieved from http://whc.unesco.org/archive/2000/whc-00-conf204-21e.pdf ; see chapter 5 for a thorough discussion of elections to the World Heritage Committee.

comparability, complementarity and universality of cultural and natural properties of universal importance" and, in a gesture that naively ignored national interests, recommended that countries should consult on properties of a comparable nature in order to harmonize approaches for the selection of sites.[93]

To establish a comparative context, countries were asked to submit preliminary lists of properties they were considering for nomination. The Convention required each State Party to submit "an inventory of property forming part of the cultural and natural heritage, situated in its territory and suitable for inclusion in the list" (article 11.1). According to the Committee, these inventories would provide a framework to "enable the Committee to have a better global idea of the form that the World Heritage List would take and thus better define the selection criteria."[94] The term "tentative list" instead of inventories was introduced in the 1980 revision to the Operational Guidelines. Evidently unaware of the work required, the Committee optimistically thought it possible that countries "would make available to the Committee before its next session a list of those properties which it intends to nominate to the World Heritage List during the next five to ten years."[95] This plea, repeated annually, went largely unheeded for years. Lack of tentative lists and inadequate comparative frameworks contributed to the growing imbalance of the World Heritage List.

The response to this situation led to one of the major achievements of the World Heritage Convention, namely the intellectual theorization of the heritage field through research on definitions and comparative frameworks. 1980 marks the beginning of this undertaking when the Committee expanded the definition of heritage place to include a series of cultural properties in different geographical locations, provided they belong to the same historico-cultural group and typology.[96] The technical advisors explored ways to develop comparative research. IUCN used its global network to prepare a worldwide inventory of natural sites using questionnaires and expert meetings followed by in-depth comparative analysis,

93   UNESCO, Report of the rapporteur on the second session of the World Heritage Committee in Washington, 5–8 September 1978, Paris, 9 October 1978, CC-78/conf.010/10 rev., para. 47. Retrieved from http://whc.unesco.org/archive/1978/cc-78-conf010-10reve. pdf

94   UNESCO, Report of the rapporteur on the third session of the World Heritage Committee in Cairo and Luxor, 22–26 October 1979, Paris, 30 November 1979, CC-79/ conf.003/13, para. 34. Retrieved from http://whc.unesco.org/archive/1979/cc-79-conf003-13e.pdf

95   UNESCO, Report of the rapporteur on the third session of the World Heritage Committee in Cairo and Luxor, 22–26 October 1979, Paris, 30 November 1979, CC-79/ conf.003/13, para. 44. Retrieved from http://whc.unesco.org/archive/1979/cc-79-conf003-13e.pdf

96   UNESCO, Report of the rapporteur on the fourth session of the World Heritage Committee in Paris, 1–5 September 1980, Paris, 29 September 1980, CC-80/conf.016/ WHC/2 rev., para. 19e. Retrieved from http://whc.unesco.org/archive/1980/cc-80-conf016-10e.pdf

in order to assist countries in the preparation of their tentative lists. At that time, ICOMOS judged it unfeasible to carry out global thematic studies on cultural properties and offered instead to do targeted comparative research as part of the evaluation of specific nominations.[97]

Faced with the obvious complexity of selecting World Heritage properties, the Committee made various attempts to create adequate frameworks. Conscious of the potential contradiction between the goals of representativeness and selectivity, it crafted a statement that attempted to bring the two concepts together: "The World Heritage List should be as representative as possible of all cultural and natural properties which meet the Convention's requirement of outstanding universal value and the cultural and natural criteria." It also requested that countries group their tentative lists according to bio-geographical provinces for natural sites, and cultural periods or areas for cultural sites, a directive that was never followed. The Committee naively appealed to countries to hold back to prevent the list from becoming increasingly imbalanced, encouraging "those countries which have several properties already inscribed on the list to exercise restraint in putting forward additional nominations (especially cultural nominations) at least for a limited period of time."[98]

In 1982, countries were scrambling to put together tentative lists, although they lacked direction about scale and parameters. For example, France and the Federal Republic of Germany had ambitious proposals to submit tentative lists with fifty properties each, eliciting a request for harmonization among countries of the same cultural region.[99] The Committee ruled that, as of 1984, it would no longer consider cultural nominations from countries who had not submitted tentative lists because they were essential to carry out "comparative and serial studies which are necessary for a methodological approach in building up the World Heritage List." To accelerate the work, countries were encouraged to consult the technical advisors, IUCN and ICOMOS, and if necessary to request financial assistance from the World Heritage Fund.[100]

---

97   UNESCO, Report of the rapporteur on the fourth session of the World Heritage Committee in Paris, 1–5 September 1980, Paris, 29 September 1980, CC-80/conf.016/ WHC/2 rev., paras. 47–50. Retrieved from http://whc.unesco.org/archive/1980/cc-80-conf016-10e.pdf

98   UNESCO, Report of the rapporteur on the fifth session of the World Heritage Committee in Sydney, 26–30 October 1981, Paris, 5 January 1982, CC-81/conf/003/6, paras. 22, 24–5. Retrieved from http://whc.unesco.org/archive/1981/cc-81-conf003-6e.pdf

99   UNESCO, Report of the rapporteur on the sixth session of the World Heritage Committee in Paris, 13–17 December 1982, Paris, 17 January 1983, CLT-82/conf.015/8, para. 15. Retrieved from http://whc.unesco.org/archive/1982/clt-82-conf015-8e.pdf

100   UNESCO, Report of the rapporteur on the seventh session of the World Heritage Committee in Florence 5–9 December 1983, Paris, January 1984, SC/83/conf.009/8, paras. 17–20. Retrieved from http://whc.unesco.org/archive/1983/sc-83-conf009-8e.pdf

Global Study: 1983–1994

One of the analytical tools that the Committee considered for the selection of cultural sites was a global study which evolved from its initial phase in 1983 until 1994 when it was discarded in favour of the global strategy. Emerging from the tension between universality and cultural diversity, the global study was intended to render the Committee's decision-making more objective and transparent.

A key figure in the early phase of this theorization of cultural heritage was French architect Michel Parent who was involved first as a member of the French delegation and then as President of ICOMOS until his retirement in 1986. In 1979 he had presented the Committee with an analysis and initial classification of over 50 cultural nomination files using three broad categories listed in the Convention (monuments, groups of buildings and sites).[101] Five years later, as President of ICOMOS, he again took stock of the situation using over 150 listed cultural properties and others in the nomination queue. In a remarkable and insightful speech to the 1983 Bureau, he laid out the key challenges to the credibility of World Heritage in a trenchant analysis that anticipated years of struggle on these issues. Concerned about the quality of Committee decisions, he argued that comparative evaluation is the "guarantee of consistency" and that without it decisions have "neither objectivity or scientific or moral foundation." He pointed to varying national interpretations of universality, representativeness, integrity and authenticity that resulted in contradictory evaluations as well as disappointment among States Parties when their sites were deferred or rejected. He recommended taking time "for a critical examination of the list as it stands and for reflection on the future that will lead to definite aims which go beyond abstract reference to a set of criteria and will eventually resolve the ambiguities observed at present." Through the harmonization of tentative lists and thematic studies, he envisaged a framework that would:

> enable us to distinguish the horizon of World Heritage, and clear both the Committee and ICOMOS of any charges of laxity or inconsistency. My basic misgiving about the practice of choosing solely on the basis of individual cases is that it might, in the end, by force of circumstances produce a list that lacked consistency and would consequently lose some of its credibility.[102]

---

101   Michel Parent, Comparative study of nominations and criteria for world cultural heritage, principles and criteria for inclusion of properties on the World Heritage List, Paris, 11 October 1979, CC-79/conf.003/11 annex. Retrieved from http://whc.unesco.org/archive/1979/cc-79-conf003-11e.pdf

102   UNESCO, Speech by Mr. Michel Parent, Chairman of ICOMOS, during the seventh session of the Bureau of the World Heritage Committee in Paris, 27–30 June 1983, Paris, 1 September 1983, SC/83/conf.009/inf. 2. Retrieved from http://whc.unesco.org/archive/1983/sc-83-conf009-inf2e.pdf

Parent's proposal led to the development of the global study, a kind of world inventory of all types of property that might be eligible for inscription on the World Heritage List. The global study was meant to be a comprehensive framework within which the World Heritage List could develop and which could guide countries in the selection of properties. Thus began years of expert meetings and research focused on solutions to ensure the credibility of the World Heritage List.

In 1983, the Committee approved research studies on three topics: historic towns, sites representing events, ideas or beliefs, and the concept of authenticity.[103] The next year, it considered the report from an ICOMOS expert meeting on historic towns and adopted, with minor amendments, the framework for selecting candidate historic towns for the World Heritage List.[104] Although approved in 1984, the framework was only published in the 1987 Operational Guidelines and has remained virtually unchanged to this day, thereby influencing heritage theory and practice for a quarter of a century in many parts of the world. In brief, three categories of historic towns were proposed along with illustrative examples. These are:

1. towns which are no longer inhabited but which provide immutable archaeological evidence of the past;
2. historic towns which are still inhabited and which, by their very nature, have developed and will continue to develop under the influence of socioeconomic and cultural change, a situation that renders the assessment of their authenticity more difficult and any conservation policy more problematical. Within this group, four sub-categories are specified: towns representing a specific period of culture, towns that have evolved over time, historic centres that are enclosed within modern cities, and isolated units which provide evidence of towns that have disappeared;
3. new towns of the twentieth century.

The Committee decided that new towns should not be considered until "traditional historic towns which represent the most vulnerable part of the human heritage, have been entered on the World Heritage List."[105]

---

103 UNESCO, Report of the rapporteur on the seventh session of the World Heritage Committee in Florence 5–9 December 1983, Paris, January 1984, SC/83/conf.009/8, para. 24. Retrieved from http://whc.unesco.org/archive/1983/sc-83-conf009-8e.pdf

104 UNESCO, Report of the rapporteur on the eighth session of the World Heritage Committee in Buenos Aires, 29 October–2 November 1984, Buenos Aires, 2 November 1984, SC/84/conf.004/9, para. 14. Retrieved from http://whc.unesco.org/archive/1984/sc-84-conf004-9e.pdf

105 UNESCO, Operational Guidelines for the implementation of the World Heritage Convention, WHC/2 rev. January 1987, paras. 23–32. Retrieved from http://whc.unesco.org/archive/opguide87.pdf

In 1985 and 1986, States Parties were again urged to complete their tentative lists so that properties could be considered in the widest possible context.[106] Despite these repeated appeals, only 30 of 98 States Parties had submitted tentative lists by late 1987, meaning that it was impossible to use them to build thematic comparative frameworks. To speed up the process, ICOMOS led several harmonization exercises in various regions including the Balkans, North Africa, French-speaking African countries, Northern Europe and Asia.[107] But even in those rare instances when the Committee had guidance, the majority of members were apparently not ready to follow it, as is demonstrated by the 1987 inscription of the twentieth-century city of Brasilia. Although the Committee had decided in 1984 that new towns "should be deferred until all the traditional historic towns ... have been entered on the World Heritage List," the revised Operational Guidelines were only distributed by the secretariat during the 1987 meeting, and not until after Brasilia had been inscribed on the list and publicly announced in Brazil. Three Committee members (Canada, India and United States of America) recorded their objection to this flagrant disregard of procedures.[108] In his interview, M'Bow sheds light on the secretariat's tardiness in distributing the guidelines, stating that in his role of Director-General he had insisted on listing Brasilia. "From the beginning, the World Heritage Committee tended to consider essentially ancient heritage. I was the one who intervened for the first time on behalf of recent heritage, namely the city of Brasilia."[109]

Faced with a high number of nominations and few comparative frameworks, the 1987 Committee created a working group to propose ways and means of ensuring rigorous application of the criteria, and hence the credibility of the system.[110] 1988 marks a significant turning point for cultural heritage in the World Heritage system. The working group proposed "a global study which might include an international

---

106   UNESCO, Report of the rapporteur on the ninth session of the World Heritage Committee in Paris, 2–6 December 1985, Paris, December 1985, SC-85/conf.008/9, para. 19. Retrieved from http://whc.unesco.org/archive/1985/sc-85-conf008-9e.pdf

107   UNESCO, Report of the rapporteur on the eleventh session of the World Heritage Committee in Paris, 7–11 December 1987, Paris, 20 January 1988, SC-87/conf.005/9, para.8. Retrieved from http://whc.unesco.org/archive/1987/sc-87-conf005-9e.pdf

108   UNESCO, Report of the rapporteur on the eleventh session of the World Heritage Committee in Paris, 7–11 December 1987, Paris, 20 January 1988, SC-87/conf.005/9, para. 10. Retrieved from http://whc.unesco.org/archive/1987/sc-87-conf005-9e.pdf

109   Canada Research Chair on Built Heritage, Université de Montréal, audio interview of Amadou-Mahtar M'Bow by Mechtild Rössler and Petra Van Den Born, Paris, 22 October 2009. "Dès le début, il y avait une tendance du Comité du patrimoine, effectivement de prendre en considération essentiellement le patrimoine ancien. Ah! Très ancien et c'est moi qui suis intervenu pour la première fois pour qu'on inscrive un patrimoine récent, c'est-à-dire la ville de Brasilia."

110   UNESCO, Report of the rapporteur on the eleventh session of the World Heritage Committee in Paris, 7–11 December 1987, Paris, 20 January 1988, SC-87/conf.005/9, para. 35. Retrieved from http://whc.unesco.org/archive/1987/sc-87-conf005-9e.pdf

tentative list of references designed to assist the States Parties in identifying their properties and the Committee in evaluating nominations."[111] In other words, instead of waiting for countries to submit tentative lists, the proposal was a top-down exercise for cultural properties, mirroring IUCN's 1982 publication on *The World's Greatest Natural Areas.*[112] In an earlier interview, Raidl had anticipated the need for such a study to show the cultural patterns and broader context: "It seems to me that these patterns are worthy of careful consideration and discussion by knowledgeable experts to investigate if such thematic ideas might provide an alternative to the present practice in inscribing the properties one by one as if they were in isolation."[113]

ICOMOS was to lead "a retrospective and prospective global reflection" for cultural properties, including those belonging to States Parties to the Convention and those countries which had not yet signed on. This research was to be framed according to "different parameters of coherence: chronological, geographical, ecological, functional, social, religious, and so forth."[114] The priority thematic studies mentioned in the working group report, covering problematic typologies such as rural landscapes, traditional villages and contemporary architecture, were deemed to be part of the global study. The first phase took the form of a thematic framework, raising methodological questions about its relationship to the other studies. Originally aimed at cultural properties, it was broadened to include natural heritage, with the addition of an IUCN global inventory of geological and fossil sites.[115]

1990 was the year when the first specific results of the global study were reported to the Committee. The initial research involved work on listed sites, tentative lists and additional research from experts. Within the secretariat, Raidl produced research with the help of two Greek experts, often working late into the night. In her view, the global study was intended to build a system of theme studies: "The idea behind it was to make, even for States that had not yet ratified, an overview of what exists and then the characteristics of those that should be

111   UNESCO, Report of the rapporteur on the twelfth session of the World Heritage Committee in Brasilia, 5–9 December 1988, Paris, 23 December 1988, SC-88/conf.001/13, para. 12. Retrieved from http://whc.unesco.org/archive/1988/sc-88-conf001-13e.pdf

112   IUCN Commission on National Parks and Protected Areas, *The World's Greatest Natural Areas: an indicative inventory of natural sites of world heritage quality* (Gland, 1982), pp. 1–69.

113   Derek Linstrum, "An alternative approach? An interview with Anne Raidl," *Momentum,* special issue (1984), p. 52. Retrieved from http://www.international.icomos. org/monumentum/vol-special/vol-special_4.pdf

114   UNESCO, Report of the rapporteur on the twelfth session of the World Heritage Committee in Brasilia, 5–9 December 1988, Paris, 23 December 1988, SC-88/conf.001/13, para. 14. Retrieved from http://whc.unesco.org/archive/1988/sc-88-conf001-13e.pdf

115   UNESCO, Report of the rapporteur on the thirteenth session of the World Heritage Committee in Paris, 11–15 December 1989, Paris, 22 December 1989, SC-89/conf.004/12, paras. 40–43. Retrieved from http://whc.unesco.org/archive/1989/sc-89-conf004-12e.pdf

selected in order to achieve a certain equilibrium."[116] Three initial studies, all well-documented areas from the European region, were relatively easy to classify: Greco-Hellenistic, Roman and Byzantine cultures. Even so, inevitably a diversity of views on geographical and cultural categories arose, revealing the futility of imposing rigid frameworks on cultural phenomena at a global level. There were concerns about overlapping cultures and the relevance of capturing so much data through a mixed temporal, cultural and thematic approach. Not surprisingly, the global study stalled because of its complexity and lack of rigorous methodology. It was an unrealistically ambitious undertaking to create an overall framework for all cultures and regions of the world.

In the words of Stovel, the global study was an "effort to find a comprehensive matrix within which you could plant all civilizations and all forms of expressions of those civilizations." In a wry comment on what he considered an ill-advised approach, he says that it was a "geo-cultural template into which you could pop all the possible World Heritage nominations. If you can have a matrix that was big enough ... then you can just take every new nomination and say 'ah, we don't have any of those' and we'd know right away what the answer should be."[117]

In spite of these difficulties, the global study represents an important milestone in fostering an international discourse on heritage. It engaged States Parties, professional organizations and independent scholars in a rich global dialogue.[118] The flurry of activity engendered by the global study was impressive, if unstructured. By the end of 1991, new research areas included aesthetic topics (Gothic, Romanesque, Baroque and Art Nouveau), cultural groupings (Hittites and Slavs), and geographic groupings (east Europe from Antiquity to the modern age). Responding to criticism that the focus was mainly on Europe, the secretariat announced projected studies for Islamic and Buddhist sites. The global study also marks an important shift in roles as the secretariat took over much of ICOMOS's responsibilities in this area, relegating to the technical advisor the thankless (and impossible) task of drafting a general global framework.[119]

---

116 Canada Research Chair, interview Raidl. "L'idée derrière était de faire, même pour les États qui n'avaient pas encore ratifié, pour avoir déjà une vue d'ensemble de ce qui existe et ensuite les caractéristiques de ceux qu'on devrait choisir pour avoir un certain équilibre."

117 Canada Research Chair on Built Heritage, Université de Montréal, audio interview of Herb Stovel by Christina Cameron, Ottawa, 16 March 2011.

118 UNESCO, Report of the rapporteur on the fourteenth session of the World Heritage Committee in Banff, 7–12 December 1990, Banff, 12 December 1990, CLT-90/conf.004/13, paras. 50–56. Retrieved from http://whc.unesco.org/archive/1990/cc-90-conf004-13e.pdf

119 UNESCO, Report of the rapporteur on the fifteenth session of the World Heritage Committee, Carthage, 9–13 December 1991, Carthage, 12 December 1991, SC-91/conf.002/15, para. 37. Retrieved from http://whc.unesco.org/archive/1991/sc-91-conf002-15e.pdf; UNESCO, Global study on cultural properties, Paris, 6 November 1991, SC-91/conf.002/5, paras. 1–6. Retrieved from http://whc.unesco.org/archive/1991/sc-91-conf002-5e.pdf

The Committee's enthusiasm for a tool to support the evaluation of cultural nominations was tempered by a growing awareness of its methodological weaknesses and its cost. Some cautioned that the study "should not result in an ossified list of the cultural values of World Heritage at a time when the very notion of heritage is undergoing rapid changes." Others insisted on practical outcomes, pointing out that the global study was not meant to be an "encyclopaedia of the history of world art" but a "reference framework to facilitate the work of the Committee."[120] Raidl expresses her disappointment at the 1991 Committee's lack of support for the global study. "At Carthage, many delegates did not really understand the value of these studies. There were countries that were so opposed and then the budget was voted in a very restrictive manner."[121]

At the same time, as part of taking stock of the Convention after 20 years of operation, the global study as a concept was seen as a key component for the agenda of the twentieth anniversary in 1992.[122] Countries saw it as a platform for future-oriented discussions on cultural diversity and equitable balance of cultures and regions on both the Committee and the list.[123] To prepare for the anniversary stock-taking, UNESCO commissioned an evaluation report in 1991 from Beschaouch who chaired the Committee that year. This important analysis exposed the system's failure to finalize a methodology for the global study. He reported that the Committee's mixed temporal, cultural and thematic approach did not meet unanimous support and noted that some UNESCO staff wondered whether it would be better to base it on chronology, geography or art history. Others considered that a socio-cultural approach might be preferable. Beschaouch's evaluation report concludes that:

> The preparation of a global study is an arduous, complex and, of necessity, multidisciplinary task [that] cannot afford to ignore the present and future evolution of ideas and attitudes and should never become a binding document.

---

120   UNESCO, Report of the rapporteur on the fifteenth session of the World Heritage Committee, Carthage, 9–13 December 1991, Carthage, 12 December 1991, SC-91/conf.002/15, para. 39. Retrieved from http://whc.unesco.org/archive/1991/sc-91-conf002-15e.pdf

121   Canada Research Chair, interview Raidl. "À Carthage … c'est là où beaucoup de délégués n'avaient pas compris vraiment de ce qui était la valeur de ces études. … Il y avait des États qui étaient tellement contre et ensuite le budget a été voté de manière extrêmement restrictif."

122   UNESCO, Report of the rapporteur on the fourteenth session of the World Heritage Committee in Banff, 7–12 December 1990, Banff, 12 December 1990, CLT-90/conf.004/13, paras. 33–5. Retrieved from http://whc.unesco.org/archive/1990/cc-90-conf004-13e.pdf

123   UNESCO, Report of the rapporteur on the fifteenth session of the World Heritage Committee, Carthage, 9–13 December 1991, Carthage, 12 December 1991, SC-91/conf.002/15, para. 49. Retrieved from http://whc.unesco.org/archive/1991/sc-91-conf002-15e.pdf

> As a mere *general frame of reference*, the global study should primarily enable the Committee better to bring out the 'outstanding universal value' of the sites nominated for inclusion and achieve a better balance in the list.[124]

During the anniversary discussions at Santa Fe in 1992, the Committee set five strategic directions. Of relevance to the global study was the second goal which focused on ensuring the representativity and credibility of the World Heritage List through refining evaluation procedures and promoting cultural inscriptions from all geo-cultural regions of the world and natural inscriptions demonstrating all geo-morphological formations and ecosystems.[125] During this meeting, frustrated with the slow progress of the global study and interested in applying computer technology to the exercise, the United States tabled an alternative framework based on a three-dimensional grid of time, culture and human achievement. The Committee referred the American framework to the working group.[126]

As work progressed, rejection of the global study became inevitable. It is true that there were successes with regard to thematic studies, including ICOMOS reports on industrial heritage and twentieth-century architecture, as well as expert meetings proposed by Spain and Canada on cultural itineraries and heritage canals respectively.[127] But key weaknesses were the global study's Euro-centric bias and overall methodology. Describing it as "a kind of misguided effort," Stovel criticizes the methodology as fundamentally flawed:

> Implicitly this involved a focus on categories of heritage as opposed to thematic typologies into which heritage properties might fit. What do I mean by that? I mean that people looked at the question of new nominations in terms of how many of this type do we have. So if we had a new French cathedral coming along, a new nomination, we'd ask "well, how many of these French cathedrals do we already have? We have six. Do we need more? Do we need seven?" ... That's a superficial question but that's where it would start and you say "okay, what are the qualities of this seventh that might allow us to distinguish it from

---

124   Azedine Beschaouch, Towards an evaluation of the implementation of the Convention, Paris, December 1991, WHC-92/conf.002/3, annex IV. Retrieved from http://whc.unesco.org/archive/1992/whc-92-conf002-3e.pdf

125   UNESCO, Final report: the evaluation of the implementation of the Convention concerning the protection of the world cultural and natural heritage and the strategic orientations, Paris, 16 November 1992, WHC-92/conf.002/4, II B, III. Retrieved from http://whc.unesco.org/archive/1992/whc-92-conf002-4e.pdf

126   UNESCO, Report of the rapporteur for the sixteenth session of the World Heritage Committee in Santa Fe, 7–14 December 1992, 14 December 1992, WHC-92/conf.002/12, para. XIII 3. Retrieved from http://whc.unesco.org/archive/1992/whc-92-conf002-12e.pdf

127   UNESCO, Report of the rapporteur for the seventeenth session of the World Heritage Committee in Cartagena, 6–11 December 1993, Paris, 4 February 1994, WHC-93/conf.002/14, paras. V.4, XVIII.2. Retrieved from http://whc.unesco.org/archive/1993/whc-93-conf002-14e.pdf

the first six? And if we are going to talk about outstanding universal value, can we actually make the claim that this has a different kind of outstanding universal value than something that's already there on the list?" The global study somehow accepted the idea that typologies were okay. Somehow those of us working with typologies were always dissatisfied with this discussion. In reality the answer to the question "how many French cathedrals is enough?" There is no answer to that ... no clear objective mathematical answer to that.[128]

As ICOMOS's World Heritage coordinator at the time, Cleere disagrees with those who criticize the global study: "I still think that the approach that was developed by ICOMOS, Greece and the US, which was bitterly attacked by Pressouyre as being a mechanistic approach, was the right one." In response to criticism that it was too rigid, he continues, "Well, it was only a framework. You can interpret it in different ways. You can take a broad brush approach to a whole group, looking at it as a whole group, or you can focus on one aspect of that group which is the sort of thing ICOMOS was doing."[129]

A new framework proposed by a 1993 expert meeting in Sri Lanka did not achieve consensus in the scientific community. According to UNESCO's report, there were disagreements on the philosophical premises and conceptual framework:

> Some specialists fear that this procedure might give too much importance to the traditional categories of traditional art history which have developed around the study of the great monuments and great civilisations, precisely at a time when the Convention's bodies ... are questioning the advisability of extending it in the future to other types of property and other cultures which, at present, are not at all or only slightly represented.[130]

This shift in understanding of the meaning of heritage heralds a change in direction. It marks a period when historical and anthropological perspectives gained a stronger foothold through the development of the World Heritage global strategy.

Global Strategy: 1994–2000

1994 marks an important moment in the life of the World Heritage Convention. During this year, a common understanding of the evolving notion of cultural heritage was advanced, leading to measures that encouraged the inscription of non-monumental sites as well as greater recognition of intangible values and cultural diversity. After a decade of striving, breakthroughs were achieved to broaden the scope of potential nominations by approval of a global strategy for

---

128   Canada Research Chair, interview Stovel, March 2011.
129   Canada Research Chair, interview Cleere.
130   UNESCO, Global study, Paris, 20 October 1993, WHC/93/conf.002/8, p. 5. Retrieved from http://whc.unesco.org/archive/1993/whc-93-conf002-8e.pdf

cultural properties, re-interpretation of the notion of authenticity and identification of innovative themes.

At the Committee's request, UNESCO and ICOMOS jointly organized a singularly important meeting of experts representing different regions and disciplines who were charged with defining a conceptual framework, a methodology and common goals for inscribing cultural properties.[131] Like a phoenix rising from the ashes, the faltering global study made a remarkable comeback as an open-ended global strategy that gained support from the Committee and the scientific community alike. The word "strategy" was deliberately chosen to indicate a dynamic, open-ended and evolutionary approach. No longer bound by the constraints of the earlier global study, this initiative called for a more flexible and adaptable approach to identifying outstanding universal value. The experts acknowledged the evolution of the concept of cultural heritage that had developed since the creation of the Convention. Arguing that in 1972 the idea of cultural heritage had been largely confined to architectural monuments and sites of great antiquity, the experts noted that current considerations included "cultural groupings that were complex and multidimensional, which demonstrated in spatial terms the social structures, ways of life, beliefs, systems of knowledge and representations of different past and present cultures." To enhance the World Heritage List, the global strategy strove to "take into account all the possibilities for extending and enriching it by means of new types of property whose value might become apparent as knowledge and ideas developed."[132]

The most significant contribution of the global strategy was its elaboration of a general anthropological framework that broadened the kinds of properties that could be considered for listing. The framework was structured under two principal themes and a number of subthemes, all formulated in general terms in an effort to encourage cultural nominations from under-represented regions and cultures. The theme of "human coexistence with the land" included the sub-themes of movement of peoples, settlement, modes of subsistence and technological evolution; the theme of "human beings in society" included human interaction, cultural coexistence as well as spirituality and creative expression. It was intended to be a process of continuous collaborative study of the development of scientific thought and world cultures. While such an open-ended approach encouraged countries to introduce new categories of nominations, thereby greatly widening the scope of outstanding universal value, further guidance would have enhanced the effect even further. While acknowledging the success of the global strategy,

---

131 UNESCO, Report of the rapporteur on the eighteenth session of the World Heritage Committee in Phuket, 12–17 December 1994, Paris, 31 January 1995, WHC-94/conf.003/16, para. X 2. Retrieved from http://whc.unesco.org/archive/1994/whc-94-conf003-16e.pdf

132 UNESCO, Expert meeting on the global strategy and thematic studies for a representative World Heritage List, Paris, 13 October 1994, WHC-94/conf.003/inf. 6, p. 3. Retrieved from http://whc.unesco.org/archive/1994/whc-94-conf003-inf6e.pdf

Stovel believes that the general themes needed to be elaborated in more detail in order to provide guidance to countries on their choice of nominations. "I think it would have changed the way nominations come forward. I think it would have helped countries better understand ... if they can see the place of their site in aiding the expression of one of those important themes. Then its link to OUV is quite clear."[133]

The strategy proposed an action plan with regional meetings, thematic studies on priority subjects and slight adjustments to inscription criteria. The report illustrated how it might be applied in operational terms by referring to twentieth-century architecture which "should not be considered solely from the point of view of 'great' architects and aesthetics, but rather as a striking transformation of multiple meanings in the use of materials, technology, work, organization of space, and, more generally, life in society."[134] Some of the recommended revisions to cultural criteria were retained and appear in the 1996 version of the Operational Guidelines. For criterion (i), the phrase "a unique artistic achievement" was deleted in order to remove an aesthetic bias (and to align with the French version) and replaced with "a masterpiece of human creative genius." To remove the sense that cultural influences only occur in one direction, seen as a hierarchical relic from the period of colonization, criterion (ii) replaced the phrase "have exerted great influence" with "exhibit an important interchange of human values." For criterion (iii), words were added to include "a civilization which is living," not just those that have disappeared.[135]

Two of the world's best heritage theorists agree on the significance of the global strategy.[136] For Jokilehto, a thematic framework "is the starting point for any nomination. We have to identify what are the themes, what is the message, what is the story, what are we talking about? And then, you will identify what is the regional historical context, cultural context, and then you can come to what is the type of heritage."[137] Stovel sees the global strategy as "a major departure for the Committee from previous ways of looking at heritage" because it moved from a typology-focused list to a thematically-focused list:

---

133   Canada Research Chair on Built Heritage, Université de Montréal, audio interview of Herb Stovel by Christina Cameron, Ottawa, 5 April 2011.

134   UNESCO, Expert meeting on the global strategy and thematic studies for a representative World Heritage List, Paris, 13 October 1994, WHC-94/conf.003/inf. 6, p. 4. Retrieved from http://whc.unesco.org/archive/1994/whc-94-conf003-inf6e.pdf

135   UNESCO, Operational Guidelines for the implementation of the World Heritage Convention, February 1996, WHC/2/revised, para. 24. Retrieved from http://whc.unesco.org/archive/opguide96.pdf

136   See also, Sophia Labadi, "A review of the Global Strategy for a Balanced, Representative and Credible World Heritage List 1994–2004," *Conservation and Management of Archaeological Sites*, 7/2 (2005), pp. 89–102.

137   Canada Research Chair, interview Jokilehto.

It's a question really of backing up and recognizing what those typologies express in terms of development on the planet. And the thematic framework offers an opportunity to say "what are the basic messages, what are the basic themes, what are the basic stories of human development on this planet that are important?" ... Once you work in a thematic framework, you are much more likely to come up with a range of inscriptions which encompass all of the major ways in which society has organized itself to produce conditions for living and development over time in the world. It's a much more sophisticated way of looking at heritage.[138]

While the global strategy originally concerned only cultural properties, natural heritage was soon added because of concerns about the uneven regional distribution of natural sites. In line with a more holistic notion of heritage, States Parties also wanted more emphasis on the role of culture in maintaining biodiversity within ecological systems. The Committee set priorities for studies that included the main types of ecosystems, the natural aspect of cultural landscapes and interpretation of the term "integrity" in natural site assessment procedures. In line with the new conceptualization of heritage, the global strategy discussion evoked the first mention of a unified set of criteria. "The Delegate from Niger expressed his hope that, eventually, separate criteria for Natural and Cultural sites could be eliminated in favour of a unified set of criteria applicable for all types of World Heritage sites."[139]

Evolving Concept of Authenticity

The year 1994 also marks an important achievement in clarifying the term "authenticity" in light of the evolution of the concept of cultural heritage. At the beginning of World Heritage implementation, Parent wrote about the difficulty of defining the term "authenticity" which is a qualifying condition for the inscription of cultural sites. He remarked that "authenticity is relative and depends on the nature of the property involved," arguing that a Japanese wooden temple "whose timbers have been replaced regularly as and when they decay – without any alteration of the architecture or the look of the material over ten centuries – remains undeniably authentic."[140] In the 1980s, the Committee vacillated between a rigorous materials-

---

138   Canada Research Chair, interview Stovel, April 2011.

139   UNESCO, Report of the rapporteur on the eighteenth session of the World Heritage Committee in Phuket, 12–17 December 1994, Paris, 31 January 1995, WHC-94/conf.003/16, paras. VII 9–10, X 6, 8–9. Retrieved from http://whc.unesco.org/archive/1994/whc-94-conf003-16e.pdf

140   Michel Parent, Comparative study of nominations and criteria for world cultural heritage, principles and criteria for inclusion of properties on the World Heritage List, Paris, 11 October 1979, CC-79/conf.003/11 annex, p. 19. Retrieved from http://whc.unesco.org/archive/1979/cc-79-conf003-11e.pdf

Figure 3.9   Participants at the 1994 Nara conference on authenticity in Nara, Japan

based interpretation of authenticity and a more flexible symbolic one, as Pressouyre points out with regard to Carcassonne and other sites.[141]

While ideas about the limitations of a materials-bound approach to authenticity had been circulating among experts for some time, the impetus for a global debate coincided with Japan's ratification of the Convention in 1992. Following a preparatory meeting in Bergen sponsored by Norway and Canada, Japan hosted a major international meeting in 1994 in Nara with 45 experts from diverse regions and cultures. Roland Silva, then President of ICOMOS, chaired the meeting. He believes it marked "a growth of an interpretation or of a principle."[142] The two general rapporteurs symbolically embodied the confrontation of old and new world approaches. Lemaire from Belgium was an author of the Venice Charter, a founding member of ICOMOS and interested in monumental architecture; Stovel from Canada represented a new generation of heritage specialists focused on vernacular buildings, site management and community participation. While the Nara conference confirmed that authenticity was a qualifying condition for outstanding universal value, the body of thought reflected in the conference

---

141   Léon Pressouyre, *The World Heritage Convention, Twenty Years Later* (Paris, 1996), pp. 11–14.

142   Canada Research Chair on Built Heritage, Université de Montréal, audio interview of Roland Silva by Christina Cameron, Victoria, 12 October 2011.

Figure 3.10   Raymond Lemaire

proceedings captures a doctrinal shift towards greater recognition of cultural and heritage diversity as well as associative values of heritage sites.[143] Matsuura supports this more flexible approach, noting that the disputed 1980 listing of the largely reconstructed Warsaw was ahead of its time: "Warsaw was inscribed based on the Nara document's concept in a way, though at the time, you know, it was inscribed as an exception."[144]

Of particular importance in the Nara Document on Authenticity is a relativist view on assessing authenticity within specific cultural contexts:

> All judgements about values attributed to cultural properties as well as the credibility of related information sources may differ from culture to culture, and even within the same culture. It is thus not possible to base judgements of value and authenticity on fixed criteria. On the contrary, the respect due to all cultures requires that heritage properties must be considered and judged within the cultural contexts to which they belong.[145]

---

143   Knut Einar Larsen (ed.), *Nara Conference on authenticity in relation to the World Heritage Convention: Proceedings, Nara, Japan, 1–6 November 1994* (Paris, 1995), pp. 1–427.

144   Canada Research Chair, interview Matsuura.

145   Knut Einar Larsen (ed.), "Nara Document on authenticity," *Nara Conference on Authenticity in relation to the World Heritage Convention: Proceedings, Nara, Japan, 1–6*

**Figure 3.11   Herb Stovel and Henry Cleere, Riomaggiore, Cinque Terre,
Italy, 19 March 2001**

Cleere recalls that at the beginning of the Nara meeting former Chairperson Beschaouch did "the best thing he ever did" by asking participants to say what authenticity meant in their language and in their countries:

> A number of people said "we haven't even got a word for it in our languages" and I thought that's it, that's it ... By doing that, not intentionally, I think he put his finger on it, the inability of us to identify an absolute authenticity. There's no such thing. There's no absolute. And of course Herb and Raymond Lemaire ... spent hours and hours and hours coming up with a three-page document which can be summed up in one sentence. Authenticity is culturally dependent. It really is as simple as that.[146]

Von Droste considers that the Nara meeting "opened the eyes of everyone that we have to change the course of our action and we have to look differently at the question of authenticity. We have to be ... more understanding of what other cultures do in order to perpetuate their heritage."[147] Luxen adds that:

November 1994 (Paris, 1995), p. xxix.

   146   Canada Research Chair, interview Cleere.

   147   Canada Research Chair on Built Heritage, Université de Montréal, audio interview of Bernd von Droste by Christina Cameron and Mechtild Rössler, Paris, 1 February 2008.

Nara was one of those moments when one realized that definitions of heritage could differ from one region to another. That is a great enrichment because one suddenly discovers that there are values that other civilizations pay more attention to than we do, and it is important that we open our eyes to these dimensions.[148]

But the response to the Nara document was not all positive. Luxen finds it difficult to interpret: "Nara allows for contradictory interpretations and that is dangerous. In fact the English and French versions were never completely coordinated because Raymond Lemaire died before Herb could finalize it. It is a shame because the two texts are not exactly the same. The French is stricter than the English."[149] Jokilehto agrees that it is not sufficiently clear. While appreciating its emphasis on the recognition of cultural diversity and heritage diversity, he adds:

> There were certain points which came out as the attributes that had to be referred to and on the basis of these I have developed my own thinking about what authenticity could be ... If we look at the works of art or the creativity, I think that's one aspect; another one is this historical aspect which is represented in the material and so on; and the third one is the social-cultural aspect which is in the society as such.[150]

He sums up by noting that "whenever we talk about authenticity we are talking about something that is genuine and true."[151] Guo Zhan from China regrets that the Nara document does not improve on the Venice Charter by providing general conservation guidance: "When I check the document, I have found almost nothing except diversity, the importance of diversity. But for the standards, for the principles, for the methods, approach of conservation, nothing ... instead of anything from the Venice Charter."[152]

---

148 Canada Research Chair, interview Luxen. "À Nara, c'est un des moments dans lequel on s'est rendu compte que les définitions du patrimoine pouvaient être différentes dans une région ou l'autre et ça c'est un grand enrichissement parce que on découvre tout d'un coup qu'il y a des choses, il y a des valeurs auxquelles d'autres civilisations sont plus attentives que nous et c'est important que nous aussi nous ouvrions aussi les yeux à ces dimensions."

149 Canada Research Chair, interview Luxen. "Le texte, je trouve, de Nara est difficile à interpréter et ce que je reproche quant il y a trop de textes doctrinaux c'est que ça permet des interprétations contradictoires ou différentes. Nara permet des interprétations contredisantes et c'est dangereux. Déjà que la version anglaise et la version française n'ont jamais été tout à fait coordonnées parce que Raymond Lemaire est décédé avant qu'avec Herb ils aient pu mettre ça au point. C'est très dommage parce que les deux textes ne sont pas exactement les mêmes. En français c'est plus strict qu'en anglais."

150 Canada Research Chair, interview Jokilehto.

151 Canada Research Chair, interview Jokilehto.

152 Canada Research Chair, interview Guo.

The Nara report to the World Heritage Committee recommended more dialogue in different regions of the world to further refine the concept and its application to cultural properties.[153] It took several years for the Committee to absorb the implications of Nara, not without open disagreement. Some argued that authenticity must remain materials-based and universal, as understood in the Venice Charter; others insisted that authenticity should be interpreted as intangible and relative to specific cultural contexts, as presented in the Nara document.[154] At the repeated urging of Japan as well as ICOMOS, which had finally adopted it at its 1998 General Assembly in Mexico, the Committee formally endorsed the Nara Document on Authenticity in 1999. It was only in 2005 that it was annexed to the World Heritage Operational Guidelines and that the attributes for the term "authenticity" were expanded from "design, material, workmanship or setting" to a longer list that included such intangible dimensions as "use and function, traditions, techniques and management systems, language and other forms of intangible heritage and spirit and feeling."[155] The expanded definition for authenticity irrevocably altered the World Heritage selection process for cultural properties.[156]

Implementation Strategies for Increased Representativity

Following the extraordinary policy gains made with cultural landscapes, the global strategy and the interpretation of authenticity, the period from 1995 to 2000 saw the consolidation and implementation of these ideas. The global strategy provided a framework and methodology for increasing the representativity of the World Heritage List, encouraging countries to ratify the Convention and calling on all States Parties to prepare nominations of properties from under-represented categories and regions. The challenge was to fill the gaps while not losing the manageability and credibility of the World Heritage List through an unreasonable number of inscriptions.

The flurry of activities unleashed by these initiatives was eventually formalized in 1998 when the Committee grouped them under the global strategy action plan. It called for regional meetings to engage countries in the World Heritage process. The Committee, having broadened the scope of the global strategy to

---

153   UNESCO, Nara document on authenticity, experts meeting, 1–6 November 1994, Paris, 31 November 1994, WHC-94/conf.003/inf.008. Retrieved from http://whc. unesco.org/archive/1994/whc-94-conf003-inf8e.pdf

154   Christina Cameron, "From Warsaw to Mostar: the World Heritage Committee and Authenticity," *Bulletin of the Association for Preservation Technology*, 39/ 2–3 (2008), pp. 19–24.

155   UNESCO, Operational Guidelines for the implementation of the World Heritage Convention, WHC.05/2, 2 February 2005, paras. 79–86. Retrieved from http://whc.unesco. org/archive/opguide05-en.pdf

156   Herb Stovel, "Origins and influence of the Nara Document on Authenticity," *Bulletin of the Association for Preservation Technology*, 39/ 2–3 (2008), pp. 9–18; Nora J. Mitchell, "Considering the Authenticity of Cultural Landscapes," *Bulletin of the Association for Preservation Technology*, 39/ 2–3 (2008), pp. 25–32.

include natural heritage, requested the participation of natural heritage specialists alongside their cultural counterparts. Many such meetings, especially in under-represented regions, explored the scope and diversity of potential candidate sites using the new anthropological themes and modified criteria from the global strategy as well as the broader interpretation of authenticity following Nara.[157] Stovel believes that the real benefit came with the mobilization of people through the regional workshops organized by the World Heritage Centre:

> For some it might have been repetitive but, by putting people together to compare notes within a cultural group or within a cultural context, it really opened the door to recognizing cultural specificities in a region.... So they'd go back home with some new energy and some new thinking. They'd go back home also as part of a kind of World Heritage network.[158]

The global strategy had specifically pointed to Africa as an under-represented region "in spite of its tremendous archaeological, technological, architectural, spiritual wealth, its ways of organizing land and space, its exchange network system for merchandise and ideas."[159] Therefore, the Committee directed UNESCO to hold discussions with African experts on the World Heritage Convention in general and the implications of the global strategy in particular. Focused on regional types of African cultural heritage, three major meeting were held in Zimbabwe, Ethiopia and Benin.[160] Dawson Munjeri, then responsible for museums and heritage sites in Zimbabwe, attributes his involvement with World Heritage to this initiative. At the meeting in Harare, he recalls that the group was charged with:

> coming up with what we thought should be the components of World Heritage, in the particularities of our region. And we came up with what we thought were

---

157 UNESCO, Progress report on the global strategy and thematic and comparative studies, Paris, 9 October 1997, WHC-97/conf.208/11. Retrieved from http://unesdoc.unesco.org/images/0011/001128/112878e.pdf

158 Canada Research Chair, interview Stovel, April 2011.

159 UNESCO, Progress report on the implementation of the global strategy and thematic studies, Paris, 2 October 1995, WHC-95/conf.203/8, para. A 1. Retrieved from http://whc.unesco.org/archive/1995/whc-95-conf203-8e.pdf

160 Dawson Munjeri et al. (eds.), African cultural heritage and the World Heritage Convention: first global strategy meeting in Harare, Zimbabwe, 11–13 October 1995. Retrieved from http://whc.unesco.org/uploads/events/documents/event-594-1.pdf; UNESCO, Synthetic report of the second meeting on global strategy of the African cultural heritage and the World Heritage in Addis Ababa, Ethiopia, 29 July – 1 August 1996, Paris, 14 October 1996, WHC-96/conf.201/inf. 7 Retrieved from http://whc.unesco.org/archive/1996/whc-96-conf201-inf7e.pdf; UNESCO, Report of the fourth global strategy meeting for West Africa, 16–19 September 1998, Porto Novo, Republic of Benin, Paris, 9 November 1998, WHC-98/conf.203/inf. 9. Retrieved from http://whc.unesco.org/archive/1998/whc-98-conf203-inf9e.pdf

important categories of heritage to be included and which had not yet been included although the framework had come out from the 1994 global strategy. But we then had to put it in our context.[161]

Other meetings focused on the harmonization of tentative lists, the pertinence of the new category of cultural landscapes and the meaning of authenticity and integrity in an African context.[162] The activities met with the approval of African delegations to World Heritage who linked them to improving the representativity of the list.[163] The sustained activity also exemplified the principle of international solidarity since support came from all parts of the World Heritage system: the World Heritage Fund, the technical advisory bodies and various States Parties, as well as from the new Nordic World Heritage Office. The reports from the African expert meetings capture the enthusiasm and excitement of the participants as they explored Africa's heritage places. Taken collectively, these African reports constitute a remarkable assemblage that makes a considerable contribution to knowledge and merits further study. They culminated in a ground-breaking international expert meeting organized by Munjeri in Zimbabwe in 2000 which further advanced the concepts of authenticity and integrity in an African context.[164]

In the five years following the adoption of the global strategy, UNESCO initiated dozens of meetings in other regions which were all listed in annual progress reports to the Committee.[165] One of particular interest was the Pacific

---

161   Canada Research Chair on Built Heritage, Université de Montréal, audio interview of Dawson Munjeri by Christina Cameron, Brasilia, 3 August 2010.

162   UNESCO, Synthetic report of the expert meeting on African cultural landscapes, Tiwi, Kenya, 9–14 March 1999, Paris, 25 May 1999, WHC-99/conf.204/inf. 4. Retrieved from http://whc.unesco.org/archive/1999/whc-99-conf204-inf4e.pdf; UNESCO, Synthetic report of the meeting on "authenticity and integrity in an African context," Great Zimbabwe National Monument, Zimbabwe, 26–29 May 2000, Paris, 9 October 2000, WHC-2000/conf.204/inf. 11. Retrieved from http://whc.unesco.org/archive/2000/whc-00-conf204-inf11e.pdf

163   UNESCO, Report of the rapporteur of the twenty-fourth session of the World Heritage Committee in Cairns, 27 November–2 December 2000, Paris, 16 February 2001, WHC-2000/conf.204/21, paras. IX 1–2. Retrieved from http://whc.unesco.org/archive/2000/whc-00-conf204-21e.pdf

164   UNESCO, *L'Authenticité et l'intégrité dans un contexte africain: réunion d'experts, Grand Zimbabwe, 26–9 mai 2000/Authenticity and integrity in an African context: expert meeting, Great Zimbabwe, 26–9 May 2000*, Galia Saouma-Forero, ed. (Paris, 2001), pp. 1–204. Retrieved from http://unesdoc.unesco.org/images/0012/001225/122598mo.pdf

165   UNESCO, Progress report, synthesis and action plan on the global strategy for a representative and credible World Heritage List, Paris, 27 October 1998, WHC-98/conf.203/12. Retrieved from http://whc.unesco.org/archive/1998/whc-98-conf203-12e.pdf; UNESCO, Progress report on the implementation of the regional actions described in the global strategy action plan adopted by the Committee at its twenty-second session, Paris, 21 October 1999, WHC-99/conf.209/8. Retrieved from http://whc.unesco.org/archive/1999/whc-99-conf209-8e.pdf

Islands meeting in Fiji. In a clear confirmation of the evolving concept of heritage, regional experts discussed the inextricable interrelationship between culture and nature. They noted the "inseparable connection between the outstanding seascapes and landscapes in the Pacific Islands region which are woven together by the rich histories, oral and life traditions of the Pacific Island peoples. These elements comprise the cultural heritage of the region ... bound through voyaging, kinship, trade and other relationships."[166] As Chairperson of the 1998 session of the World Heritage Committee, Matsuura takes pride in his leadership in recognizing traditional management for the nomination of East Rennell (Solomon Islands). At the time, the Committee's guidelines only accepted traditional protection and management mechanisms for cultural sites, not for natural ones.[167] Faced with IUCN's advice to adhere to the rules as set forth in the Committee's Operational Guidelines, Matsuura proposed first to change the rules on the spot and secondly to inscribe the site: "This is the kind of thing we have to do in the context of the global strategy because in many countries ... management plans are based on traditional law, not necessarily on legislation passed by parliaments. So we should make the requirement more inclusive."[168]

The new category of cultural landscapes proved to be particularly attractive for countries with non-monumental sites. In an unprecedented collaboration, UNESCO, ICOMOS and IUCN prepared a series of regional meetings to explore the implications of this new category in specific natural and cultural contexts. At a meeting in the Philippines, experts studied Asian rice terraces that result from the application of pond-field agriculture in mountainous terrain. They concluded that terrace landscapes are valuable not only for their intrinsic value but also for what they reveal about traditional cultures that have developed systems of cultural, socio-economic, ecological, agricultural and hydraulic practices:

> They are monuments to life itself. These landscapes celebrate the traditional lifestyle of the Asian people. It is this particular regional culture's special imprint on and relationship with nature manifested with significant aesthetic and harmonic values. It is a landscape that is being renewed daily and will continue its existence for as long as the unbroken line of this lifestyle continues.[169]

---

166   UNESCO, Findings and recommendations of the third global strategy meeting, Suva, Fiji, 15–19 July 1997, Paris, 17 October 1997, WHC-97/conf.208/inf. 8, para. II. Retrieved from http://whc.unesco.org/archive/1997/whc-97-conf208-inf8e.pdf

167   UNESCO, Operational Guidelines for the implementation of the World Heritage Convention, WHC-97/2, February 1997, para. 24.b.ii. Retrieved from http://whc.unesco.org/archive/opguide97.pdf

168   Canada Research Chair, interview Matsuura.

169   UNESCO, Report of the regional thematic study meeting "Asian rice culture and its terraced landscapes," Manila, Philippines, 28 March–4 April 1995, 25 September 1995, WHC-95/conf.203/inf. 8, 2, 4. Retrieved from http://whc.unesco.org/archive/1995/whc-95-conf203-inf8e.pdf

Other meetings further explored the potential of a cultural landscape approach to identify properties of outstanding universal value. At a meeting in Australia, experts discussed the intangible dimension of sites in the Asia-Pacific region where a diversity of traditional cultures both depend on and influence the landscape.[170] In Europe, meetings in Austria, Poland and Slovakia assessed the region's cultural landscapes, contributing as well to the drafting of the European Landscape Convention.[171] At a meeting in the Andean region, experts attributed the variety of cultural landscapes to the "rich diversity of the Andean mountain range, produced by a North-South and East-West variety of climatic zones," noting that these landscapes evolved through the influence of biological diversity and human creativity in managing the agricultural systems. In addition, experts acknowledged the sacred value of Andean landscapes often associated with ancient belief systems.[172]

As implementation of the global strategy unfolded, there was a move towards decentralization of the World Heritage system. Regional nodes emerged to play a leadership role in strengthening the implementation of the Convention. The first example was the creation in 1995 of the Nordic World Heritage Office, a joint effort of the Nordic countries to offer tangible support for the preparation of nominations and other conservation activities in under-represented regions like Africa.[173] In 1997 Japan expressed its interesting in playing a similar role in Asia and asked for clarification of the Committee's policy on the regionalization of its work.[174] Another model was the creation in 1999 of the Asia-Pacific focal point for

170   UNESCO, Report of the Asia-Pacific workshop on associative cultural landscapes Australia, 27–29 April 1995, Paris, 25 September 1995, WHC-95/conf.203/inf. 9. Retrieved from http://whc.unesco.org/archive/1995/whc-95-conf203-inf9e.pdf

171   UNESCO, Report of the expert meeting on European cultural landscapes of outstanding universal value, Poland, Austria, 21 April 1996, Paris, 30 April 1996. Retrieved from http://whc.unesco.org/archive/1996/whc-96-conf202-inf10e.pdf

UNESCO, Synthesis report of the expert meeting on management guidelines for cultural landscapes, Banská Stiavnica, Slovakia, 1–4 June 1999, Paris, 22 June 1999, WHC-99/conf.204/inf. 16. Retrieved from http://whc.unesco.org/archive/1999/whc-99-conf204-inf16e.pdf

UNESCO, Report of the regional thematic expert meeting on cultural landscapes in eastern Europe, 29 September – 3 October 1999, Bialystok, Poland, Paris, 20 October 1999, WHC-99/conf.209/inf. 14. Retrieved from http://whc.unesco.org/archive/1999/whc-99-conf209-inf14e.pdf

172   UNESCO, Report of the regional thematic meeting on cultural landscapes in the Andes, Arquipa/Chivay, Peru, 17–22 May 1998, Paris, 26 October 1998, WHC-98/conf.203/inf. 5, p. 6. Retrieved from http://whc.unesco.org/archive/1998/whc-98-conf203-inf8e.pdf

173   Further detail on the creation of the Nordic World Heritage Office is found in chapter 5 in the section on the World Heritage Centre.

174   UNESCO, Report of the rapporteur on the twenty-first session of the World Heritage Committee in Naples, 1–6 December 1997, Paris, 27 February 1998, WHC-97/conf.208/17, para. III.12. Retrieved from http://whc.unesco.org/archive/1997/whc-97-conf208-17e.pdf

south-east Asia, the Pacific, Australia and New Zealand. Located in Australia, it aimed at sharing information and experience, developing networks and facilitating training in support of World Heritage.[175]

In terms of intellectual development, the global strategy emphasized the need for comprehensive overarching frameworks to help guide the process. It inspired studies which reflected an enlarged concept of what constitutes heritage. With its call for thematic studies on little known aspects of heritage, the strategy opened a path for consideration of other kinds of heritage.[176]

Canals for example came under the global strategy theme of "Human co-existence with the land," including "industrial technology," "water management" and "routes for people and goods." A 1994 international experts meeting on heritage canals held in Canada recommended definitions and tackled the question of how to consider authenticity in the case of operating canals. The experts concluded that a distinctive feature of heritage canals is their evolution over time as they adapt to changing uses and new technologies, making it difficult to meet the test of authenticity as originally set out in the Operational Guidelines. The four original attributes of authenticity did not accommodate technological sites very well, but the new ones from the Nara discussions relating to use, function and techniques were better suited to the evolving nature of canals.[177] Proposals from the canals meeting to add "technological" to criteria (ii) and (iv) became part of the 1996 Operational Guidelines.[178]

Heritage corridors also fit within the expanded thematic framework. A 1994 expert meeting in Madrid explicitly refers to the global strategy in identifying heritage routes as a new open, dynamic and evocative concept. The proposed definition reveals the influence of heritage discourses on cultural landscapes and authenticity as well as the global strategy: "a heritage route is composed of tangible elements of which the cultural significance comes from exchanges and a multi-dimensional dialogue across countries or regions, and that illustrate the interaction of movement, along the route, in space and time."[179]

---

175   UNESCO, Report of the rapporteur on the twenty-third session of the World Heritage Committee in Marrakesh, 29 November to 4 December 1999, Paris, 2 March 2000, WHC-99/conf.209/22, para. VI 13. Retrieved from http://whc.unesco.org/archive/1999/whc-99-conf209-22e.pdf

176   An interesting analysis of World Heritage practice and how dispersed local phenomena are organised into a World Heritage of global relevance is found in Jan Turtinen, *Globalising Heritage: On UNESCO and the Transnational Construction of a World Heritage* (Stockholm, 2000), pp. 1–25.

177   UNESCO, Report on the expert meeting on heritage canals, Paris, 31 October 1994, WHC-94/conf.003/inf. 10, III B. Retrieved from http://whc.unesco.org/archive/1994/whc-94-conf003-inf10e.pdf; UNESCO, Operational Guidelines for the implementation of the World Heritage Convention, http://whc.unesco.org/archive/opguide11-en.pdf

178   When the Operational Guidelines were completely revised between 1999 and 2005, the canals text was integrated into annex 3.

179   UNESCO, Report of the expert meeting on routes as a part of our cultural heritage in Madrid, November 1994, Paris, 30 November 1994, WHC-94/conf.003/inf. 13, pp. 1–3.

The advisory bodies also contributed to this explosion of research. Partly supported by the World Heritage Fund, they initiated thematic studies, often based on typologies, using their own professional networks and databases. During the period, ICOMOS undertook studies on Roman theatres, fossil hominid sites, Iberian colonial towns, Gothic cathedrals, Islamic military sites in central and south Asia, castles of the Teutonic order in central and east Europe, cultural landscapes in southern Africa, and prehistoric sites in West Africa. In addition ICOMOS worked in partnership with the International Committee for Documentation and Conservation of Buildings, Sites and Neighbourhoods of the Modern Movement (DoCoMoMo) on a study of twentieth-century architecture and with the International Committee for the Conservation of Industrial Heritage (TICCIH) on studies of heritage canals, bridges, railways and so forth. During the same period, IUCN carried out research studies on geological and fossil sites, wetland and marine ecosystems, forests and sites of exceptional biodiversity.[180] To build up data for this research, IUCN proposed that tentative lists, previously only required for cultural properties, be obligatory for the consideration of natural nominations.[181]

As implementation of the global strategy action plan steamed ahead, work continued on procedures and rules for inscribing properties on the list. An irritant for many countries was IUCN's rigorous interpretation of natural criteria for site evaluations although it was in line with the Committee's Operational Guidelines "to be as strict as possible."[182] In 1996, at the meeting at Parc national de la Vanoise in France, an impressive group of international experts from the natural heritage community proposed to re-frame the term "natural" to recognize human influence. In ground-breaking wording that made its way into the Operational Guidelines, the experts proposed a definition that shifted towards ecological sustainability:

Retrieved from http://whc.unesco.org/archive/1994/whc-94-conf003-infl3e.pdf

180   UNESCO, Progress report, synthesis and action plan on the global strategy for a representative and credible World Heritage List, Paris, 27 October 1998, WHC-98/conf.203/12, pp. 8–11. Retrieved from http://whc.unesco.org/archive/1998/whc-98-conf203-12e.pdf; UNESCO, Progress report on the implementation of the regional actions described in the global strategy action plan adopted by the Committee at its twenty-second session, Paris, 21 October 1999, WHC-99/conf.209/8, pp. 31–4. Retrieved from http://whc.unesco.org/archive/1999/whc-99-conf209-8e.pdf; Roderick T. Wells, *Earth's Geological History: a contextual framework for assessment of World Heritage fossil site nominations* (Gland: IUCN, 1996), pp. 1–43. Retrieved from http://www.unep-wcmc.org/medialibrary/2010/10/11/e51b9f6e/wh_fossil_sites.pdf

181   UNESCO, Report of the expert meeting on evaluation of general principles and criteria for nominations of natural World Heritage Sites, Parc national de la Vanoise, France, 22–24 March 1996, Paris, 15 April 1996, WHC-96/conf.202/inf. 9, para. 3. Retrieved from http://whc.unesco.org/archive/1996/whc-96-conf202-inf9e.pdf

182   UNESCO, Operational Guidelines for the implementation of the World Heritage Convention, February 1996, WHC/2/rev, para. 61. Retrieved from http://whc.unesco.org/archive/opguide96.pdf

A natural area is one where bio-physical processes and landform features are still relatively intact and where a primary management goal of the area is to ensure that natural values are protected. The term "natural" is a relative one. It is recognized that no area is totally pristine and that all natural areas are in a dynamic state. Human activities in natural areas often occur and when sustainable may complement the natural values of the area.[183]

To recognize the unique character of the Convention's comprehensive approach to culture and nature, the experts made some startling recommendations, especially the merging of natural and cultural heritage criteria for World Heritage nominations in order to promote a unified identity for the outstanding heritage of humankind. This idea had first been raised by Niger in 1994. It was these same experts who recommended a common approach for assessing integrity by proposing that it apply to cultural as well as to natural properties. Noting that the Operational Guidelines had been in a state of constant revision and had accumulated numerous inconsistencies, the experts proposed a thorough review and restructuring of this document, including the preparation of a glossary of terms which the Committee had already requested in 1995.[184]

After inconclusive discussions of the Vanoise report at the Committee and further debates on a differentiated cultural approach to authenticity,[185] these ideas were eventually pursued at a 1998 meeting in Amsterdam with experts in both natural and cultural heritage. They studied the proposal for a harmonized set of criteria, issues related to authenticity and integrity as well as the notion of outstanding universal value and its application in different regional and cultural contexts. Experts at the Amsterdam meeting concluded that a unified set of criteria would better reflect the continuum between culture and nature as clearly expressed in the Convention. They also recommended that integrity should be a requirement for all ten criteria and that specific explanatory notes should be developed for each one. To better demonstrate the link between authenticity and cultural value, the group recommended that authenticity provisions should be

---

183   UNESCO, Report of the expert meeting on evaluation of general principles and criteria for nominations of natural World Heritage Sites, Parc national de la Vanoise, France, 22–24 March 1996, Paris, 15 April 1996, WHC-96/conf.202/inf. 9, para. 2c. Retrieved from http://whc.unesco.org/archive/1996/whc-96-conf202-inf9e.pdf

184   UNESCO, Report of the expert meeting on evaluation of general principles and criteria for nominations of natural World Heritage Sites, Parc national de la Vanoise, France, 22–24 March 1996, Paris, 15 April 1996, WHC-96/conf.202/inf. 9. Retrieved from http://whc.unesco.org/archive/1996/whc-96-conf202-inf9e.pdf

185   UNESCO, Report of the rapporteur on the twenty-first session of the World Heritage Committee in Naples, 1–6 December 1997, Paris, 27 February 1998, WHC-97/conf.208/17, para. VIII 11. Retrieved from http://whc.unesco.org/archive/1997/whc-97-conf208-17e.pdf

defined for each of the cultural criteria with reference to geo-cultural contexts.[186] As for outstanding universal value, the Amsterdam meeting recommended that it be interpreted as an outstanding response to issues of a universal nature, thereby encouraging a more regional and thematic approach.[187] Jokilehto considers that this guidance appropriately set the stage for a more culturally diverse selection of properties.

While acknowledging the progress that had been achieved in the implementation of the global strategy, the experts pushed for greater effort to fill gaps in the World Heritage List.[188] The Committee endorsed this position, finding that the Amsterdam report did not go far enough to quell dissatisfaction with the uneven representation on the World Heritage List. During an open debate at the following Committee session, States Parties acknowledged that the question of balance on the list was less about numbers of properties and more about expressions of cultural and natural diversity and themes from different regions.[189] A number of delegates at the 1998 meeting criticized their Committee colleagues, regretting that decisions concerning nominations were "sometimes disconnected from the implementation of the global strategy as had been seen by the high number of European sites the Committee had inscribed on the World Heritage List at its twenty-second session."[190] As for the specific recommendations from the Amsterdam meeting, the Committee referred them to the group tasked with revising the Operational Guidelines.

With regard to the concept of representativity, there was a tacit assumption that it would be evaluated through the lens of existing nation states. Not everyone agrees with this view. Cleere considers it a fundamental flaw of the Convention.

------

186   UNESCO, Report of the rapporteur on the twenty-second session of the World Heritage Committee in Kyoto, 30 November–5 December 1998, Paris, 29 January 1999, WHC-98/conf.203/18, paras. IX 8–12. Retrieved from http://whc.unesco.org/archive/1998/whc-98-conf203-18e.pdf

187   Jukka Jokilehto, "World heritage: defining the outstanding universal value," *City and Time*, 2/2 (2006), p. 3. Retrieved from http://www.ct.ceci-br.org

188   UNESCO, Report of the World Heritage global strategy natural and cultural heritage expert meeting, 25 to 29 March 1998, Amsterdam, the Netherlands, Paris, 20 October 1998, WHC-98/conf.203/inf. 7. Retrieved from http://whc.unesco.org/archive/1998/whc-98-conf203-inf7e.pdf; Bernd von Droste et al. eds, *Linking Nature and Culture: Report of the Global Strategy Natural and Cultural Heritage Expert Meeting 25 to 29 March 1998, Amsterdam, Netherlands* (The Hague: UNESCO/Ministry of Foreign Affairs, 1998), pp. 1–237.

189   For a thoughtful analysis of the tension between universality and cultural diversity, see Jean Musitelli, "Opinion: World Heritage, between Universalism and Globalization, *International Journal of Cultural Property*, 11/2 (2002), pp. 327–30.

190   UNESCO, Report of the rapporteur on the twenty-second session of the World Heritage Committee in Kyoto, 30 November–5 December 1998, Paris, 29 January 1999, WHC-98/conf.203/18, para. IX 16. Retrieved from http://whc.unesco.org/archive/1998/whc-98-conf203-18e.pdf

"It has to be based on modern states, modern countries, the frontiers of which in many cases bear no relationship to the cultural movements."[191] While acknowledging the dramatic imbalance of European sites, Ray Wanner, an American diplomat who attended many Committee sessions, voices his concern:

> One principle that I hear articulated that I had always been very uncomfortable with and I've heard Director-General Matsuura articulate it regularly, namely that ... every country has the right ...to inscriptions, I will put it that way, if not a certain number of inscriptions. I have never felt comfortable with that. At the end of the day, maybe the principles of hazard or something will end up where there will be in a sense a so-called equitable distribution of sites. But I think that should be at the end of the day and not as a principle ... I don't feel that every country has a right to a certain number of sites.[192]

To respond to the need to edit and reorganize the Operational Guidelines, an expert meeting finally took place in spring 2000 and began the lengthy process of restructuring the guidelines in an effort to create a more user-friendly version. All the pending revisions that the Committee had approved over time were lumped into this exercise which took five years to complete.[193]

In spite of the many efforts to implement the global strategy, the degree of dissatisfaction with the slow pace of change boiled over at the end of the decade. At the 1998 meeting, the outgoing Director of the World Heritage Centre warned that "the problem of regional imbalance will aggravate if the present trend continues. Of the 35 nominations to be reviewed by the Committee at this session, there is not a single property in Africa and only one site in the Arab States. In fact, the vast majority is from Europe."[194] Outraged, the Committee demanded that an item be placed on the agenda of the next General Assembly of States Parties to examine ways and means to ensure a representative World Heritage List. The fact that geographical distribution of World Heritage Sites had in fact deteriorated since 1994 contributed to an intense and emotionally charged debate. Stovel notes that "nothing has changed in spite of the best efforts, in spite of the concentration of funds and workshops in those regions to generate more nominations, in spite

---

191   Canada Research Chair, interview Cleere.

192   Canada Research Chair on Built Heritage, Université de Montréal, audio interview of Ray Wanner by Christina Cameron, Springfield, 18 May 2011.

193   UNESCO, Report of the rapporteur on the twenty-third session of the World Heritage Committee in Marrakesh, 29 November to 4 December 1999, Paris, 2 March 2000, WHC-99/conf.209/22, paras. XIIL 16–20. Retrieved from http://whc.unesco.org/archive/1999/whc-99-conf209-22e.pdf

194   UNESCO, Report of the rapporteur on the twenty-second session of the World Heritage Committee in Kyoto, 30 November–5 December 1998, Paris, 29 January 1999, WHC-98/conf.203/18, para. IV 3. Retrieved from http://whc.unesco.org/archive/1998/whc-98-conf203-18e.pdf

of the fact that ICCROM, for example, tried very hard for a decade to get African countries to produce more nominations by working with them."[195]

Frustration was fuelled by the disconnect between unanimous commitment to the global strategy and the fact that in 1997, for example, half the new nominations still came from Europe, with Italy leading the pack with an unprecedented 12 proposed sites.[196] There was also an issue of capacity. A participant at this session, Lopez Morales recalls a sense of alarm, "a red warning light that if there were going to be inscriptions of 60 or 70 cultural or natural properties a year, the future was really troubling because at that rate neither the secretariat nor the Committee could guarantee in a rigorous way the proper conservation of sites already inscribed."[197] UNESCO attributed the lack of nominations from under-represented regions to a lack of capacity of some countries "to prepare nominations at a sufficiently sustained rhythm" and urged the General Assembly to take action "so that in the future the List will not only be associated with limited categories of properties mainly situated in States with a solid conservation record to the exclusion of those States which devote an important part of their resources to health, education and the fight against poverty."[198]

The 1999 General Assembly of States Parties unanimously adopted a resolution that identified the responsibilities of each of the actors and laid out measures to reach an equitable representation on the World Heritage List. These measures included accelerated implementation of the global strategy action plan as well as greater bilateral and multi-lateral cooperation. Well-represented countries were encouraged to propose only properties in under-represented categories and to link their nominations with other countries whose heritage was under-represented. The most controversial measure was an appeal to countries with a substantial number of sites in the list to voluntarily slow down or even suspend future nominations. In the ensuing debate, many countries observed that the resolution would only

195   Canada Research Chair, interview Stovel, February 2011.

196   UNESCO, Examination of nominations of natural and cultural properties to the World Heritage List and the List of World Heritage in Danger, Paris, 7 May 1997, WHC-97/conf.204/3B. Retrieved from http://whc.unesco.org/archive/1997/whc-97-conf204-3be.pdf Originally Italy nominated 17 sites that year but was persuaded to reduce the number to 12.

197   Canada Research Chair on Built Heritage, Université de Montréal, audio interview of Francisco Lopez Morales by Christina Cameron, Brasilia, 3 August 2010. "Ce qui mettait déjà une espèce de feu rouge d'alarme de voir que si on commençait à inscrire une quantité à peu près en moyenne par année de soixante, soixante-dix biens culturels ou naturels, le futur était vraiment très inquiétant parce qu'avec ce rythme, le Secrétariat et le Comité ne pouvaient pas assurer d'une manière rigoureuse la bonne conservation des biens qui étaient déjà inscrits."

198   UNESCO, Summary record of the twelfth General Assembly of States Parties in Paris, 28–29 October 1999, Paris, 8 November 1999, WHC-99/conf.206/7, paras. 7-8. Retrieved from http://whc.unesco.org/archive/1999/whc-99-conf206-7e.pdf

succeed if "supported by the political will of the States."[199] To implement the resolution, yet another working group was created to prepare a report for the reform discussions in 2000.

In concluding this overview of populating the World Heritage List from its inception to 2000, one remarks that one of the positive aspects of the Convention is the lack of definition of outstanding universal value in the legal text. This lack of definition allows for a flexible and nuanced approach to its implementation. The general nature of the Convention text gradually became more specific as a result of decisions on listing criteria, authenticity and integrity as well as thematic studies. From the beginning to the year 2000, the requirement to identify sites of outstanding universal value fostered a rich global dialogue about heritage theory and practice.[200] As Stovel observes, "Our views of what constitutes heritage have evolved and continue to evolve and so what any given country might have thought appropriate to put on the list for itself 30 years ago is now different."[201] The high point occurred in the middle of the 1990s with ground-breaking studies and decisions on the nature of heritage. Munjeri believes that these efforts have been helpful. "If you look at the sites that are being inscribed from Africa, most of them fall in the categories of the global strategy: cultural landscapes, cultural itineraries, spiritual sites and the like. So if we go by that, then I see the positive dimension."[202]

The Convention's evolution could be characterized as the disappearance of the notion of artistic masterpiece, the emergence of an anthropological concept of cultural heritage, the re-interpretation of the concept of authenticity and the articulation of cultural landscapes as the connecting tissue between culture and nature. Near the end of the century, Pressouyre observed that:

> In the space of one generation, opposition between cultural heritage and natural heritage lost much of its validity. Coded terms such as landscapes, properties, ensembles, monuments, do not have the same meaning today as in 1972. The concept of authenticity, recently discussed at the Nara Conference in Japan, has evolved in a spectacular manner.[203]

---

199   UNESCO, Summary record of the twelfth General Assembly of States Parties in Paris, 28–29 October 1999, Paris, 8 November 1999, WHC-99/conf.206/7, paras. 38–9, 41.Retrieved from http://whc.unesco.org/archive/1999/whc-99-conf206-7e.pdf

200   Christina Cameron, "The Evolution of the Concept of Outstanding Universal Value," *Conserving the Authentic: Essays in Honour of Jukka Jokilehto*, Nicholas Stanley-Price and Joseph King eds (Rome, 2009), pp. 127–36.

201   Canada Research Chair, interview Stovel, February 2011.

202   Canada Research Chair, interview Munjeri.

203   UNESCO, Synthetic report of the second meeting on global strategy of the African cultural heritage and the World Heritage in Addis Ababa, Ethiopia, 29 July– 1 August 1996, Paris, 14 October 1996, WHC-96/conf.201/inf. 7, II. Retrieved from http://whc.unesco.org/archive/1996/whc-96-conf201-inf7e.pdf

This rich dialogue on the evolution of the concept of heritage is an outstanding achievement of this period. Despite Pressouyre's regret that the global strategy meeting was held behind closed doors and did not publish its rich debates, he judges that the reflections on authenticity and the global strategy make important epistemological contributions.[204]

Nonetheless at the turn of the century, dissatisfaction remained. The regional and thematic imbalances had not diminished. The majority of new proposals for inscription on the World Heritage List continued to originate from the northern hemisphere. While many extraordinary properties had been listed, some countries were passionate in their conviction that implementation was unfair with regard to inscriptions. This issue gnawed away at the credibility of the entire World Heritage system. The statistics speak for themselves. In the year 2000, three-quarters of World Heritage Sites were cultural and half came from Europe and North America. The issue of equitable distribution boiled over amid demands for fairness. As part of the World Heritage reform agenda at the turn of the millennium, a task force tackled the issue of representativity of the list, deemed by the Committee to be "the most difficult of the reform issues."[205] The monumental efforts to develop and implement a global strategy for a representative, credible and balanced World Heritage List had not yet succeeded. By the year 2000, mounting dissatisfaction with the listing process made it a high priority for the Committee's reform agenda.

---

204   Canada Research Chair, interview Pressouyre.

205   UNESCO, Report of the rapporteur of the twenty-fourth session of the World Heritage Committee in Cairns, 27 November–2 December 2000, Paris, 16 February 2001, WHC-2000/conf.204/21, para. VI.2.3. Retrieved from http://whc.unesco.org/archive/2000/whc-00-conf204-21e.pdf

Chapter 4

Conserving World Heritage Sites

Conserving places of outstanding universal value is the ultimate goal of the World Heritage Convention. This visionary treaty paved the way for significant cooperation among countries with the common purpose of safeguarding natural and cultural heritage properties of international significance. In the early years, the World Heritage Committee focused on the nominations process and the development of the World Heritage List. This chapter deals with the post-inscription obligation to conserve such properties for the benefit of present and future generations. It reviews how ideas about conservation and monitoring evolved and how key actors advanced them through policy development and practical experience. The chapter begins with the Committee's informal conservation efforts linked to the nominations process. It then examines how monitoring and management tools such as informal and systematic reporting, Danger listing and international assistance were gradually developed to support the conservation of World Heritage Sites.

Conservation and World Heritage

Over time, the Convention has arguably become the most influential instrument in heritage conservation globally.[1] However, in the early days of its implementation not all participants were well versed in conservation theory and practice. Prior to the 1972 Convention, UNESCO and the three advisory bodies had distinctly different experiences in the field of heritage conservation.

For natural sites, both UNESCO and IUCN were active in the field of conservation of natural resources long before 1972. International conservation efforts gained momentum through several important conferences including the 1949 United Nations Conference on the Conservation and Utilization of Resources (UNSCCUR) at Lake Success, New York, the 1968 UNESCO biosphere conference in Paris and the 1972 United Nations Conference on the Human Environment in Stockholm.[2] Instrumental in establishing nature conservation as a discipline, the three conferences also confirmed UNESCO in the lead role for natural heritage conservation within the United Nations system.

In the late 1940s, several inter-twined initiatives reveal the gathering strength of the environmental movement. In advance of UNSCCUR, preparatory work was

---

1   See chapter 6 for a fuller discussion of the significance of the World Heritage Convention.

2   See chapter 1 for greater detail on the 1968 UNESCO biosphere conference.

underway to establish the International Union for the Protection of Nature (IUPN, later IUCN).[3] A 1948 document links the two initiatives:

> During its second session, in November-December, 1947 in Mexico City, the General Conference of UNESCO instructed the Director-General of UNESCO to convene ... an International Conference on the Protection of Nature. While waiting for this resolution to be carried out, as now planned for 1949 in conjunction with the Scientific Conference on the Conservation and Utilization of Resources, the French Government judged it advisable to convene, this year, an International Conference for the purpose of adopting a final constitution for the International Union of the Protection of Nature.[4]

IUPN was established at Fontainebleau in 1948 with the support of UNESCO. The next year, the two organizations prepared the International Conference on the Protection of Nature, also at Lake Success, with an agenda focused on human ecology and resolutions that included new concepts like environmental impact assessments.

For cultural sites, efforts were mainly focused on the theory and practice of restoring monumental architecture and archaeological sites.[5] UNESCO gained significant operational experience through large scale projects starting with the 1960s campaign to move and restore the Nubian monuments in Egypt, followed by work at threatened sites at Borobudur (Indonesia), Venice (Italy) and elsewhere. At the same time, ICCROM developed hands-on capacity through restoration projects and international site missions.[6] ICOMOS, as an association of professionals, focused mainly on doctrine based on the landmark 1964 code for conservation practice, the International Charter for the Conservation and Restoration of Monuments and Sites (Venice Charter).

The World Heritage Convention captures the essence of these earlier efforts at preservation and conservation. In its preamble, it points to a global context of continuous threat, "noting that the cultural heritage and the natural heritage are increasingly threatened with destruction not only by the traditional causes of decay, but also by changing social and economic conditions which aggravate the situation with even more formidable phenomena of damage or destruction." While the specific words "monitoring" and "state of conservation" do not appear in the Convention text, they are implied in its stated purpose and in the

---

3   See chapter 5 for the establishment of IUPN, later IUCN.

4   UNESCO, Conference for the establishment of the International Union for the Protection of Nature, Paris, 20 July 1948, NS/UIPN, p. 3. Retrieved from http://unesdoc. unesco.org/images/0015/001547/154739eb.pdf

5   See chapter 1 for the development of UNESCO's cultural initiatives and chapter 5 for the work of ICOMOS and ICCROM.

6   Jukka Jokilehto, *ICCROM and the Conservation of Cultural Heritage. A History of the Organization's First 50 Years, 1959–2009* (Rome, 2011), pp. 1–174.

obligations of States Parties and the international community. The Convention requires countries to ensure "the identification, protection, conservation, presentation and transmission to future generations of the cultural and natural heritage" (article 4). Furthermore, it emphasizes "that such heritage constitutes a world heritage for whose protection it is the duty of the international community as a whole to co-operate" and obligates each State Party "not to take any deliberate measures which might damage directly or indirectly the cultural and natural heritage" (article 6.3). A requirement to report on World Heritage matters was included in the text (article 29), but remained dormant until it was activated in the 1990s when the issue of systematic monitoring and reporting emerged.

Nominations and Informal Conservation Oversight

While the World Heritage Convention gives primacy to conservation and protection, the first session of the World Heritage Committee did not directly discuss such matters because the sites to preserve were yet to be inscribed. In the early years, issues of conservation, management and monitoring inadvertently entered through the back door, either during the nomination and inscription process or through allocations of international assistance.

Throughout the first decade, the Committee and its Bureau sometimes made site-specific conservation recommendations during the inscription process. Two examples serve to illustrate the kinds of concerns that emerged for cultural heritage. First, the 1980 debate on the nomination of Palmyra (Syrian Arab Republic) raised issues about the boundaries of the nomination and threats from tourist development: "The necropolises and the remains of the Roman aqueduct which are situated outside the fortified walls should be included in the protected zone. The Committee draws attention to the hotel facilities on the site which should not, in its opinion, be further extended."[7] It is interesting to note that two years earlier UNESCO had dispatched a mission to the Syrian Arab Republic to advise authorities on potential sites for World Heritage listing. During this visit, the UNESCO specialist made a number of specific conservation observations and identified threats to the values, authenticity and integrity of various archaeological zones and ensembles, including Palmyra: "The integrity of the archaeological site is threatened on the one hand by an increasing and insufficiently planned urban

---

7   UNESCO, Report of the rapporteur on the fourth session of the World Heritage Committee in Paris, 1–5 September 1980, Paris, 29 September 1980, CC-80/conf. 016/10, para. 12. Retrieved from http://whc.unesco.org/archive/1980/cc-80-conf016-10e. pdf; ICOMOS evaluation of the site of Palmyra, Paris, May 1980, Retrieved from http://whc.unesco.org/archive/advisory_body_evaluation/023.pdf

scheme for the 'new village' (25,000 inhabitants) and by the spatial planning of
the region which does not take into account the importance of the place."[8]

The second example is the 1988 nomination of the archaeological site at Mystras
(Greece) which opened a bitter year-long debate on appropriate conservation policy
for ruins. The Greeks proposed to rebuild the roof and top storey of the Byzantine
palace to protect the building and open it to tourists. ICOMOS was fiercely opposed
on doctrinal grounds. It recommended against inscription, calling for retention of the
ruins on the grounds of authenticity and advising that "such extensive, irreversible
restoration … would seriously modify the aesthetic and historic links of the site whose
authenticity is specifically linked to its state as a ruin."[9] After the Greek delegation
threatened to withdraw from the meeting, the Committee ignored the advisory body's
advice and listed the property as a World Heritage Site in December 1989.

On the natural side, the case of Ichkeul National Park (Tunisia) demonstrates
the World Heritage Committee's vigilance in using the nomination process and
financial assistance to identify and overcome conservation problems. At its third
session in 1979, the Committee deferred the nomination of Ichkeul "until the
Tunisian Government has contacted the other States concerned to ensure adequate
protection of summering and wintering areas of major migratory species found in
Ichkeul."[10] The next year Tunisia made a commitment to "implement a plan for
corrective measures, as described in documents submitted to the Secretariat, so
that the integrity of Ichkeul National Park will be maintained in the future."[11] On
the basis of the State Party's response, the Committee inscribed the property on the
World Heritage List. The following year it approved $30,000 (US) for technical
cooperation at the site.[12] The matter came back to the ninth Committee session in
1985 when IUCN was so concerned about the critical situation at Ichkeul that it
proposed the site for the List of World Heritage in Danger, one of the conservation
tools created in the Convention (article 11.4):

> IUCN noted that if compensatory measures to re-establish the water regime of this
> Park were not taken very soon, the property would lose its international importance

---

8   UNESCO, Arno Heinz, République arabe syrienne, compte-rendu de mission,
13.3.-27.3.1978, p. 3.

9   ICOMOS evaluation for Mystras, Greece, 15 November 1989. Retrieved from
http://whc.unesco.org/archive/advisory_body_evaluation/511.pdf

10   UNESCO, Report of the rapporteur on the third session of the World Heritage
Committee in Cairo and Luxor, 22–26 October 1979, Paris, 30 November 1979, CC-79/
conf.003/13, para. 47. Retrieved from http://whc.unesco.org/archive/1979/cc-79-conf003-
13e.pdf

11   UNESCO, Report of the rapporteur on the fourth session of the World Heritage
Committee in Paris, 1–5 September 1980, Paris, 29 September 1980, CC-80/conf.016/10,
para. 12. Retrieved from http://whc.unesco.org/archive/1980/cc-80-conf016-10e.pdf

12   UNESCO, Report of the rapporteur on the fifth session of the World Heritage
Committee in Sydney, 26–30 October 1981, Paris, 5 January 1982, CC-81/conf/003/6, para.
32. Retrieved from http://whc.unesco.org/archive/1981/cc-81-conf003-6e.pdf

Figure 4.1   Ichkeul National Park, Tunisia

for migratory wildfowl. IUCN therefore strongly recommended the inclusion of this Park in the List of World Heritage in Danger ... The representative of Tunisia informed the Committee that he would take up this matter at the highest level and he would inform the Secretariat and IUCN of the results of this enquiry."[13]

When IUCN reiterated its concerns the following year, the pressure was kept on.[14] At the eleventh session in 1987 the State Party finally acknowledged the gravity of the situation, informing the Committee "that a UNESCO/World Heritage consultant was currently reviewing the situation of Ichkeul National Park which had been mentioned in the IUCN document: he stated that his country would certainly nominate this site to the List of World Heritage in Danger if this was recommended in the consultant's report."[15]

---

13   UNESCO, Report of the rapporteur on the ninth session of the World Heritage Committee in Paris, 2–6 December 1985, Paris, December 1985, SC-85/conf.008/9, para. 37. Retrieved from http://whc.unesco.org/archive/1985/sc-85-conf008-9e.pdf

14   UNESCO, Report of the rapporteur on the tenth session of the World Heritage Committee in Paris, 24–28 November 1986, Paris, 5 December 1986, CC-86/conf.003/10, para. 15. Retrieved from http://whc.unesco.org/archive/1986/cc-86-conf003-10e.pdf

15   UNESCO, Report of the rapporteur on the eleventh session of the World Heritage Committee in Paris, 7–11 December 1987, Paris, 20 January 1988, SC-87/conf.005/9, para

Although the integrity of this wetland site remained a major concern to the World Heritage Committee for years, the State Party did not request its inclusion on the List of World Heritage in Danger. It was almost ten years later when the Committee took a strong stand at its 1996 session:

> The Committee was informed that the Bureau at its twentieth extraordinary session recalled debates held concerning inclusion of the site on the List of World Heritage in Danger beginning in 1985 and considered the possibility of an eventual deletion of this property from the World Heritage List. The Bureau discussed if a rehabilitation of the site is at all possible and requested the Secretariat to write immediately to the Tunisian authorities to (a) inform them about the Bureau's concerns, (b) to inform them about the Bureau's recommendation to include the site on the List of World Heritage in Danger, and (c) to inform them of the possible deletion of Lake Ichkeul from the World Heritage List if the integrity of the site is lost.[16]

In an interview, Hal Eidsvik, who worked for IUCN at that time, looks back on the inscription of Ichkeul and its conservation problems:

> Ichkeul was marginally qualified to meet the criteria by migratory species in effect. But even at the time when it was put up on the list, there were dams, irrigation dams, the lake was being drawn down and you could see the threat that was there. So Ichkeul is one that I still feel very uncomfortable about.[17]

The 1996 session proceeded to inscribe the property on the Danger List and asked Tunisia to provide a programme of corrective measures to reverse the degradation. Moreover, the Committee informed the authorities "of the possibilities of the deletion of the property from the World Heritage List if rehabilitation of the site would not be possible."[18] The site was only removed from the List of World Heritage in Danger in 2006.

What the case of Ichkeul National Park illustrates is the gradual development of the Committee's conservation processes. From the 1980s to the early 1990s, it articulated concepts for conservation, state of conservation reporting, Danger listing including specific reasons and corrective measures, and potential delisting.

---

19. Retrieved from http://whc.unesco.org/archive/1987/sc-87-conf005-9e.pdf

16 UNESCO, Report of the rapporteur on the twentieth session of the World Heritage Committee in Merida, 2–7 December 1996, Paris, 10 March 1997, WHC-96/conf.201/21, para. VII.36. Retrieved from http://whc.unesco.org/archive/1996/whc-96-conf201-21e.pdf

17 Canada Research Chair on Built Heritage, Université de Montréal, audio interview of Hal Eidsvik by Christina Cameron, Ottawa, 3 July 2009.

18 UNESCO, Report of the rapporteur on the twentieth session of the World Heritage Committee in Merida, 2–7 December 1996, Paris, 10 March 1997, WHC-96/conf.201/21, para. VII.36. Retrieved from http://whc.unesco.org/archive/1996/whc-96-conf201-21e.pdf

This case is also instructive as a demonstration of the collaboration among different international instruments and programmes. Ichkeul became a UNESCO Biosphere Reserve in 1977 and a wetland of international importance in 1980 under the Ramsar Convention. As a general practice, UNESCO coordinated exchanges of information among these programmes on critical conservation issues at specific natural sites designated under each of the programmes and conventions.[19]

In the early years, no description of conservation and reporting requirements existed in the Committee's Operational Guidelines. Such direction as there was came from Committee comments during the nomination process on conservation issues at specific properties. Until 1984, no site visits were carried out by the advisory bodies as part of the nomination process; inscriptions were based on written documentation rather than on the verification of the actual state of conservation. Informally, information on some properties came from IUCN through its field network and from UNESCO through its missions and safeguarding campaigns.

Monitoring World Heritage Sites

Developing Reactive Monitoring: Ad Hoc State of Conservation Reports

The roots of reactive monitoring go back to 1982 when the World Heritage Committee supported the idea of being regularly informed about the state of conservation of World Heritage Sites, the measures taken to protect and manage them, and the activities undertaken with assistance from the World Heritage Fund. "Although the principle of yearly reporting was considered to be highly desirable," the Committee recognized that this question would require careful study and asked IUCN, ICOMOS and ICCROM to prepare advice on the contents of such reports and the procedures required.[20]

Only IUCN responded the following year to the Committee's request by presenting a theoretical framework on monitoring natural World Heritage properties.[21] This marks the unofficial launch of monitoring the state of conservation of World Heritage Sites. While the IUCN document was robust and forward-looking, the Committee response was tentative:

---

19   This collaboration only works for natural sites because there are no other place-based international instruments for cultural heritage.

20   UNESCO, Report of the rapporteur on the sixth session of the World Heritage Committee in Paris, 13–17 December 1982, Paris, 17 January 1983, CLT-82/conf.015/8, paras. 47–8. Retrieved from http://whc.unesco.org/archive/1982/clt-82-conf015-8e.pdf

21   IUCN, Monitoring natural world heritage properties, Gland, 11 April 1983, SC-83/conf.009/6, paras. 1–4. Retrieved from http://whc.unesco.org/archive/1983/sc-83-conf009-6e.pdf; UNESCO, Report of the rapporteur on the seventh session of the World Heritage Committee in Florence 5–9 December 1983, Paris, January 1984, SC/83/conf.009/8, para. 40. Retrieved from http://whc.unesco.org/archive/1983/sc-83-conf009-8e.pdf

The Committee considered that it was highly desirable to be regularly informed of the state of conservation of World Heritage properties, particularly on measures undertaken to protect and manage these properties and on the way in which the funds allocated under the World Heritage Fund are used. However, the Committee preferred not to establish a formal reporting system at the present time and rather encouraged IUCN, ICOMOS and ICCROM to collect information through their experts. The Committee will continue to seek information from States Parties on an *ad hoc* basis whenever this is necessary for making its decisions.[22]

This statement is remarkable in that it confirms the primary focus of the Convention on conservation and the need for accurate information as a basis for decision making. It also highlights the role of the advisory bodies in collecting relevant information through their expert networks. Significantly, it anticipates two major future developments: ad hoc state of conservation information which eventually became "reactive monitoring" and a formal monitoring system that only materialized in the late 1990s as "periodic reporting."

At this same 1983 meeting, the first state of conservation report on a specific site was presented orally to the Committee by Jim Thorsell, newly arrived at IUCN. He recalls the Florence session:

I would say that a significant thing happened at that meeting, however. I gave what I think was the very first monitoring report for any site ever on the World Heritage List. And that took place only because I was living in Tanzania and I had just come from a two-week study tour with a bunch of my African students to Ngorongoro and I was quite disturbed at what was going on there. So at the meeting, there somehow came up an opportunity to introduce this whole idea of monitoring and I gave them probably a 15 or 20 minute summary of what I thought the issues were at Ngorongoro. I remember there was a one or two minute total silence after I gave this report. The Committee members couldn't believe that one of their sites could be in so much trouble.[23]

After his presentation, the Committee asked the State Party to "initiate the procedure for including this property in the List of World Heritage in Danger."[24] Ngorongoro was duly inscribed on the Danger List in 1984.

---

22   UNESCO, Report of the rapporteur on the seventh session of the World Heritage Committee in Florence, 5–9 December 1983, Paris, January 1984, SC/83/conf.009/8, para. 41. Retrieved from http://whc.unesco.org/archive/1983/sc-83-conf009-8e.pdf

23   Canada Research Chair on Built Heritage, Université de Montréal, audio interview of Jim Thorsell by Christina Cameron, Banff, 11 August 2010.

24   UNESCO, Report of the rapporteur on the seventh session of the World Heritage Committee in Florence, 5–9 December 1983, Paris, January 1984, SC/83/conf.009/8, para. 43. Retrieved from http://whc.unesco.org/archive/1983/sc-83-conf009-8e.pdf

Having been encouraged by the Committee to collect information through its expert networks, IUCN reported the next year on what it considered to be critical conservation matters.[25] It presented the state of conservation of four natural sites: Simien National Park (Ethiopia), Mount Nimba (Côte d'Ivoire and Guinea), Tai National Park (Côte d'Ivoire) and Durmitor National Park (Yugoslavia, now Montenegro).[26] At the same session, IUCN recommended and the Committee adopted Danger listing for three natural sites: Djoudj National Bird Sanctuary (Senegal), Ngorongoro Conservation Area (Tanzania) and Garamba National Park (Zaire, now Democratic Republic of the Congo).[27] With regard to Wieliczka Salt Mine (Poland), the only cultural property under consideration for inscription on the Danger List, ICOMOS played a less active role. It did not present any specific assessment but merely transmitted a report from the national government:

> ICOMOS provided the Committee with the information which the Polish authorities had given for this property. The Committee considered that there was insufficient geological information at present to evaluate the dangers facing this property. The Committee therefore decided to defer a decision on this nomination until more information had been obtained and expressed the wish that in the meantime the national authorities concerned ensure the necessary protection.[28]

The following year, IUCN institutionalized its work by reporting on the state of conservation of twelve World Heritage properties, organized by priority into three groups: sites on the Danger List, sites for possible inclusion on the Danger List, and other natural properties.[29] IUCN was able to take on this role because

---

25 UNESCO, Report of the rapporteur on the eighth session of the World Heritage Committee in Buenos Aires, 29 October–2 November 1984, Buenos Aires, 2 November 1984, SC/84/conf.004/9, para. 40. Retrieved from http://whc.unesco.org/archive/1984/sc-84-conf004-9e.pdf

26 UNESCO, Report of the rapporteur on the eighth session of the World Heritage Committee in Buenos Aires, 29 October–2 November 1984, Buenos Aires, 2 November 1984, SC/84/conf.004/9, para. 40. Retrieved from http://whc.unesco.org/archive/1984/sc-84-conf004-9e.pdf

27 UNESCO, Report of the rapporteur on the eighth session of the World Heritage Committee in Buenos Aires, 29 October–2 November 1984, Buenos Aires, 2 November 1984, SC/84/conf.004/9, para. 26. Retrieved from http://whc.unesco.org/archive/1984/sc-84-conf004-9e.pdf

28 UNESCO, Report of the rapporteur on the eighth session of the World Heritage Committee in Buenos Aires, 29 October–2 November 1984, Buenos Aires, 2 November 1984, SC/84/conf.004/9, para. 26. Retrieved from http://whc.unesco.org/archive/1984/sc-84-conf004-9e.pdf

29 UNESCO, Report of the rapporteur on the ninth session of the World Heritage Committee in Paris, 2–6 December 1985, Paris, December 1985, SC-85/conf.008/9, para. 37. Retrieved from http://whc.unesco.org/archive/1985/sc-85-conf008-9e.pdf

it had access to reliable and current information on natural World Heritage Sites through its field offices as well as broad membership and collaboration with other non-governmental organizations. In recognition of its leadership, a new and separate agenda item on the state of conservation was introduced in 1985 as the "conservation status of natural properties inscribed on the World Heritage List and List of World Heritage in Danger."[30]

It took longer to develop ad hoc monitoring approaches for cultural sites. In 1983 UNESCO worked with ICOMOS and ICCROM on a handbook for managing World Heritage that dealt with "the preservation of these properties, ranging from general principles and legal considerations to practical means for carrying out a management programme The outline was elaborated by an international group of experts during a meeting organized by ICCROM and ICOMOS in 1983 at the suggestion of the Secretariat."[31] Despite the invitation from the 1983 Committee, ICOMOS and ICCROM did not immediately present monitoring reports on cultural sites. Even for seriously threatened properties like the Royal Palaces of Abomey (Benin), information reached the Committee from other sources: "Taking account in particular of the considerable damage caused by the 1984 tornado and the urgency of the work needed to preserve the site, the Committee decided to include the Royal palaces of Abomey (Benin) on the List of World Heritage in Danger."[32] The contrast in capacity between ICOMOS and IUCN at this time is striking in terms of people and networks to gather information. At this same 1985 session, the Committee specifically thanked IUCN "for these comprehensive reports and for regularly providing information on the status of natural properties. It furthermore welcomed the proposal of ICOMOS to submit similar reports, as far as its means would allow, in the near future."[33]

By 1987, cultural heritage reporting was carried out more by the secretariat than by ICOMOS. At the 1987 Committee meeting, Raidl from UNESCO's Cultural Heritage Division reported on the state of conservation of cultural sites[34] and in

---

30   UNESCO, Report of the rapporteur on the ninth session of the World Heritage Committee in Paris, 2–6 December 1985, Paris, December 1985, SC-85/conf.008/9, para. 37. Retrieved from http://whc.unesco.org/archive/1985/sc-85-conf008-9e.pdf

31   UNESCO, Report of the rapporteur on the eighth session of the World Heritage Committee in Buenos Aires, 29 October–2 November 1984, Buenos Aires, 2 November 1984, SC/84/conf.004/9, para. 37. Retrieved from http://whc.unesco.org/archive/1984/sc-84-conf004-9e.pdf

32   UNESCO, Report of the rapporteur on the ninth session of the World Heritage Committee in Paris, 2–6 December 1985, Paris, December 1985, SC-85/conf.008/9, para. 30. Retrieved from http://whc.unesco.org/archive/1985/sc-85-conf008-9e.pdf

33   UNESCO, Report of the rapporteur on the ninth session of the World Heritage Committee in Paris, 2–6 December 1985, Paris, December 1985, SC-85/conf.008/9, para. 38. Retrieved from http://whc.unesco.org/archive/1985/sc-85-conf008-9e.pdf

34   UNESCO, Report of the rapporteur on the eleventh session of the World Heritage Committee in Paris, 7–11 December 1987, Paris, 20 January 1988, SC-87/conf.005/9, para 14. Retrieved from http://whc.unesco.org/archive/1987/sc-87-conf005-9e.pdf

1989 on an earthquake in Tipasa (Algeria).[35] But it was a Committee member from Canada who intervened about the potential impact of an upcoming universal exhibition in Venice, prompting the Committee to immediately express its "grave concerns about the new threats to Venice" noting that "a universal exhibition, which would attract several hundreds of thousands of visitors ... risks threatening the integrity of this heritage."[36] In 1990, UNESCO staff continued to report on the state of conservation of cultural properties.[37] It is quite telling that seventeen years after the adoption of the Convention and eleven years after the listing of the first sites, no adequate system was in place for monitoring cultural heritage properties. Ad hoc monitoring was truly ad hoc.

At the same 1990 session, the Committee took a key decision that affected future implementation of the Convention. It was a response to the proliferation of infrastructure projects at or near many World Heritage Sites. As a result of a debate concerning major new construction at the Monastery of the Hieronymites and the Tower of Belém (Portugal), the Committee added a new requirement to its Operational Guidelines. In future, States Parties were asked "to inform the Committee, through the UNESCO Secretariat, of their intention to undertake or to authorize in an area protected under the Convention major restorations or new constructions which may affect the World Heritage value of the property." Further, countries were directed to give early notice "before making any decisions that would be difficult to reverse, so that the Committee may assist in seeking appropriate solutions to ensure that the world heritage value of the site is fully preserved."[38] This important paragraph has been used ever since to remind States Parties of their obligation to inform the Committee of large–scale building and infrastructure developments which might impact on the outstanding universal value of sites. It serves as a major tool in a process of exchange within the World Heritage system. While it is not the origin of reactive monitoring as such, it is an important building block for this activity. The wording in the Operational Guidelines remains essentially unchanged to this day.[39]

---

35   UNESCO, Report of the rapporteur on the thirteenth session of the World Heritage Committee in Paris, 11–15 December 1989, Paris, 22 December 1989, SC-89/conf.004/12, paras. 19–20. Retrieved from http://whc.unesco.org/archive/1989/sc-89-conf004-12e.pdf

36   UNESCO, Report of the rapporteur on the thirteenth session of the World Heritage Committee in Paris, 11–15 December 1989, Paris, 22 December 1989, SC-89/conf.004/12, para. 22. Retrieved from http://whc.unesco.org/archive/1989/sc-89-conf004-12e.pdf

37   UNESCO, Monitoring of the state of conservation of World Heritage cultural properties, Paris, 15 October 1990, CC-90/conf.004/03. Retrieved from http://whc.unesco.org/archive/1990/cc-90-conf004-3e.pdf

38   UNESCO, Report of the rapporteur on the fourteenth session of the World Heritage Committee in Banff, 7–12 December 1990, Banff, 12 December 1990, CLT-90/conf.004/13, para. 24. Retrieved from http://whc.unesco.org/archive/1990/cc-90-conf004-13e.pdf

39   UNESCO, Operational Guidelines for the implementation of the World Heritage Convention, WHC.11/01 rev. November 2011, para. 172. Retrieved from http://whc.unesco.org/archive/opguide11-en.pdf

With his long experience with World Heritage, Pressouyre assesses the usefulness of this alert system. In an interview, he wryly remarks that some countries see inscription as the end of the process: "There was a time when States Parties considered that's good, that's it, it's been inscribed, very good. It was one more small distinction, but it did not lead to further reflection."[40] He explains why an early warning system is needed:

> There have been significant cases where, after inscription, terrible things have happened that should not have happened and there are also cases where, through a failure to react, States Parties have acted as if the World Heritage Convention does not exist and have not had the courtesy to warn that they were going to undertake some sort of urban development ... that would change the nature of the property. ... I'm not sure it is won. There are so many examples of interventions that were made without prior reference to the World Heritage Centre or the Committee.[41]

By the twentieth anniversary of the World Heritage Convention in 1992, there was still no systematic monitoring system in place despite many discussions on concepts and methodologies. For more than ten years, the Committee had received reports from IUCN on the state of conservation of natural World Heritage Sites but for cultural properties, the different approaches and overlapping reports from the secretariat and ICOMOS, working with its limited means, meant that the Committee was not informed in a systematic way. The 1992 session illustrates the situation. UNESCO reported on the methodology and results of the UNDP-UNESCO regional exercise in Latin America and the Caribbean as well as on nine cultural properties. Then ICOMOS presented its own monitoring report with some theoretical and methodological reflections and fourteen site-specific reports. IUCN followed with a presentation on its seven-step process and reports on twenty sites.[42]

---

40 Canada Research Chair on Built Heritage, Université de Montréal, audio interview of Léon Pressouyre by Christina Cameron and Mechtild Rössler, Paris, 18 November 2008. "Il y a eu un moment où les États parties considéraient que bon, ça y est, c'était inscrit, très bien, ça faisait une petite distinction de plus, mais ça n'engageait pas de réflexion supplémentaire."

41 Canada Research Chair, interview Pressouyre. "Il y a eu des grands cas où après l'inscription, il y a eu des choses terribles qui se sont produites et qui n'auraient pas dû se produire et il y a eu aussi des cas où, par manque de réactivité des États parties ont fait comme si la Convention du patrimoine mondial n'existait pas et n'ont pas simplement eu la courtoisie d'avertir qu'ils allaient faire telle opération d'urbanisme ... qu'elle est changée, la nature du bien ... Je ne suis pas sûr que ça soit gagné et il y a tant d'exemples d'intervention qui ont été faites sans référence préalable au Centre du patrimoine mondial ni au Comité."

42 UNESCO, Report of the rapporteur for the sixteenth session of the World Heritage Committee in Santa Fe, 7–14 December 1992, 14 December 1992, WHC-92/conf.002/12, paras. VIII.3–13. Retrieved from http://whc.unesco.org/archive/1992/whc-92-conf002-12e.pdf

By the mid-1990s, the Committee took significant steps to formalize the monitoring process. The 1994 Operational Guidelines added to the Committee's responsibilities a fourth essential function to "monitor the state of conservation of properties inscribed on the World Heritage List."[43] In 1996, a whole new section on monitoring appeared in the Operational Guidelines subdivided into a section on systematic monitoring and reporting, and another on reactive monitoring.[44] The latter was defined as "reporting by the World Heritage Centre, other sectors of UNESCO and the advisory bodies to the Bureau and the Committee on the state of conservation of specific World Heritage sites that are under threat."[45] There followed a period during which the Bureau played the role of clearing house for reactive monitoring reports prior to Committee sessions. For example the Bureau at its twenty-first extraordinary session in November 1997 examined reports of fifty-one sites, of which the Committee only reviewed twenty-one directly and took note of the others.[46] This role of the Bureau was later discontinued as part of the 2000 reform agenda.

Proof that better reactive monitoring could produce positive results may be seen in two cases from the late 1990s: the group of sites from the Democratic Republic of the Congo and the Mexican site of El Vizcaino. Four sites from Zaire, now the Democratic Republic of the Congo, were inscribed on the World Heritage List between 1979 and 1984: Virunga National Park, Garamba National Park, Kahuzi-Biega National Park and Salonga National Park. Reactive monitoring reports were presented from time to time individually on these sites. Okapi Wildlife Reserve from the same country was added to the list in 1996. With civil unrest spreading in the great lakes region, four of these sites were inscribed on the List of World Heritage in Danger between 1994 and 1997 and the fifth, Salonga National Park, in 1999. It was a unique situation to have all five World Heritage Sites from a single country inscribed as endangered and therefore requiring priority action. The Committee's decisions led to significant investments of funding and expertise from UNESCO, governments, non-governmental organizations and the United Nations Foundation (UNF). The Committee was especially proud of the UNF commitment:

---

43   UNESCO, Operational Guidelines for the implementation of the World Heritage Convention, WHC/2/rev. February 1994, para 3. Retrieved from http://whc.unesco.org/archive/opguide94.pdf

44   UNESCO, Operational Guidelines for the implementation of the World Heritage Convention, WHC/2/rev. February 1996, paras 68–75. Retrieved from http://whc.unesco.org/archive/opguide96.pdf

45   UNESCO, Operational Guidelines for the implementation of the World Heritage Convention, WHC/2/rev. February 1996, para 75. Retrieved from http://whc.unesco.org/archive/opguide96.pdf

46   UNESCO, Report of the rapporteur on the twenty-first session of the World Heritage Committee in Naples, 1–6 December 1997, Paris, 27 February 1998, WHC-97/conf.208/17, para. VII.31. Retrieved from http://whc.unesco.org/archive/1997/whc-97-conf208-17e.pdf

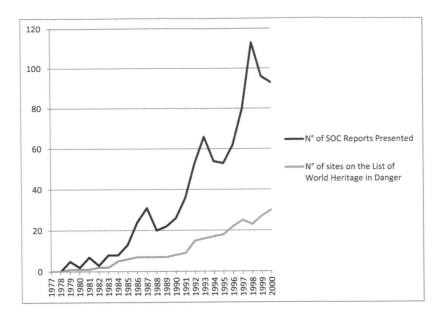

(Source: UNESCO, M. Rössler/C. Bigio)

Figure 4.2   Evolution of the number of state of conservation reports and the List of World Heritage in Danger 1978–2000

the support ... will be extended over a period of 4 years through a project, approved by the United Nations Foundation (UNF), for a sum of US$ 4,186,600 and entitled: "Biodiversity Conservation in Regions of Armed Conflict: Protecting World Natural Heritage in the Democratic Republic of the Congo." UNF will provide US$ 2,895,912 as an outright contribution and co-operate with UNESCO and its partners to raise the balance of US$ 1,290,688 from alternative sources.[47]

Attracting substantial funding for the conservation of severely threatened World Heritage Sites in the Democratic Republic of the Congo was a major achievement for the Convention.

A second victory for World Heritage conservation efforts was the case of the Whale Sanctuary of El Vizcaino (Mexico) where a mission had reviewed the potential impacts of a proposed salt production operation on the outstanding

---

47   UNESCO, Report of the rapporteur on the twenty-third session of the World Heritage Committee in Marrakesh, 29 November–4 December 1999, Paris, 2 March 2000, WHC-99/conf.209/22, para. X.4. Retrieved from http://whc.unesco.org/archive/1999/whc-99-conf209-22e.pdf

Figure 4.3   Kahuzi-Biega National Park, Democratic Republic of the Congo

universal value of this key site for the migratory species of the grey whale population. The 1999 Committee reviewed the mission's findings which proposed Danger listing in the event that the salt works proceeded. The Committee judged that the site was not in danger at that time, noting that "scientific data show that the whale population is not endangered and continues to increase." However it sent a strong signal to the State Party: "If any significant change to the present situation should occur, documented by appropriate evidence, the conclusion concerning the site's status under the World Heritage Convention should be promptly re-evaluated in co-operation and co-ordination with the State Party."[48] Following high level discussions involving World Heritage representatives and the State Party, the President of Mexico announced in March 2000 that development of the proposed salt works at San Ignacio would not proceed in order to protect the unique features of the World Heritage Site.[49]

---

48   UNESCO, Report of the rapporteur on the twenty-third session of the World Heritage Committee in Marrakesh, 29 November–4 December 1999, Paris, 2 March 2000, WHC-99/conf.209/22, para. X.25. Retrieved from http://whc.unesco.org/archive/1999/whc-99-conf209-22e.pdf

49   Pedro Rosabal and Mechtild Rössler, "A model of teamwork: El Vizcaino," *World Conservation*, 2 (2001), 21.

Figure 4.4   Whale Sanctuary of El Vizcaino, Mexico

With such positive results, Matsuura as Chairperson of the World Heritage Committee reported to the 1999 General Assembly of States Parties on the success of reactive monitoring but warned about the increased workload:

> During the past two years, over 200 state of conservation reports have been presented to the Bureau and to the Committee. The increasing number of cases being brought to the attention of the Committee is an indication of the growing reach of the Convention and the authority of the Committee. The Committee's knowledge of conservation problems at sites is essential in mobilizing international co-operation for their safeguard. However, given the time constraint inherent in the Committee's current method of work, how seriously can the Committee treat each case?[50]

Management planning
The World Heritage Convention provided a framework for developing principles and tools of good practice. The emergence of a common standard for site management and a concept for management effectiveness are important

---

50   UNESCO, Report of the rapporteur for the twelfth session of the General Assembly of the States Parties to the Convention Concerning the Protection of the World Cultural and Natural Heritage , 28–29 October 1999, 8 November 1999, WHC-99/conf.206/7, annex I. Retrieved from http://whc.unesco.org/archive/1999/whc-99-conf206-7e.pdf

components of a global conservation system. The idea of management planning first appeared in the 1980 Operational Guidelines, hidden in brackets under the nomination section.[51] In practical terms, it appeared in 1981 during a discussion of the Willandra Lakes Region (Australia), when the Committee expressed its desire "to see a management plan rapidly established for the whole area."[52]

As seen earlier with regard to monitoring methods, management planning was more advanced among professionals working with natural properties. While management plans had long been prepared for national parks and natural sites in some parts of the world, such plans were not often used for cultural sites. Given the inclusion of both natural and cultural heritage under the Convention, the Committee in 1981 requested that, for cultural sites: "the Secretariat examine with ICCROM and ICOMOS the question of protection and management of listed properties and report back to the Committee."[53] At the same time, noting that for natural properties the greatest conservation problems were related to integrity and management, it decided to "encourage States Parties to prepare a management plan appropriate to the capacity of the country concerned for each property nominated and to make such plans available when technical co-operation is requested."[54]

In his interview, McNeely, IUCN's representative at that time, recalls the emphasis on management:

> The other thing that IUCN has been really strict about, or has tried to be, is to look at the management effectiveness of the area; and so, if an area physically qualified or seemed to qualify, but was not sufficiently well managed then we would defer, recommend deferral, well, it was never our decision, it was always the Committee's decision, our recommendation would be to defer ... we would say "defer because" and give them reasons and tell them what they needed to do to make this qualify in terms of management and I think that this has been, in many ways, ..I won't say more important than being on the list, but it's been a way the list has been used to improve the management of the site.[55]

---

51   UNESCO, Operational Guidelines for the implementation of the World Heritage Convention, Report of the rapporteur on the fourth session of the World Heritage Committee in Paris, 1–5 September 1980, Paris, 29 September 1980, CC-80/conf.016/WHC/2 rev., para. 33. Retrieved from http://whc.unesco.org/archive/1980/opguide80.pdf

52   UNESCO, Report of the rapporteur on the fifth session of the World Heritage Committee in Sydney, 26–30 October 1981, Paris, 5 January 1982, CC-81/conf/003/6, para. 15. Retrieved from http://whc.unesco.org/archive/1981/cc-81-conf003-6e.pdf

53   UNESCO, Report of the rapporteur on the fifth session of the World Heritage Committee in Sydney, 26–30 October 1981, Paris, 5 January 1982, CC-81/conf/003/6, para. 28. Retrieved from http://whc.unesco.org/archive/1981/cc-81-conf003-6e.pdf

54   UNESCO, Report of the rapporteur on the fifth session of the World Heritage Committee in Sydney, 26–30 October 1981, Paris, 5 January 1982, CC-81/conf/003/6, para. 29 (b). Retrieved from http://whc.unesco.org/archive/1981/cc-81-conf003-6e.pdf

55   Canada Research Chair on Built Heritage, Université de Montréal, audio interview of Jeff McNeely by Mechtild Rössler, Gland, 17 September 2010.

As early as 1977, the Committee recommended that "when setting the boundary of a property to be nominated to the List, the concept of a buffer zone around the property may be applied where appropriate and feasible."[56] The requirement for management plans first appeared in 1980 and then more explicitly in 1983 when the Operational Guidelines called on States Parties to prepare plans "for the safeguarding of each cultural property nominated".[57] Yet it would seem that these directives were largely ignored by cultural heritage experts. Robertson Vernhes describes a seminal moment during a working session with the advisory bodies in the mid-1980s:

> I remember that it was very interesting because we were talking about protection of natural heritage and we were talking about the need for a management plan, adequate protection and things like that, and I do remember that the cultural heritage people who were in the same room said, "Oh, that's a good idea. We haven't asked that for cultural heritage." And they got the idea. They said, "Yes, we have never actually asked that the properties themselves for the cultural heritage actually have a protected status." In other words it could have been a church or some building but had there been a planning permission given for a car park to be placed right next to it, that was nothing against the Convention. I think that was rather seminal in that it did sort of make the cultural heritage people click and realize that there was an idea of buffer zones and things like that. I think that was important.[58]

Feilden recalls UNESCO's initiative at this time to improve management effectiveness for cultural sites:

> In the mid-1980s, Anne Raidl of UNESCO convened an international meeting of experts, to consider the management of World Cultural Heritage Sites and write the detailed Guidelines ... I was nominated to carry out this task, and after some delay, produced a first draft of the chapters. This was circulated for comment.

---

56 UNESCO, Operational Guidelines for the implementation of the World Heritage Convention, Final report of the first session of the intergovernmental committee for the protection of the world cultural and natural heritage in Paris, 27 June–1 July 1977, Paris, 20 October 1977, CC-77/conf. 001/8 rev., para. 25. Retrieved from http://whc.unesco.org/archive/1977/cc-77-conf001-8reve.pdf

57 UNESCO, Operational Guidelines for the implementation of the World Heritage Convention, Report of the rapporteur on the fourth session of the World Heritage Committee in Paris, 1–5 September 1980, Paris, 29 September 1980, CC-80/conf.016/WHC/2 rev., para. 33. Retrieved from http://whc.unesco.org/archive/1980/opguide80.pdf; UNESCO, Operational Guidelines for the implementation of the World Heritage Convention, WHC/2 rev. November 1983, para. 18. Retrieved from http://whc.unesco.org/archive/opguide83.pdf

58 Canada Research Chair on Built Heritage, Université de Montréal, audio interview of Jane Robertson Vernhes by Christina Cameron and Mechtild Rössler, Paris, 24 November 2009.

I then asked Dr Jukka Jokilehto to co-author it. UNESCO did not respond to our draft. During each of about 3 or 4 years on my annual stint of lecturing at ICCROM, I re-read the draft and improved it. As UNESCO had not responded with comments or approval, ICCROM decided to publish it as "Management Guidelines for World Cultural Heritage Sites" in 1993.[59]

These guidelines proposed an expert management committee for each World Heritage Site whose aim, according to Feilden, was "to separate management from political manipulation, manage the care and maintenance of the Site, and to manage the impact of tourism."[60]

Commenting on the challenges of bridging different cultures and regional approaches to site management, former Chairperson Collinson notes the example of Sichuan Giant Panda Sanctuaries (China) that illustrates a positive impact of the Convention:

> It's one of the difficulties in trying to have a Convention involving a lot of individual countries who have their own internal perspective of what's important in terms of heritage that may or may not correspond to some international definition and including how it's managed. And I can give an example, another example because they've changed their views and methods drastically since that time. If I remember correctly, China had recently signed the Convention and put forward nine panda reserves for nomination ... Yet the management of natural areas at that time in China ... wasn't based on what we would call an ecological systems approach but on a species approach, I guess would be the way to put it. And so, by banning the hunting of pandas, China honestly believed that they had done the right thing and that was going to solve the problem.

But Collinson points out that Chinese representatives did not realize the impact of nearby roads and bamboo harvesting that were chasing the pandas away and reducing their habitat: "When it was pointed out to them, they changed their management plans considerably and that made a difference. That was a positive example of a Member State having a management approach that wasn't appropriate for what the Convention described and being willing to make adjustments to that."[61]

From 1980 on, the Committee encouraged countries to prepare plans for managing and safeguarding World Heritage Sites. The aim was to ensure that the properties were protected and conserved through effective management. In his statement to the World Heritage Committee on the occasion of the twentieth anniversary, Train recognized the achievements of the Convention in this regard:

---

59   Canada Research Chair on Built Heritage, Université de Montréal, written response to interview questions by Bernard Feilden, Bawburgh, Norwich, December 2007.

60   Canada Research Chair, interview Feilden.

61   Canada Research Chair on Built Heritage, Université de Montréal, audio interview of Jim Collinson by Christina Cameron, Windsor, 12 July 2010.

"It has raised management standards and, most importantly has provided technical training opportunities, particularly on a regional basis. World Heritage status has become an important bulwark against actions which threaten the integrity of listed areas and sites."[62]

Traditional management

Progress in standard setting through World Heritage discussions also touched traditional management systems. In this case, cultural properties paved the way in 1992, to be followed at the end of the decade by natural sites. It was the development of the cultural landscapes category that brought into focus the involvement of communities and traditional management systems emanating from those who live on the land. In the early years, the Committee's Operational Guidelines actively discouraged community participation during the nomination process: "So as to maintain the objectivity of the evaluation process and to avoid possible embarrassment to those concerned, States Parties should refrain from giving undue publicity to the fact that a property has been nominated for inscription pending the final decisions of the Committee on the nomination in question."[63] The Committee made a fundamental change in 1992 with regard to cultural landscapes by adding in the guidelines that "participation of local people in the nomination process is essential to make them feel a shared responsibility with the State Party in the maintenance of the site, but should not prejudge future decision-making by the Committee."[64] This reversal acknowledged the contribution of local communities and indigenous peoples towards the stewardship of cultural properties and their responsibility in the transmission of this heritage.

With regard to traditional management systems, the Committee changed its Operational Guidelines as part of the cultural landscapes debates to allow for customary law and traditional management. The new wording stated that sites must "have adequate legal and/or contractual and/or traditional protection and management mechanisms to ensure the conservation of the nominated cultural properties or cultural landscapes."[65] This revision had substantive implications for World Heritage practices, including the documentation of such forms of management, which may be based on oral traditions, or in the case of the first

---

62   UNESCO, Report of the rapporteur for the sixteenth session of the World Heritage Committee in Santa Fe, 7–14 December 1992, 14 December 1992, WHC-92/conf.002/12, annex VII. Retrieved from http://whc.unesco.org/archive/1992/whc-92-conf002-12e.pdf

63   UNESCO, Operational Guidelines for the implementation of the World Heritage Convention, WHC/2 rev. 27 March 1992, para. 14. Retrieved from http://whc.unesco.org/archive/opguide92.pdf

64   UNESCO, Operational Guidelines for the implementation of the World Heritage Convention, WHC/2/rev. February 1994, para. 14. Retrieved from http://whc.unesco.org/archive/opguide94.pdf

65   UNESCO, Operational Guidelines for the implementation of the World Heritage Convention, WHC/2/rev. February 1994, para. 24. Retrieved from http://whc.unesco.org/archive/opguide94.pdf

cultural landscape inscribed on the World Heritage List in 1993, Tongariro National Park (New Zealand), on songs.

These two modifications concerning community involvement and customary law had far-reaching consequences. For cultural properties, people were now integrated as major actors and partners in site management and their traditional knowledge systems were recognized as an effective way to look after heritage properties. This is the first acceptance of customary law and management in an international legal instrument in the heritage field. It opened the Convention to regions and traditional cultures particularly from sub-Saharan Africa, the Pacific and the Caribbean. It also in a sense paved the way for UNESCO's 2003 Convention on the Safeguarding of the Intangible Cultural Heritage which specifically recognizes "practices, representations, expressions, knowledge, skills – as well as the instruments, objects, artefacts and cultural spaces associated therewith – that communities, groups and, in some cases, individuals recognize as part of their cultural heritage."[66]

It was not until 1998 that traditional management systems were accepted for natural properties. At its session in Kyoto, the nomination of East Rennell (Solomon Islands) led to the inclusion of traditional management practices for natural sites. Home to approximately 1,200 people of Polynesian origin living mainly by subsistence gardening, hunting and fishing, the property is under customary ownership and management. It was Matsuura, Chairperson of the 1998 meeting, who championed the inscription of this site. At his suggestion, first the Operational Guidelines were amended to add "traditional protection" to the requirement for natural sites to "have adequate long-term legislative, regulatory, institutional or traditional protection."[67] Then East Rennell was inscribed on the World Heritage List. It was a landmark decision to allow for customary law and traditional protection for natural World Heritage properties. Matsuura sums up his motivation: "Management plans are based on traditional law, not necessarily on legislation passed by parliaments. So we should make the requirement more inclusive."[68]

The Emergence of Systematic Monitoring and Periodic Reporting

There is no provision for systematic monitoring of sites in the World Heritage Convention. It is noteworthy that the concept was discussed but not retained by the experts preparing the final draft of the Convention. According to Professor Ito of

---

66   UNESCO, 2003, Convention on the Safeguarding of the Intangible Cultural Heritage, art. 2, Retrieved from http://www.unesco.org/culture/ich/index.php?lg=en&pg=00006

67   UNESCO, Operational Guidelines for the implementation of the World Heritage Convention, WHC-99/2 March 1999, para. 44. Retrieved from http://whc.unesco.org/archive/opguide99.pdf

68   Canada Research Chair on Built Heritage, Université de Montréal, audio interview of Koïchiro Matsuura by Christina Cameron and Mechtild Rössler, Paris, 24 November 2009.

Japan, a participant at the 1972 expert meeting, the word "periodic" was used with respect to monitoring.[69] This recollection probably refers to the American draft convention that required the Committee to conduct "periodic surveys and give notification of the need for corrective action when warranted." [70]

The word "monitoring" does not appear in the text; however, a "reporting process" is included that requires States Parties to submit reports to the General Conference of UNESCO "on the legislative and administrative provisions which they have adopted and other action which they have taken for the application of this Convention, together with details of the experience acquired in this field." In addition, the reports "shall be brought to the attention of the World Heritage Committee" (article 29). It was only in 1997 that this article was activated to underpin a systematic monitoring system even though the idea of systematic monitoring, including periodic (five year) reports to the Committee on a regional basis, was included in the 1996 Operational Guidelines.[71]

The earliest appearance of systematic monitoring in World Heritage discourse was the American proposal to the 1982 Bureau for a formal monitoring system, modelled on its experience with national parks. In a letter to the Committee, the American delegation explained the reasons for the proposal:

> the World Heritage List has grown and diversified over the past few years to the point where it is not possible for the World Heritage Committee to monitor the condition of World Heritage properties through informal contacts and communications alone. One of the important responsibilities of the Committee is to ensure that properties inscribed on the List retain those values that initially qualified them for inscription.[72]

Monitoring was to be based on a brief, standardized form for use by each country in a system of regular self-reporting every two to three years. Von Droste recalls the details of the proposal:

---

69   Canada Research Chair on Built Heritage, Université de Montréal, audio interview of Nobuo Ito by Christina Cameron, Mechtild Rössler and Nobuko Inaba, Kyoto. 8 November 2012.

70   UNESCO, Comparative table of the provisions of the revised draft convention concerning the protection of monuments, groups of buildings and sites of universal value, submitted by the Director-General of UNESCO, and the provisions of the World Heritage Trust draft convention concerning the preservation and protection of natural areas and cultural sites of universal value, submitted by the United States of America, SHC-72/conf.37/inf. 3, p.4. Retrieved from http://whc.unesco.org/archive/1972/shc-72-conf37-inf3e.pdf

71   UNESCO, Operational Guidelines for the implementation of the World Heritage Convention, WHC/2/rev. February 1996, paras. 68–74. Retrieved from http://whc.unesco. org/archive/opguide96.pdf

72   UNESCO, 1982, Examination of a proposal to establish a programme for monitoring the conditions of sites inscribed on the World Heritage List, Paris, CLT-82/CH/conf.014/2, paras.1. Retrieved from http://whc.unesco.org/archive/1982/clt-82-conf014-2e.pdf

The Americans came up with the initiative that World Heritage should adopt a procedure for assessing the state of conservation, monitoring as they called it, and they tabled, as an example, a document on monitoring the state of conservation of Yellowstone in which they said eighty deficiencies are stated and the monitoring would therefore prescribe a management strategy how to overcome these deficiencies and World Heritage should do the same because this would be in line with the Convention. We need to monitor in order to know what's going on and in order to give direction to the World Heritage Fund and to a policy choice and so on.[73]

The Bureau rejected this proposal on the grounds that it was premature "given the current state of infrastructures in the majority of countries concerned."[74] Von Droste gives a different explanation for the Bureau's reaction:

Now, this document on monitoring and Yellowstone was translated into French and unfortunately the word "monitoring" was translated in French by the word "control". You cannot imagine ... and you can see that story in the records, that they [Committee members] say: it's out of the question, we don't want to be controlled and who is paying for the control? We are autonomous. We know what we do. We need that everyone trusts us that we fulfil the Convention and it's our business and we do not submit ourselves to the international discussion and so on. So there was a violent reaction against the proposal which nevertheless for the first time raised the issue which came back and back to the Committee.[75]

Although this monitoring proposal was not accepted, the Committee nonetheless added a new item on its agenda to cover the protection and management

---

73   Canada Research Chair on Built Heritage, Université de Montréal, audio interview of Bernd von Droste by Christina Cameron and Mechtild Rössler, Paris, 5 April 2007.

74   UNESCO, Report of the rapporteur on the sixth session of the Bureau of the World Heritage Committee in Paris, 21–24 June 1982, Paris, 20 August 1982, CLT-82/conf.014/6, paras. 17–18. Retrieved from UNESCO, http://whc.unesco.org/archive/1982/clt-82-conf014-6e.pdf

75   Canada Research Chair, interview von Droste, 2007; see also UNESCO, Examination of a proposal to establish a programme for monitoring the conditions of sites inscribed on the World Heritage List, Paris, 21–24 June 1982, CLT-82/CH/conf.014/2, para. 3. Retrieved from http://whc.unesco.org/archive/1982/clt-82-conf014-2e.pdf; UNESCO, Examen d'une proposition visant à l'établissement d'un programme de rapports sur l'état de préservation des sites inscrits sur la Liste du Patrimoine mondial, Paris, 21–24 juin 1982, CLT-82/CH/conf.014/2, paras. 3. Retrieved from http://whc.unesco.org/archive/1982/clt-82-conf014-2f.pdf. Bernd von Droste, "A Gift from the Past to the Future. Natural and cultural world heritage," *Sixty Years of Science at UNESCO 1945–2005*, (Paris, 2006), pp. 397–8, refers to the translation error: "I remember for example that the debate in 1982 was profoundly influenced by a translation error in our documents: in the French-language version, 'monitoring' had been (mis)translated as 'controlling', which ruined the whole discussion."

of properties inscribed on the World Heritage List and reports on their condition. It also supported the desirability of yearly reporting and called for the establishment of conservation guidelines.[76]

After the rejection of the American proposal, there was another attempt in 1986 to set up a more systematic approach. The Committee "agreed that a more encompassing monitoring-reporting system was required as an integral part of the process of maintaining a World Heritage List" and set up a working group to study principles and procedures for a potential monitoring system.[77] At the 1986 Bureau, ICOMOS presented an ambitious though unfunded monitoring plan, estimating that it could report on the state of conservation of about twenty properties a year by creating "a monitoring committee which would work on the basis of reports from States Parties, information provided by ICOMOS National Committees and information from other sources. The data thus collected could be computerized at ICOMOS Headquarters."[78] Modelled to some extent on IUCN's approach, this was a comprehensive and intelligent proposal that anticipates later developments in World Heritage monitoring.

Stovel recalls this proposal which he attributes to two Canadians, François Leblanc and Jacques Dalibard:

> ICOMOS was encouraged in the middle 80s to develop monitoring systems. Francois LeBlanc, Jacques Dalibard actually developed a system which was submitted to the Committee in about 1986. It was rejected because people weren't sure why they needed monitoring. Mrs. Raidl may have understood the need for that, but many countries interpreted that as ICOMOS policing what was sovereign, sovereign decision-making. Nevertheless, the concern for monitoring remained in place, appropriate monitoring.[79]

---

76   UNESCO, Report of the rapporteur on the sixth session of the World Heritage Committee in Paris, 1–17 December 1982, Paris, 17 January 1983, CLT-82/conf.015/8, paras.18–19 Retrieved from http://whc.unesco.org/archive/1982/clt-82-conf015-8e.pdf

77   UNESCO, Report of the rapporteur on the tenth session of the World Heritage Committee in Paris, 24–28 November 1986, Paris, 5 December 1986, CC-86/conf.003/10, para. 30. Retrieved from http://whc.unesco.org/archive/1986/cc-86-conf003-10e.pdf; UNESCO, Monitoring the status of conservation of properties inscribed on the World Heritage List; report of the working group on cultural properties, SC-87/conf.004/5, para. II 6. Retrieved from http://whc.unesco.org/archive/1987/sc-87-conf004-5e.pdf

78   UNESCO, Report of the rapporteur on the tenth session of the Bureau of the World Heritage Committee in Paris, 16–19 June 1986, Paris, 15 September 1986, CC-86/conf.001/11, para. 13. Retrieved from http://whc.unesco.org/archive/1986/cc-86-conf001-11e.pdf; ICOMOS, Monitoring procedures for properties included on the World Heritage List, Paris, May 1986, CC-86/conf.001/5, pp. 1–16. Retrieved from http://whc.unesco.org/archive/1986/cc-86-conf001-5e.pdf

79   Canada Research Chair on Built Heritage, Université de Montréal, audio interview of Herb Stovel by Christina Cameron, Ottawa, 16 March 2011.

The Bureau was unable to formulate a recommendation on this proposal and directed the secretariat to study it further in collaboration with ICOMOS. A different proposal was presented to the 1986 Committee. This new programme was designed to monitor forty cultural sites per year in the chronological order of their inscription:

> The main purpose of this system, which would be based on questionnaires sent to States Parties, would be to help the States concerned to identify the conservation problems of the sites and the assistance that they may need. The monitoring of such a number of sites would presuppose the establishment of a formal system of data collection and an important increase in the financial and man-power resources allocated to the Secretariat and to ICOMOS, but more flexible solutions could also be envisaged.[80]

Interestingly, this text anticipates a number of challenging issues including the increasing number of sites, the need for funding, especially for ICOMOS as a non-governmental organization, and the lack of a methodology for collecting base-line information on specific sites. Although the proposal received only partial support, the principles reappeared in the subsequent development of periodic reporting.

The next year, the Committee considered the results of a working group set up the previous year "to examine the problems raised by the establishment of a system to monitor the state of conservation of cultural properties included in the World Heritage List." The group proposed principles, procedures and two questionnaires, one addressed to States Parties and a second for use by UNESCO when further details were required.[81] The Committee however viewed this complex proposal with caution, concerned with preserving State Party authority and uneasy about a possibly shift in control from the advisory body to UNESCO:

> Emphasis was placed on the need to ensure that States were the primary source of information, on the need for the Committee to have objective information at its disposal and on the fact that the system should be considered by the States as an incentive to conserve their listed sites and not of a means of control …It was furthermore suggested that ICOMOS should be more closely associated with the proposed system.[82]

---

80   UNESCO, Report of the rapporteur on the tenth session of the World Heritage Committee in Paris, 24–28 November 1986, Paris, 5 December 1986, CC-86/conf.003/10, para. 20. Retrieved from http://whc.unesco.org/archive/1986/cc-86-conf003-10e.pdf

81   UNESCO, Report of the rapporteur on the eleventh session of the World Heritage Committee in Paris, 7–11 December 1987, Paris, 20 January 1988, SC-87/conf.005/9, para 12. Retrieved from http://whc.unesco.org/archive/1987/sc-87-conf005-9e.pdf

82   UNESCO, Report of the rapporteur on the eleventh session of the World Heritage Committee in Paris, 7–11 December 1987, Paris, 20 January 1988, SC-87/conf.005/9, para 13. Retrieved from http://whc.unesco.org/archive/1987/sc-87-conf005-9e.pdf

In the end the Committee decided to implement the system but on a limited experimental basis.

Many problems were reported to the following Committee. States Parties had difficulty meeting deadlines. The secretariat and ICOMOS needed more time to review the replies, requesting an extension of the timeline. ICOMOS complained that the "exceedingly brief answers provided little information on the extent of danger referred to and very succinct analyses prevented any serious evaluation of problems raised" and concluded "that procedures could no doubt be improved, either by reformulating the questionnaire or by encouraging States to answer in greater detail."[83] In the ensuing debate, in which several States Parties participated as well as representatives from the United Nations Development Programme (UNDP) and the International Council of Museums (ICOM), different views and opinions were expressed. Despite the problems, there was consensus on a desire to see an efficient and effective monitoring system. Therefore it was agreed that the experiment should be continued and the next fifty properties should be examined in 1989.

By that time, the Committee "considered that the system underway was both cumbersome and not fully satisfactory, especially in comparison to the system for monitoring natural sites, and that the system did not enable the Committee to assume this important function efficiently." Moreover, the Committee was critical of the performance of ICOMOS and ICCROM, suggesting that "more use be made of non-governmental organizations specialized in cultural heritage conservation, not forgetting to use structures which already exist in the field, as well as the expertise of the members of the Secretariat."[84] In 1990, the experimental monitoring project with ICOMOS kept going although the third series of questionnaires was postponed in order to analyse the ones already received.

To complicate matters, UNESCO proposed a different approach, "a programme for the systematic diagnosis of World Heritage cultural sites" using a UNDP project already under way in Latin America and the Caribbean as the basis for an experiment in that region.[85] The next year UNESCO expanded this initiative to a second region in collaboration with the United Nations Environment Programme (UNEP) for the protection of sites in the Mediterranean.

In a later article, von Droste analysed the slow emergence of a more systematic approach to monitoring: "Despite the evident advantages of knowing more about the state of conservation of World Heritage properties, commitments to monitor

---

83   UNESCO, Report of the rapporteur on the twelfth session of the World Heritage Committee in Brasilia, 5–9 December 1988, Paris, 23 December 1988, SC-88/conf.001/13, para. 34. Retrieved from http://whc.unesco.org/archive/1988/sc-88-conf001-13e.pdf

84   UNESCO, Report of the rapporteur on the thirteenth session of the World Heritage Committee in Paris, 11–15 December 1989, Paris, 22 December 1989, SC-89/conf.004/12, para. 18. Retrieved from http://whc.unesco.org/archive/1989/sc-89-conf004-12e.pdf

85   UNESCO, Report of the rapporteur on the fourteenth session of the World Heritage Committee in Banff, 7–12 December 1990, Banff, 12 December 1990, CLT-90/conf.004/13, para. 21. Retrieved from http://whc.unesco.org/archive/1990/cc-90-conf004-13e.pdf

Figure 4.5   Twentieth anniversary experts meeting in Paris (October 1992)
from left to right: Azedine Beschaouch (rapporteur), Bernd von
Droste (Director, World Heritage Centre), Christina Cameron
(Chairperson) and Laurent Levi-Strauss (UNESCO)

have been slow to arrive in the World Heritage community. Consensus around
the need for, and value of, monitoring has been difficult to achieve – primarily
because of different perceptions of the purpose of monitoring."[86]

The turning point for more systematic monitoring came on the twentieth
anniversary of the Convention. In preparation for the 1992 celebrations, an
evaluation was carried out and two expert meetings took place, one in Washington
(June 1992) and one at UNESCO headquarters in Paris (October 1992), which
prepared strategic goals and objectives for the Committee's consideration.[87] The
fourth strategic goal to "pursue more systematic monitoring of World Heritage
Sites" was reinforced the following year by the Assistant Director-General for
Culture when he addressed the General Assembly of States Parties. Recalling
the previous discussions, he reminded countries of "their obligation to preserve
properties inscribed on the World Heritage List, and of the World Heritage
Committee's increasing efforts to ensure regular monitoring of the state of these

---

86   Von Droste, Gift from the Past, pp. 397–8 Retrieved from http://unesdoc.unesco.
org/images/0014/001481/148187e.pdf#xml=http://www.unesco.org/ulis/cgi-bin/ulis.pl?da
tabase=&set=4F96A404_3_209&hits_rec=10&hits_lng=eng

87   UNESCO, Report of the rapporteur for the sixteenth session of the World Heritage
Committee in Santa Fe, 7–14 December 1992, 14 December 1992, WHC-92/conf.002/12,
annex II, para. II.B. Retrieved from http://whc.unesco.org/archive/1992/whc-92-conf002-
12e.pdf

properties, with the assistance of ICOMOS, IUCN and ICCROM, and also, on a broader scale, with competence drawn from the different regions."[88]

To implement the fourth strategic goal, an important expert meeting was held at the World Conservation Monitoring Centre (WCMC) in Cambridge[89] involving key players with previous experience in the various monitoring experiments for both natural and cultural sites. The choice of Cambridge was deliberate since the WCMC held the pre-eminent position globally in gathering and analysing data on natural properties world-wide. This meeting dealt exclusively with systematic monitoring. It proposed a definition for monitoring that spelled out basic concepts that eventually made their way into the periodic reporting framework:

> By "monitoring" we mean, therefore, a process of continuous co-operation between site managers, States Parties and the World Heritage Convention and its partners involving the continuous/repeated observation of the condition(s) of the site, identification of issues that threaten the conservation and World Heritage characteristics of the site and the identification of decisions to be taken; and reporting the results of monitoring ... Monitoring in this sense is predicated on the existence of a base of information that describes the heritage properties, their use and management as well as their characteristics, qualities and significance. It is a process of repeated comparison of the current state of a site against the original baseline information.[90]

The Committee examined the methodological aspects of the experts' report and fully endorsed the findings which proposed three types of monitoring:

> **Systematic monitoring**: the continuous process of monitoring the conditions of World Heritage sites with periodic reporting on its state of conservation;

> **Administrative monitoring**: follow-up actions by the World Heritage Centre to ensure the implementation of recommendations and decisions of the World Heritage Committee and Bureau at the time of the inscription or at a later date;

---

88   UNESCO, Report of the rapporteur for the ninth session of the General Assembly of the States Parties to the Convention Concerning the Protection of the World Cultural and Natural Heritage , 29–30 October 1993, 2 November 1993, WHC-93/conf.003/6, para. 6. Retrieved from http://whc.unesco.org/archive/1993/whc-93-conf003-6e.pdf

89   The World Conservation Monitoring Centre (WCMC) was founded in 1979 by IUCN, and in 1988 became an independent institution with the support of IUCN, WWF and UNEP. Since 2000, the UNEP World Conservation Monitoring Centre (UNEP-WCMC) has been UNEP's specialist biodiversity assessment arm. WCMC has supported IUCN's World Heritage work since the early 1980s.

90   UNESCO, Report of the expert meeting on "Approaches to the monitoring of World Heritage properties: exploring ways and means," Cambridge, U.K. (1 to 4 November 1993, Paris, 23 November 1993, WHC-93/conf.002/inf. 5, para. 2. Retrieved from http://whc.unesco.org/archive/1993/whc-93-conf002-inf5e.pdf

Ad hoc monitoring: the reporting by the Centre or other sectors by UNESCO and the advisory bodies to the Bureau and the Committee on the state of conservation of specific World Heritage sites that are under threat. Ad hoc reports and impact studies are necessary each time exceptional circumstances occur or work is undertaken which may have an affect [sic] on the state of conservation of the sites.[91]

It is noteworthy that the term "periodic reporting" appears for the first time as part of systematic monitoring. The Committee debate identified other aspects including the involvement of States Parties in the further development of the concept, the importance of impact studies and the need for comprehensive baseline information as a requirement at the time of inscription.[92] It also asked for a "format for periodic reporting" and the establishment of a "small unit to oversee the implementation of a systematic monitoring and reporting system."[93] Although such a unit was not developed, responsibility for monitoring was assigned to staff member Herman van Hooff who had worked with Sylvio Mutal on the UNDP/UNESCO project to review cultural properties in Latin America and the Caribbean. Van Hooff and Mutal were both participants at the Cambridge meeting. Mutal's regional project stands as a precursor to periodic reporting.

Herb Stovel who represented ICOMOS at the Cambridge meeting, recalls this moment:

By 1993, there was a very good expert meeting in Cambridge, cultural heritage and natural heritage together, which came up with some recommendations for monitoring systems. The recommendations were well linked to understanding the values of the site, the OUV of the site and to ensuring that all decisions were built around respect for that central theme or central idea of OUV. Those recommendations did not fly right away though, because as I remember, one of the State Parties, this was India I think, raised once again this specter that: "well, how can we involve ourselves with this? We are sovereign countries and we don't need ICOMOS or anybody to come into our country and tell us how to do our job." And they said it with sincerity and they said it with

---

91   UNESCO, Report of the rapporteur for the seventeenth session of the World Heritage Committee in Cartagena, 6–11 December 1993, Paris, 4 February 1994, WHC-93/conf.002/14, para. IX.2. Retrieved from http://whc.unesco.org/archive/1993/whc-93-conf002-14e.pdf

92   UNESCO, Report of the rapporteur for the seventeenth session of the World Heritage Committee in Cartagena, 6–11 December 1993, Paris, 4 February 1994, WHC-93/conf.002/14, paras. IX.3–4. Retrieved from http://whc.unesco.org/archive/1993/whc-93-conf002-14e.pdf

93   UNESCO, Report of the rapporteur for the seventeenth session of the World Heritage Committee in Cartagena, 6–11 December 1993, Paris, 4 February 1994, WHC-93/conf.002/14, para. IX.6. Retrieved from http://whc.unesco.org/archive/1993/whc-93-conf002-14e.pdf

enough force that it stalled the monitoring initiative for a while. It was picked up toward the end of the 90s.[94]

The 1994 Committee session consolidated the foundation of a systematic monitoring system, commending the secretariat for progress on developing a framework for periodic reporting and monitoring and noting that "one of the principal aims of monitoring was to assess if the values, on the basis of which the site was inscribed on the World Heritage List, have remained intact." The Committee:

> also stressed that a monitoring methodology should be flexible and adaptable to regional and national characteristics, as well as to the natural and cultural specificities of the sites. Furthermore, it expressed the need to involve external advice in the periodic reporting through the non-governmental advisory bodies and/or the existing decentralized UNESCO structures.[95]

Italy was particularly insistent on the involvement of experts to ensure better management, while India emphasized the need for States Party consent in any monitoring process. The Committee brought the pieces together by inviting countries to put monitoring structures in place, to include it in training courses and to develop work plans for regional monitoring programmes.[96] More significantly, it approved the text for "systematic monitoring and reporting" for the next edition of the Operational Guidelines.[97] Stovel welcomed this major development:

> The commitment by the World Heritage Committee in December 1994 to ensure the continuous monitoring of the properties inscribed on the World Heritage List concluded a decade-long series of exploratory discussions on the subject. It also offered a number of significant opportunities to ICOMOS to increase its involvement in meaningful conservation activity at the international level.[98]

---

94 Canada Research Chair, interview Stovel, April 2011.

95 UNESCO, Report of the rapporteur on the eighteenth session of the World Heritage Committee in Phuket, 12–17 December 1994, Paris, 31 January 1995, WHC-94/conf.003/16, para. IX.5. Retrieved from http://whc.unesco.org/archive/1994/whc-94-conf003-16e.pdf

96 UNESCO, Report of the rapporteur on the eighteenth session of the World Heritage Committee in Phuket, 12–17 December 1994, Paris, 31 January 1995, WHC-94/conf.003/16, paras. IX.10–11. Retrieved from http://whc.unesco.org/archive/1994/whc-94-conf003-16e.pdf

97 UNESCO, Report of the rapporteur on the eighteenth session of the World Heritage Committee in Phuket, 12–17 December 1994, Paris, 31 January 1995, WHC-94/conf.003/16, para. IX.9. Retrieved from http://whc.unesco.org/archive/1994/whc-94-conf003-16e.pdf

98 Herb Stovel, "Monitoring world cultural heritage sites," *ICOMOS Canada Bulletin*, 4/3 (1995), p. 15.

In spite of this progress, some countries continued to question the authority of the Committee to put in place any kind of monitoring system. The issue led to the first significant disagreement between the Committee and the General Assembly of States Parties. The 1995 General Assembly was particularly acrimonious, with several draft resolutions in play over the issue of state sovereignty and monitoring.[99] The debate began when the Committee Chairperson, Thailand's Adul Wichiencharoen, tabled a draft Committee-approved proposal to amend the Operational Guidelines by adding a text on monitoring and reporting, stating that the Committee recognized these as essential functions. While the amendment acknowledged that monitoring is the "responsibility of the State Party concerned and is part of the site management," it also pointed to the Committee's responsibility to receive information on the state of conservation of World Heritage Sites and to "define the form, nature and extent of the regular reporting in respect of the principles of State sovereignty."[100] He explained that the Committee was motivated to bring forward this motion by a need for information to set priorities for international assistance and a belief that "conditions and circumstances that constitute serious dangers threatening World Heritage properties as to require inclusion in the List of World Heritage in Danger ... can be rescued from such an eventuality if the earlier trends have been monitored and remedial measures have been taken in time to prevent the deterioration."[101]

Some countries, led by India, disagreed on the grounds of protecting state sovereignty. India expressed concern that "reports from States Parties can only be required by the General Conference of UNESCO and not by a 'select body' such as the World Heritage Committee."[102] Australia intervened to respond to the naysayers by observing that:

> the Assembly did not seem to be close to a consensus on the matter of monitoring and reporting. In response to the fear he felt among delegates for excessive bureaucracy and an intrusion on the sovereignty of States Parties, the Delegate states that the World Heritage Committee's decisions on monitoring

---

99 The political debate is covered more fully in chapter 5 in the section on States Parties.

100 UNESCO, Summary record of the eleventh General Assembly of States Parties to the Convention concerning the protection of the world cultural and natural heritage, Paris, 27–28 October 1997, Paris, 18 December 1997, WHC-97/conf-205/7, para. 23. Retrieved from http://whc.unesco.org/archive/1997/whc-97-conf205-7e.pdf

101 UNESCO, Summary record of the tenth General Assembly of States Parties to the Convention concerning the protection of the world cultural and natural heritage, Paris, 2–3 November 1995, Paris, 22 November 1995, WHC-95/conf-204/8, annex 1. Retrieved from http://whc.unesco.org/archive/1995/whc-95-conf204-8e.pdf

102 UNESCO, Summary record of the tenth General Assembly of States Parties to the Convention concerning the protection of the world cultural and natural heritage, Paris, 2–3 November 1995, Paris, 22 November 1995, WHC-95/conf-204/8, para.16 Retrieved from http://whc.unesco.org/archive/1995/whc-95-conf204-8e.pdf

and reporting indeed strengthen the role of the Convention and the Committee but that these are in no way intrusive. Given the fact that the Convention as such, of course cannot reflect the experiences gained since 1972, he felt that there is an important role to play for UNESCO in setting standards in this field.[103]

Faced with a lack of consensus, the General Assembly deferred discussion to the following meeting.[104]

In response to this impasse, the 1995 Committee session established yet another working group to deal with reporting on specific properties as well as systematic monitoring and reporting.[105] Under the memorable leadership of Australian delegate Barry Jones, the fractious working group eventually achieved a draft resolution which proposed the activation of article 29 of the Convention.[106] The 1997 General Assembly of States Parties supported this approach and concluded that "monitoring is the responsibility of the State Party concerned and that the commitment to provide periodic reports on the state of the site is consistent with the principles set out in the Convention." It adopted a resolution to suggest to the General Conference of UNESCO to activate procedures under article 29 and "refer to the World Heritage Committee the responsibility to respond to the reports."[107] Periodic reporting thus became an official monitoring tool for the World Heritage system.

In 1998 a new chapter was added to the Operational Guidelines to draw out the distinction between reactive monitoring and periodic reporting.[108] The Committee

---

103   UNESCO, Summary record of the tenth General Assembly of States Parties to the Convention concerning the protection of the world cultural and natural heritage, Paris, 2–3 November 1995, Paris, 22 November 1995, WHC-95/conf-204/8, para. 26. Retrieved from http://whc.unesco.org/archive/1995/whc-95-conf204-8e.pdf

104   It is interesting to note that, despite the lack of agreement and deferral of discussions to 1997, the framework for a systematic reporting process was actually included in the 1996 Operational Guidelines on the basis of a Committee decision in 1994. This marks the first major disagreement between the World Heritage Committee and the General Assembly of States Parties.

105   UNESCO, Report of the rapporteur on the nineteenth session of the World Heritage Committee in Berlin, 4–9 December 1995, Paris, 31 January 1996, WHC-95/conf.203/16, para. VII.1. Retrieved from http://whc.unesco.org/archive/1995/whc-95-conf203-16e.pdf

106   UNESCO, Report of the rapporteur on the nineteenth session of the World Heritage Committee in Berlin, 4–9 December 1995, Paris, 31 January 1996, WHC-95/conf.203/16, paras. VII.52–3. Retrieved from http://whc.unesco.org/archive/1995/whc-95-conf203-16e.pdf

107   UNESCO, Summary record of the eleventh General Assembly of States Parties to the Convention concerning the protection of the world cultural and natural heritage, Paris, 27–28 October 1997, Paris, 18 December 1997, WHC-97/conf-205/7, para. 24.14. Retrieved from http://whc.unesco.org/archive/1997/whc-97-conf205-7e.pdf

108   UNESCO, Operational Guidelines for the implementation of the World Heritage Convention, WHC-99/2 December 1998, paras. 68–78. Retrieved from http://whc.unesco.org/archive/opguide99.pdf

initiated the first cycle of periodic reporting, beginning with the Arab region in the year 2000.[109] Involved from the outset, van Hooff recognized that such a process would be dynamic:

> Undoubtedly, the introduction of the principles of monitoring and reporting will not be the final step in this process. Practical experiences, the further development of the World Heritage concept and the evolution in the interpretation and application of the World Heritage Convention will require a continuous reflection on the need and the principles of monitoring and reporting. Continuing discussions by the statutory bodies of UNESCO and the World Heritage Convention are a clear indication of its dynamic and at times controversial nature.[110]

Threatened World Heritage Sites

The List of World Heritage in Danger

A key component of the World Heritage system is the List of World Heritage in Danger. The Convention requires the Committee to create and publish "a list of the property appearing in the World Heritage List for the conservation of which major operations are necessary and for which assistance has been requested under this Convention. This list shall contain an estimate of the cost of such operations" (article 11.4). The Convention text gives examples of what constitutes a serious and specific danger:

> threat of disappearance caused by accelerated deterioration, large-scale public or private projects or rapid urban or tourist development projects; destruction caused by changes in the use or ownership of the land; major alterations due to unknown causes; abandonment for any reason whatsoever; the outbreak or the threat of an armed conflict; calamities and cataclysms; serious fires, earthquakes, landslides; volcanic eruptions; changes in water level, floods and tidal waves.

At the beginning in 1978, the Committee was preoccupied with nominations and deferred consideration of Danger listing to the following year.[111] Following a 1979 earthquake, the Committee listed the Natural and Culturo-Historical Region

---

109   Herman van Hooff, "Monitoring and Reporting in the Context of the World Heritage Convention and its Application in Latin America and the Caribbean," in *Monitoring World Heritage, World Heritage 2002, Shared Legacy, Common Responsibility, Associated Workshops 11–12 November 2002, Vicenza, Italy* (Paris, 2004), p. 34.

110   Herman van Hooff, "The monitoring and reporting of the state of properties inscribed on the World Heritage List," *ICOMOS Canada Bulletin*, 4/3 (1995), p. 14.

111   UNESCO, Final report of the first session of the intergovernmental committee for the protection of the world cultural and natural heritage in Paris, 27 June–1 July 1977,

of Kotor (Yugoslavia, now Montenegro) as the first site on the World Heritage List in Danger at the request of the State Party.[112] The second site to enter the Danger List was Jerusalem in 1982. It is no coincidence that natural sites only began to be listed as endangered in 1984, the same year that IUCN began its regular reports on the state of conservation of natural properties. Three natural sites were Danger listed that year.[113] The main threats that led to inscribing sites on the Danger List during the period of this study were natural disasters, desertification and human actions.

Particularly difficult was damage resulting from military conflict. In the early 1990s, a new stage was reached when two properties affected by the Balkan Wars and the dissolution of former Yugoslavia were placed on the Danger List in 1992. Serious damage came from shelling at two Croatian sites, Plitvice Lakes National Park and the Old City of Dubrovnik. Eidsvik provides an account of a 1993 monitoring mission to the area around Plitvice, made possible through cooperation with the United Nations Protection Force (UNPROFOR). In his report, he details the vandalism and shelling around the park. He specifically suggested the creation of a "Green Beret facility" similar to the United Nations Blue Berets to safeguard precious World Heritage properties under threat, an idea that resurfaced on several occasions at World Heritage sessions.[114] Years later, he recalls the positive enabling effect of the World Heritage Convention:

> I think Plitvice in Croatia is an excellent example where during the ongoing war between Croatia-Serbia and all that area, the Serbian government had literally occupied the park. And the Croatian managers who had been there, it was actually very mixed management being both Croatians and Serbians, and I'd been there I guess four times. The fact that it was a World Heritage Site enabled UNESCO to come in, enabled us to inspect the site, enabled us to say no, the Serbs are not planting bombs on the various dams, etc, etc., enabled us to get about, over time, I think a hundred thousand dollars which is not a lot of money, but in that situation was a lot of money to assist in the rehabilitation of the park."[115]

Paris, 20 October 1977, CC-77/conf.001/9, para. 53. Retrieved from http://whc.unesco.org/archive/1977/cc-77-conf001-9e.pdf

112   UNESCO, Report of the Rapporteur on the third session of the World Heritage Committee in Cairo and Luxor, 22–26 October 1979, Paris, 30 November 1979, CC-79/conf.003/13, para. 46. Retrieved from http://whc.unesco.org/archive/1979/cc-79-conf003-13e.pdf and http://whc.unesco.org/archive/repcom79.htm#125

113   UNESCO, Report of the rapporteur on the eighth session of the World Heritage Committee in Buenos Aires, 29 October–2 November 1984, Buenos Aires, 2 November 1984, SC/84/conf.004/9, para. 26. Retrieved from http://whc.unesco.org/archive/1984/sc-84-conf004-9e.pdf

114   Harold K. Eidsvik, "Guest Comment: Plitvice National Park, World Heritage Site and the Wars in the Former Yugoslavia," *Environmental Conservation*, 20 (1993), p. 293, Retrieved from http://journals.cambridge.org/action/displayAbstract?fromPage=online&a id=5944072

115   Canada Research Chair, interview Eidsvik.

The Danger listing tool was used sparingly in the early years of implementation and conservation results were uneven. In the first decade there were fewer than eight sites listed. These numbers jumped to fifteen sites in 1992, closing out the decade with thirty properties formally recognized as endangered. A close observer, von Droste offers his analysis of the early years. "The World Heritage in Danger listing has been handled by the World Heritage Committee in a rather inconsistent way during the last thirty years." He differentiates among those requested by the State Party such as Rwenzori Mountains National Park (Uganda) or Garamba National Park (Zaire, now Democratic Republic of the Congo); those listed with State Party passive acquiescence such as Mount Nimba Strict Nature Reserve (Côte d'Ivoire/Guinea); and those listed without State Party consent such as Manas Wildlife Sanctuary (India).[116]

The Manas site was a particularly difficult case due to damage from encroachment by militants from the Bodo tribe in Assam. After several years of attempted dialogue, the 1992 Committee exercised its responsibility for the property by putting it on the List of World Heritage in Danger without the consent of the State Party. This action, which probably influenced the position of India in the debates on monitoring and periodic reporting, came after several Committee requests for information went unanswered: "The Committee noted with regret that the Indian authorities have not provided a report on the status of conservation of Manas, despite repeated requests over the last three years, and therefore decided to include Manas Wildlife Sanctuary on the List of World Heritage in Danger, in accordance with the provisions of Article 11, paragraph 4 of the Convention."[117] As the State Party had weak control over the site due to the on-going conflict, this lack of information was a continuous source of tension between India and the World Heritage Committee. A fruitful dialogue was eventually established in the late 1990s. Although a 1997 mission accessed the site and reported on rehabilitation activities, the 1998 Committee held its position, noting "that while security conditions in and around Manas have improved, the threat of insurgency still prevails in the State of Assam and militants often traversed the Sanctuary. Nevertheless, the Committee was informed that the Indian authorities were of the view that conditions for site protection and the relationship with local villagers were gradually improving."[118] Manas Wildlife Sanctuary remained on the Danger List for almost twenty years, from 1992 to 2011.

---

116   Von Droste, *Gift from the Past*, pp. 395–7.

117   UNESCO, Report of the rapporteur for the sixteenth session of the World Heritage Committee in Santa Fe, 7–14 December 1992, 14 December 1992, WHC-92/ conf.002/12, annex II, para. VII.13. Retrieved from http://whc.unesco.org/archive/1992/ whc-92-conf002-12e.pdf

118   UNESCO, Report of the rapporteur on the twenty-second session of the World Heritage Committee in Kyoto, 30 November–5 December 1998, Paris, 29 January 1999, WHC-98/conf.203/18, para. VII.9. Retrieved from http://whc.unesco.org/archive/1998/ whc-98-conf203-18e.pdf

Figure 4.6   Angkor, Cambodia

Figure 4.7   Galapagos Islands, Ecuador

Two emotional and controversial cases in this period, Angkor (Cambodia) and Galapagos Islands (Ecuador) demonstrate the positive and beneficial effects of applying Danger listing or using the mechanism of potential Danger listing to draw global attention and support. The case of Angkor came to the Committee's attention in 1992 in the context of Cambodia's civil war. The controversy arose from the political manoeuvring that led to its inscription. The request came straight from Director-General Mayor who grew tired of waiting for the Committee to do something about the imperilled site. Reacting to reports of looting and black market sales of Angkor's antiquities, Mayor recalls his impatience:

> I have enough elements to tell you that either the Committee declares Angkor Wat or Angkor in general as World Heritage or I as Director-General will declare that this heritage is in danger because I am able do it, because I think we can no longer wait for the professionals.[119]

The Bureau in July 1992 had advised the Committee "to initiate the procedure for inscription of the monuments of Angkor."[120] Several months later, on the eve of the 1992 Committee session in Santa Fe, Angkor was discussed at an informal meeting of the Bureau for which no written record was kept. In its technical evaluation of the nomination, ICOMOS judged that the magnificent ruins of the Khmer empire at Angkor clearly had outstanding universal value but recommended that its inscription be deferred until adequate protective measures were put in place. Cleere, who represented ICOMOS for the first time at this meeting, describes how the recommendation was determined:

> I got in touch with the people from the École française de l'Extrême-Orient who provided me with all their files. I did a mission there ... with guns going off in the background ... It was clear, there was nothing, nothing, nothing that is required. There was no law, there was no antiquities administration, there was no conservation, there was ziltch.[121]

---

119 Canada Research Chair on Built Heritage, Université de Montréal, audio interview of Federico Mayor by Christina Cameron, Madrid, 18 June 2009. "J'ai les éléments suffisants pour vous dire que, ou bien le Comité déclarent patrimoine mondial Angkor Vat ou Angkor en général ou bien sera moi directement le directeur général qui déclarera ce patrimoine en danger parce que j'ai la possibilité de le faire parce que je pense que nous ne pouvons plus attendre que les professionnels."

120 UNESCO, Report of the sixteenth session of the Bureau of the World Heritage Committee in Paris, 6–10 July 1992, Paris, 10 July 1992, WHC-92/conf.003/2, para. 76. Retrieved from http://whc.unesco.org/archive/1992/clt-92-conf003-12e.pdf

121 Canada Research Chair on Built Heritage, Université de Montréal, audio interview of Henry Cleere by Christina Cameron, London, 24 January 2008.

Although the informal group was divided, the Chairperson Beschaouch exercised his prerogative by declaring that the site would be inscribed. Cleere describes the situation. "I presented what ICOMOS was recommending and the chairman said: No, no, it's not possible. No. It must go on the list." Eventually Beschaouch asked each Bureau member to express an opinion. "The first of them said, Oh no, it must go on the list. The last two, Thailand and US, said no, we agree with ICOMOS. But anyway, okay, says the chairman, it will be on the List."[122]

Following intense behind the scenes consultations, during which the host country switched its position, the Committee took a decision that ignored its own rules, insisting that it should not be taken as a precedent:

> Given the unique situation in Cambodia, which, in accordance with the Paris Accords, has been placed under the temporary administration of the United Nations since July 1991, the Committee has decided to waive some conditions required under the Operational Guidelines and, on the basis of criteria (i), (ii), (iii) and (iv), has inscribed the Angkor site, together with its monuments and its archaeological zones as described in the 'Périmètre de Protection' accompanying the ICOMOS report, on the World Heritage List.

While the outstanding universal value of the site was unquestionable, the Committee recognized that other conditions had not yet been met. It requested special studies and regular reports on the status of Angkor for the next three years "in order to guarantee protection of the site."[123] What is unusual is the simultaneous decision to inscribe Angkor on the World Heritage List and the Danger List in order to deal with urgent conservation problems related to legislation, monitoring, boundaries and buffer zones.

The United States felt the need to explain publicly its reversal. The representative declared that:

> Although the United States has voted in the Bureau to inscribe the site only subject to the conditions identified by ICOMOS, that position was now to support the compromise consensus to inscribe Angkor immediately. He noted that the United States hope that inscription would in fact lead to stronger protection of this site of unquestioned international value and ... the position of the United States that this inscription not be understood as a precedent, and congratulated ICOMOS for the integrity of their position and advice to the Committee.[124]

---

122   Canada Research Chair, interview Cleere.
123   UNESCO, Report of the rapporteur for the sixteenth session of the World Heritage Committee in Santa Fe, 7–14 December 1992, 14 December 1992, WHC-92/conf.002/12, para. X.1. Retrieved from http://whc.unesco.org/archive/1992/whc-92-conf002-12e.pdf
124   UNESCO, Report of the rapporteur for the sixteenth session of the World Heritage Committee in Santa Fe, 7–14 December 1992, 14 December 1992, WHC-92/conf.002/12, annex. Retrieved from http://whc.unesco.org/archive/1992/whc-92-conf002-12e.pdf

Despite the unorthodox way the Angkor received formal World Heritage recognition, the appropriate use of the Danger listing mechanism contributed significantly to the conservation of the property. Inscription was seen as a step towards encouraging international cooperation. In this sense, the case illustrates the positive functioning of basic processes set out in the Convention "concerning the identification, protection, conservation, presentation and transmission to future generations" (article 4) as well as the principle "that such heritage constitutes a world heritage for whose protection it is the duty of the international community as a whole to co-operate" (article 6.1). After years of support from UNESCO as well as bilateral cooperation and extensive international funding, Angkor was finally removed from the World Heritage List in Danger in 2004.

The second case, Galapagos Islands, illustrates the power of productive dialogue that can arise from consideration of Danger listing. Located a thousand kilometres from the South American continent in the Pacific, the volcanic islands with their rich biodiversity are regarded as the inspiration for Charles Darwin's theory of evolution. Galapagos Islands was the first nomination officially received by UNESCO and therefore was assigned number one in the recording system. This iconic site was one of the twelve listed in 1978.[125]

For many years, the World Heritage Fund and the international community provided support for conservation and capacity-building activities at Galapagos. In 1979, funds were allocated for a training seminar and technical equipment.[126] In 1985, emergency assistance and bilateral assistance from the United Kingdom, Canada and the United States helped extinguish a major fire on Isabela Island which burned over 30,000 hectares.[127] In 1990, the Committee drew attention to threats from other countries related to elevated levels of tourism and natural resource exploitation, recalling the Convention's principle "not to take any deliberate measures which might damage directly or indirectly the cultural and natural heritage ... situated on the territory of other States Parties to this Convention" (article 6.3). Specifically the Committee noted that "Japanese, Korean and Taiwanese fishermen last year captured some 40,000 sharks in the waters adjacent to the site; this intensive fishing was halted following protests by international

---

125   UNESCO, Report of the rapporteur on the second session of the World Heritage Committee in Washington, 5–8 September 1978, Paris, 9 October 1978, CC-78/conf.010/10 rev., para. 38. Retrieved from http://whc.unesco.org/archive/1978/cc-78-conf010-10reve.pdf

126   UNESCO, Report of the rapporteur on the third session of the World Heritage Committee in Cairo and Luxor, 22–26 October 1979, Paris, 30 November 1979, CC-79/conf.003/13, para. 50. Retrieved from http://whc.unesco.org/archive/1979/cc-79-conf003-13e.pdf

127   UNESCO, Report of the rapporteur on the ninth session of the World Heritage Committee in Paris, 2-6 December 1985, Paris, December 1985, SC-85/conf.008/9, para. 37. Retrieved from http://whc.unesco.org/archive/1985/sc-85-conf008-9e.pdf

organizations, but the effectiveness of the ban was uncertain."[128] In 1992 the Committee welcomed new tourism and conservation management plans.[129]

By 1995 the situation escalated to a new level of threat, leading to a discussion about Danger listing. IUCN detailed the situation, reporting that there were:

(a) threats to the terrestrial biodiversity with the introduction of species of vertebrate animals endangering endemic flora and fauna, as well as the growing human population, which has severe impacts for example for solid waste disposal, (b) threats to the marine biodiversity with illegal and increasing export fisheries (lobsters, sea cucumbers, sharks, tuna, etc.).

The advisory body recommended several corrective measures, concluding that "in light of the serious threat of species introduction and increasing population" the Committee might want to consider placing the site on the List of World Heritage in Danger.[130]

The State Party vehemently opposed this suggestion as the intervention by the delegate from Ecuador makes clear. While he acknowledged the situation described by IUCN related to "inadequate legal and administrative structure, the population growth, the illegal fishing in the Marine Resources Reserve of the Galapagos, unbalanced tourist activities and the impact of foreign species introduced to the island;" the delegate underlined:

that Galapagos - according to scientists - continues to be an exceptional treasure of the world from which no species has been lost; ... that numerous measures to safeguard the Galapagos have been taken, including constitutional reforms, management plans and international assistance projects by GEF, UNDP, USAID and others. He concluded that the site should not be placed on the List of World Heritage in Danger.[131]

---

128   UNESCO, Report of the rapporteur on the fourteenth session of the World Heritage Committee in Banff, 7–12 December 1990, Banff, 12 December 1990, CLT-90/conf.004/13, para. 25. Retrieved from http://whc.unesco.org/archive/1990/cc-90-conf004-13e.pdf

129   UNESCO, Report of the rapporteur for the sixteenth session of the World Heritage Committee in Santa Fe, 7–14 December 1992, 14 December 1992, WHC-92/conf.002/12, annex II, para. VII.12. Retrieved from http://whc.unesco.org/archive/1992/whc-92-conf002-12e.pdf

130   UNESCO, Report of the rapporteur on the nineteenth session of the World Heritage Committee in Berlin, 4–9 December 1995, Paris, 31 January 1996, WHC-95/conf.203/16, para. VII.13. Retrieved from http://whc.unesco.org/archive/1995/whc-95-conf203-16e.pdf

131   UNESCO, Report of the rapporteur on the nineteenth session of the World Heritage Committee in Berlin, 4–9 December 1995, Paris, 31 January 1996, WHC-95/conf.203/16, para. VII.13. Retrieved from http://whc.unesco.org/archive/1995/whc-95-conf203-16e.pdf

In this case the use of Danger listing was seen by the State Party as unhelpful. The Committee was divided between those who believed that the site met the threshold for Danger listing according to the Operational Guidelines and those who sought a more flexible solution to foster a productive dialogue. The 1995 Committee chose the latter by accepting the country's invitation for a high-level mission to Ecuador, including the Committee Chairperson and the Director of the World Heritage Centre, "to discuss the pressures on and present condition of the World Heritage site and to identify steps to overcome the problems."[132]

Following discussions of the mission report, the 1996 Bureau recommended Danger listing for the Galapagos Islands, noting in its report to the Committee, "that this List should not to be considered as a 'black list', but as a signal to take emergency actions for safeguarding and protection."[133] In an emotional debate, the 1996 Committee session in Merida considered "different options, including inscription of the site on the List of World Heritage in Danger or giving more time to the Government to implement actions." American delegate Reynolds recalls the closed door meeting that tried to find common ground:

> I remember sitting in that meeting and listening to the various viewpoints and suddenly I realized that the Ecuadorians were not arguing that the criteria were not being met ... They were arguing that in their culture, taking the action of a Western-devised system would have the opposite effect of what the Committee was trying to say ... It suddenly dawned on me that there was a potential for agreement if the wording was written correctly.[134]

He describes his pleasure when a consensus was reached under the leadership of Maria-Teresa Franco, the Mexican Chairperson: "If you could devise something that took care of everybody's cultural concern, you could get to the solution as opposed to just following rules. That to me was the most satisfying, one of the very most satisfying things that ever happened to me."[135] Following this informal consultation, Germany which had pushed for strict enforcement of the Operational Guidelines showed solidarity by proposing a compromise text that was innovative and prospective:

---

132   UNESCO, Report of the rapporteur on the nineteenth session of the World Heritage Committee in Berlin, 4–9 December 1995, Paris, 31 January 1996, WHC-95/conf.203/16, para. VII.13. Retrieved from http://whc.unesco.org/archive/1995/whc-95-conf203-16e.pdf

133   UNESCO, Report of the rapporteur on the twentieth session of the World Heritage Committee in Merida, 2–7 December 1996, Paris, 10 March 1997, WHC-96/conf.201/21, para. VII.31. Retrieved from http://whc.unesco.org/archive/1996/whc-96-conf201-21e.pdf

134   Canada Research Chair on Built Heritage, Université de Montréal, audio interview of John Reynolds by Christina Cameron, Springfield, 18 May 2011.

135   Canada Research Chair, interview Reynolds.

The Committee decided to include the Galapagos National Park on the List of World Heritage in Danger effective 15 November 1997, unless a substantive written reply by Ecuador is received by 1 May 1997 and the Bureau, at its twenty-first session, determines that effective actions have been taken. The Delegate of France asked the Committee to put on record that this decision was taken on an exceptional basis, as such a decision would normally be beyond the prerogative of the Bureau.[136]

Adopted by consensus, this was an exceptional decision unlike any the Committee had ever taken. While accommodating the State Party's requests for more time, the impending Danger listing was an effective tool to draw the attention of the State Party to the urgent need for legal enforcement and improved conservation measures.

In 1997, the Committee noted the progress made and commended the commitment of the Ecuadorian government to pass a special law for Galapagos as "the centrepiece of an effective conservation strategy for the site." Noting that the draft legislation had been approved in first reading, the Committee decided not to inscribe the site on the List of World Heritage in Danger "at this time."[137] The Special Law on the Galapagos was eventually adopted on 18 March 1998. The international community had succeeded through constructive dialogue in protecting an exceptional site. It was not included in the List of World Heritage in Danger during the period of this study but much later, from 2007 until 2010.

Eidsvik, working with the World Heritage Centre at that time, reflects on the implications:

> I would say one of the good examples would be the nomination of Galapagos to the World Heritage in Danger List. There was no question when it was nominated that Galapagos obviously was going to be on the List. That was never a question. And subsequently, of course, overfishing, big increase in the population on the island, put a great deal of pressure on the site and the net result was IUCN on its own volition wanted to place it on the World Heritage in Danger List. That was totally opposed by the ambassador from Ecuador, and I must say it was a very difficult time. It went on in meeting after meeting after meeting.[138]

---

136   UNESCO, Report of the rapporteur on the twentieth session of the World Heritage Committee in Merida, 2–7 December 1996, Paris, 10 March 1997, WHC-96/conf.201/21, para. VII.31. Retrieved from http://whc.unesco.org/archive/1996/whc-96-conf201-21e.pdf

137   UNESCO, Report of the rapporteur on the twenty-first session of the World Heritage Committee in Naples, 1–6 December 1997, Paris, 27 February 1998, WHC-97/conf.208/17, para. VII.38. Retrieved from http://whc.unesco.org/archive/1997/whc-97-conf208-17e.pdf

138   Canada Research Chair, interview Eidsvik.

These two examples illustrate the effectiveness of the Convention in establishing a collective system of support for conservation and in using the List of World Heritage in Danger as a tool to engage State Parties and the international community. In the cases of Angkor and Galapagos Islands, Danger listing served as a positive measure to help these two countries in their efforts to protect, restore and rehabilitate these properties.

World Heritage and Mining

In the 1990s, threats from mining operations began to crop up on a regular basis. At first the World Heritage Committee dealt with them on a site-specific basis but near the end of the millennium a comprehensive policy was developed in dialogue with international mining organizations. Four cases serve to illustrate the complexity of balancing heritage conservation and mining: Mount Nimba Strict Nature Reserve (Côte d'Ivoire/Guinea), Yellowstone National Park (USA), Doñana National Park (Spain) and Kakadu National Park (Australia).

The earliest example is Mount Nimba which was placed on the List of World Heritage in Danger in 1992. The Committee decided that a proposed iron-ore mining project threatened the integrity of the property as defined in the Operational Guidelines. Eidsvik, then working at the World Heritage Centre, describes his mission to Mount Nimba and the proposal to extract the mining area from the World Heritage Site:

> There is a huge iron mine ... we took it out of the World Heritage Site that was nominated and adjusted the boundaries during that mission that we were down there for two weeks ... That was agreed with everybody. Greenpeace was there, non-governmental people were there, government was there, French conservationists. And we modified the boundaries, partly on the basis that the government said this was never their intention that this should be included and we accepted that ... But it is still threatened by that mining potential in the long run and the level of management it receives is pretty marginal, in that basically it is being protected by the mining company that eventually hopes to do the mining.[139]

The next case, the proposed New World gold mine outside the northeast boundary of Yellowstone National Park, came to the Bureau's attention in 1995 when the Assistant Secretary for the United States Department of Interior asked that the site be considered for Danger listing and requested a mission to assess the mining proposal.[140] There followed a highly publicized site visit in September

---

139   Canada Research Chair, interview Eidsvik.
140   UNESCO, State of conservation of properties inscribed on the World Heritage List, Paris, 10 May 1995, WHC-95/conf.201/4, p. 16. Retrieved from http://whc.unesco.org/archive/1995/whc-95-conf201-4e.pdf

by the Committee Chairperson Wichiencharoen, representatives from IUCN and the Director of the World Heritage Centre. Eidsvik, a consultant with the World Heritage Centre at the time, describes the tense atmosphere:

> It was an example also of where the politics became so heated that there was actually a public hearing on it in Yellowstone, in which the representative of Senator Baucus at the time told the chairman of the meeting, Dr. von Droste, that he should go home and they were not welcome in the State of Montana to review whether Yellowstone should be on the in Danger List or not. And Dr. von Droste advised him that he had been invited by the State Department and he was going to stay.[141]

McNeely, a staff member of IUCN, recalls how its report on the negative impacts of the mine fuelled political debate within the United States:

> We submitted the report to the US National Park Service. We didn't publicise it. It was a contract we had to give them advice. So we gave it to them. And, the US Congress went berserk about this. "This is the United Nations trying to determine our policy for us." ... and they threatened to put a grandfather clause on all the World Heritage Sites, an incredible mess ... Whenever you have a gold mine you end up with a lot of effluents from mining and it would have been a big problem. So I think it was good advice, but it shows that politics can affect virtually any World Heritage Site.[142]

The United States updated the Committee in December on the positive steps underway, including a detailed environmental impact study of the mine proposal and further public consultation. More significant is the government's letter which opened the possibility of Danger listing by saying that "the State Party does not consider action by the Committee to be an intervention in domestic law or policy."[143] In 1995 Yellowstone was put on the Danger List "on the basis of both ascertained dangers and potential dangers."[144] The following year, declaring that "Yellowstone is more precious than gold," President Clinton halted the mine by announcing "an agreement in principle with the Canadian owners of the New

---

141   Canada Research Chair, interview Eidsvik.

142   Canada Research Chair, interview McNeely.

143   UNESCO, Report of the rapporteur on the nineteenth session of the World Heritage Committee in Berlin, 4–9 December 1995, Paris, 31 January 1996, WHC-95/conf.203/16, para. VII.22. Retrieved from http://whc.unesco.org/archive/1995/whc-95-conf203-16e.pdf

144   UNESCO, Report of the rapporteur on the nineteenth session of the World Heritage Committee in Berlin, 4–9 December 1995, Paris, 31 January 1996, WHC-95/conf.203/16, para. VII.22. Retrieved from http://whc.unesco.org/archive/1995/whc-95-conf203-16e.pdf

World Mine in which the Government will swap $65 million worth of federal land in exchange for the company's dropping its claim to some $650 million worth of gold deposits upstream from the park's northeastern corner in Montana."[145] As a result of this and other measures, the property was removed from the World Heritage List in Danger in 2003. Eidsvik celebrates the successful outcome of the Yellowstone case, noting that it "brought great public attention to the Committee and World Heritage as a tool to address mining issues."[146]

The Kakadu example involved a proposal to activate uranium mining at Jabiluka, one of the enclaves within the national park that had been excised from the World Heritage designation from the outset. At the time of inscription, the Committee commended Australia "for having taken appropriate legislative measures to prohibit mineral exploration and mining, and for their efforts to restore the natural ecosystems of the site."[147] However, the proposed activation of the mine led the Committee in the late 1990s to send its Chairperson Francioni to the site. His report concluded that "there are severe ascertained and potential dangers to the cultural and natural values of Kakadu National Park posed primarily by the proposal for uranium mining and milling at Jabiluka" and recommended that the mine not proceed.[148] It raised such concern that the Committee, for the first time in its history, decided to hold an extraordinary session exclusively devoted to a single conservation issue.[149] While the July 1999 session involved detailed scientific advice as well as high-level political negotiations, the Committee decided not to inscribe Kakadu National Park on the List of World Heritage in Danger. Nonetheless it expressed "deep regret that the voluntary suspension of construction of the mine decline at Jabiluka ... has not taken place," noting its grave concerns "about the serious impacts to the living cultural values of Kakadu National Park posed by the proposal to mine and mill uranium at Jabiluka."[150] To this day, the mine has not yet been activated.

---

145   Todd S. Purdum, "Clinton unveils plan to halt gold mine near Yellowstone," *The New York Times*, 13 August 1996. Retrieved from http://www.nytimes.com/1996/08/13/us/clinton-unveils-plan-to-halt-gold-mine-near-yellowstone.html

146   Canada Research Chair, interview Eidsvik.

147   UNESCO, Report of the rapporteur on the eleventh session of the World Heritage Committee in Paris, 7–11 December 1987, Paris, 20 January 1988, SC-87/conf.005/9, para. 9. Retrieved from http://whc.unesco.org/archive/1987/sc-87-conf005-9e.pdf

148   UNESCO, Report of the rapporteur on the twenty-second session of the World Heritage Committee in Kyoto, 30 November–5 December 1998, Paris, 29 January 1999, WHC-98/conf.203/18, para.VII.28. Retrieved from http://whc.unesco.org/archive/1998/whc-98-conf203-18e.pdf

149   See chapter 6 for a discussion on Kakadu National Park and State Party consent for Danger listing.

150   UNESCO, Report of the rapporteur on the third extraordinary session of the World Heritage Committee in Paris, 12 July 1999, Paris, 19 November 1999, WHC-99/conf.205/5rev, para. XI. Retrieved from http://whc.unesco.org/archive/1999/whc-99-conf205-5reve.pdf

At Doñana National Park in Spain, an environmental disaster brought the issue of mining in or near World Heritage Sites to a head. In 1997 the collapse of a giant holding pool of the Aznalcollar mine, owned by the Canadian-Swedish Boliden Apirsa Company, resulted in a toxic flow of tailing water downstream. The dramatic images of the spill were widely publicized although it was determined the next year that "the World Heritage site and the Biosphere Reserve are currently little affected, whereas the Natural Park around the site has been impacted by the toxic spill." Spain took the matter seriously, developing the "Doñana 2005" project to purify the waters and restore the marsh dynamics and ecological systems at an estimated cost of "approximately US$ 120,000.000."[151]

It was arguably the global attention and financial implications of cases like Yellowstone, Kakadu and Doñana that led to discussions with the mining industry on conservation and World Heritage. This is the context that led to a 1998 invitation from a global mining association, the International Council on Metals and the Environment (ICME),[152] for a meeting in London to discuss strategies to prevent threats and mitigate impacts of mining in the vicinity of World Heritage properties. A representative of ICME subsequently participated in the 1999 Committee session in Marrakesh during which discussions were held on threats and potential threats from mining operations at specific World Heritage sites as well as the general issue of mining and World Heritage:

> The Committee noted that a dialogue with the mining industry had commenced and that the Centre, IUCN and ICOMOS had been invited by the International Council on Metals and the Environment (ICME) to a working session on "Mining and Protected Areas and other Ecologically Sensitive Sites" on 20 October 1998 in London (UK).[153]

In addition, the Committee was informed of the position statement of IUCN's WCPA that discouraged mining and associated activities near or in protected areas. Views were divided. While Canada and France supported the WCPA position statement, others had concerns, including the United States who asked for clarification of its status:

---

151   UNESCO, Report of the rapporteur on the twenty-second session of the World Heritage Committee in Kyoto, 30 November–5 December 1998, Paris, 29 January 1999, WHC-98/conf.203/18, para. VII.25. Retrieved from http://whc.unesco.org/archive/1998/whc-98-conf203-18e.pdf

152   ICME was later renamed as International Council on Mining and Metals (ICMM) in 2001.

153   UNESCO, Report of the rapporteur on the twenty-third session of the World Heritage Committee in Marrakesh, 29 November–4 December 1999, Paris, 2 March 2000, WHC-99/conf.209/22, para X.49. Retrieved from http://whc.unesco.org/archive/1999/whc-99-conf209-22e.pdf

The WCPA draft statement was the subject of a recent hearing before the United States Congress, because the impression had been conveyed that it would be proposed to the World Heritage Committee in Marrakesh, to adopt a policy that would ban mining outside World Heritage sites ... It is the understanding of the United States that this document was tabled for information purposes only.[154]

In light of these concerns, further study was recommended. In a gesture towards fostering a broad dialogue, the Committee asked that all interested parties be involved, including the World Heritage Centre, other UNESCO units, the advisory bodies, UN agencies, States Parties and representatives of the mining industry. It called for "a technical meeting to analyse case studies on World Heritage and mining during global events already planned for the year 2000 ... and develop recommendations for review and discussion by the twenty-fourth session of the Committee."[155]

The workshop on World Heritage and mining, organized jointly by IUCN and ICME in collaboration with the World Heritage Centre, was held in September 2000 in Gland, Switzerland. Cases studies presented by site managers and mining representatives were analysed in order to develop guidance and recommendations. With an eye to their public image, the mining industry representatives expressed interest in poverty alleviation and collaboration with local communities at mining operations near World Heritage Sites. Conservation was obviously not their first objective and the dialogue was not always easy. However agreement was reached on transparent processes and early information-sharing among concerned stakeholders, including data on potential World Heritage Sites on national tentative lists:

Importantly, the workshop agreed on a set of 10 principles that should underpin the relationship between mining and World Heritage interests. In addition, a series of recommendations were specifically targeted at three stakeholder groupings: the World Heritage Committee and State Parties; World Heritage Management Agencies; and the Mining Industry. A key recommendation of the workshop was the establishment of a joint Working Group on World Heritage and Mining.[156]

---

154 UNESCO, Report of the rapporteur on the twenty-third session of the World Heritage Committee in Marrakesh, 29 November–4 December 1999, Paris, 2 March 2000, WHC-99/conf.209/22, para X.54, annex IX. Retrieved from http://whc.unesco.org/archive/1999/whc-99-conf209-22e.pdf

155 UNESCO, Report of the rapporteur on the twenty-third session of the World Heritage Committee in Marrakesh, 29 November–4 December 1999, Paris, 2 March 2000, WHC-99/conf.209/22, para X.60. Retrieved from http://whc.unesco.org/archive/1999/whc-99-conf209-22e.pdf

156 UNESCO, IUCN and ICME, *Technical Workshop: World Heritage and Mining, Technical Workshop, 21 –23 September 2000 Gland, Switzerland* (Paris: UNESCO, 2001). Retrieved from http://unesdoc.unesco.org/images/0012/001231/123112e.pdf

At its 2000 session, the World Heritage Committee adopted the conclusions and recommendations of the Gland meeting.[157] This was a significant achievement that eventually led to the extraordinarily important pledge from the mining association in 2003 to treat World Heritage as "no go areas".[158] The process of dialogue with the mining companies proved to be the catalyst for a forward-looking strategy beyond the monitoring of individual sites to a general policy on mining and World Heritage.

Deletion from the World Heritage List

Right from the start, the Committee discussed the possibility of removing properties from the World Heritage List. At its 1979 session it explored the "procedure for the eventual deletion from the World Heritage List of properties in case of deterioration leading to the loss of characteristics which determined their inclusion."[159] It discussed the stages of a deletion process and appropriate sources of information on the damage or deterioration of a property. For delisting "the Committee retained the proposal that decisions such as the sending out of fact-finding missions should be taken by the Committee, except in the case where emergency action was necessary, when the Bureau should be authorized to request the Secretariat to take such measures."[160] In 1980 the procedure for delisting was included in the Operational Guidelines, although the Committee was clear that this would be applied on an exceptional basis: "In adopting the above procedure, the Committee was particularly concerned that all possible measures should be taken to prevent the deletion of any property from the List and was ready to offer

---

157   UNESCO, Report of the rapporteur of the twenty-fourth session of the World Heritage Committee in Cairns, 27 November–2 December 2000, Paris, 16 February 2001, WHC-2000/conf.204/21, para. III. Retrieved from http://whc.unesco.org/archive/2000/whc-00-conf204-21e.pdf

158   This was done on the occasion of the World Parks Congress (WPC) held in 2003 in South Africa and was welcomed by UNESCO in a press release dated 22 August 2003, Retrieved   from   http://portal.unesco.org/en/ev.php-URL_ID=14151&URL_DO=DO_TOPIC&URL_SECTION=201.html. Shell made a similar commitment the same year. See IUCN's press release, Shell's Commitment to Biodiversity, World Heritage sites announced as "no go" areas for its oil and gas exploration and development, 27 August 2003. Retrieved from http://liveassets.iucn.getunik.net/downloads/shell_biodiversity_commitment.pdf

159   UNESCO, Report of the rapporteur on the third session of the World Heritage Committee in Cairo and Luxor, 22–26 October 1979, Paris, 30 November 1979, CC-79/conf.003/13, para. VIII. Retrieved from http://whc.unesco.org/archive/1979/cc-79-conf003-13e.pdf

160   UNESCO, Report of the rapporteur on the third session of the World Heritage Committee in Cairo and Luxor, 22–26 October 1979, Paris, 30 November 1979, CC-79/conf.003/13, para. 22. Retrieved from http://whc.unesco.org/archive/1979/cc-79-conf003-13e.pdf

technical co-operation as far as possible to States Parties in this connection."[161] During the period covered by this research, not a single property was deleted from the World Heritage List although the possibility was discussed on several occasions.[162]

International Assistance

In the years leading up to the creation of the World Heritage Convention, international cooperation and financial assistance consistently emerged as key objectives. It is worth recalling that the 1965 proposal for a World Heritage Trust materialized as part of the White House Conference on International Cooperation. The Convention's provision for international assistance and the establishment of the World Heritage Fund are concrete expressions of the spirit of international collaboration to safeguard the world's most precious sites. The Convention charges the Committee with receiving and studying requests for international assistance in order "to secure the protection, conservation, presentation or rehabilitation of such property" (article 13), thereby linking assistance directly with the state of conservation of World Heritage Sites. While the Convention defined the process in general terms (articles 19–26), the Committee set out the parameters of international assistance in its Operational Guidelines.

International assistance was immediately on the World Heritage agenda. At its first session in 1977, the Committee defined the format of requests for international assistance, the kinds of assistance including emergency help and technical cooperation, procedures, the order of priorities and the type of agreements to be concluded with States Parties.[163] At its next session, the Committee considered Ethiopia's appeal to support Simien National Park but decided not to allocate funding until it received "a more comprehensive technical assistance request."[164] It was only at its third session in 1979 that the Committee first allocated assistance from the World Heritage Fund. On reviewing an emergency request from

---

161   UNESCO, Operational Guidelines for the implementation of the World Heritage Convention, Report of the rapporteur on the fourth session of the World Heritage Committee in Paris, 1–5 September 1980, Paris, 29 September 1980, CC-80/conf.016/WHC/2 rev., paras. 24–32. Retrieved from http://whc.unesco.org/archive/1980/opguide80.pdf

162   The removal of properties from the World Heritage List occurred only later with the delisting in 2007 of the Arabian Oryx Sanctuary (Oman) and in 2009 the Dresden Elbe Valley (Germany).

163   UNESCO, Final report of the first session of the intergovernmental committee for the protection of the world cultural and natural heritage in Paris, 27 June–1 July 1977, Paris, 20 October 1977, CC-77/conf.001/9, paras. 42–53. Retrieved from http://whc.unesco.org/archive/1977/cc-77-conf001-9e.pdf

164   UNESCO, Report of the rapporteur on the second session of the World Heritage Committee in Washington, 5–8 September 1978, Paris, 9 October 1978, CC-78/conf.010/10 rev., para. 48. Retrieved from http://whc.unesco.org/archive/1978/cc-78-conf010-10reve.pdf

Yugoslavia to deal with earthquake damage at Kotor, the first site to enter the List of World Heritage in Danger, the Committee granted $20,000 (US) for consultant services and invited further information on the equipment needed.[165]

After that, the number of requests grew rapidly, soon demonstrating the inadequacy of the level of funds available. In 1980, the Committee approved six requests, including four from Africa and an additional $7,000 (US) for the preparation of a management plan at Ngorongoro Conservation Area (Tanzania), for a total of $357,700 (US).[166] In 1981, twelve requests were approved for a total of $608,400 (US).[167]

At the same time, a general shortage of qualified personnel in cultural heritage conservation led the Committee to support a "large-scale world training programme at both the regional and national levels for specialists in the conservation of cultural property," led by UNESCO in collaboration with ICCROM. For training of specialists in natural site conservation, the Committee observed that UNESCO's Man and the Biosphere programme already provided such assistance but stood ready to supplement it with specific training for "on-the-spot" specialists like rangers, managers and scientists. The Committee believed that cooperation for training "provided one of the most effective means to attain the objectives of the Convention."[168]

An analysis of the first two decades of international assistance illustrates the steady growth in the system:

> Although the amount allocated annually for International Assistance shows an uneven progression since 1978, the overall trend is towards an increase in both the number of requests approved and their amount, with the latter standing at an average of US$20,000 per request since 1992. This increase can be explained by the growth of the World Heritage Fund, while the rise in the number of requests is largely attributed to the increasing awareness of application procedures, and the Centre's role in assisting countries in this exercise.[169]

---

165   UNESCO, Report of the rapporteur on the third session of the World Heritage Committee in Cairo and Luxor, 22–26 October 1979, Paris, 30 November 1979, CC-79/conf.003/13, para. 60. Retrieved from http://whc.unesco.org/archive/1979/cc-79-conf003-13e.pdf

166   UNESCO, Report of the rapporteur on the fourth session of the World Heritage Committee in Paris, 1–5 September 1980, Paris, 29 September 1980, CC-80/conf.016/10, para. 35. Retrieved from http://whc.unesco.org/archive/1980/cc-80-conf016-10e.pdf

167   UNESCO, Report of the rapporteur on the fifth session of the World Heritage Committee in Sydney, 26–30 October 1981, Paris, 5 January 1982, CC-81/conf/003/6, para. 32. Retrieved from http://whc.unesco.org/archive/1981/cc-81-conf003-6e.pdf

168   UNESCO, Report of the rapporteur on the fifth session of the World Heritage Committee in Sydney, 26–30 October 1981, Paris, 5 January 1982, CC-81/conf/003/6, para. 35. Retrieved from http://whc.unesco.org/archive/1981/cc-81-conf003-6e.pdf

169   UNESCO, *Investing in World Heritage: past achievements, future ambitions. A guide to International Assistance* (Paris, 2002), p. 20. Retrieved from http://whc.unesco.org/documents/publi_wh_papers_02_en.pdf

The study also notes that over time bigger amounts were requested as a result of larger sites and a more integrated approach to protection. It demonstrated the growing discrepancy between available funding and the number of requests approved, concluding that the Fund no longer kept up with urgent conservation needs, management planning and other issues related to the safeguarding of World Heritage Sites. It argued that the way forward would require tapping into potential bilateral and multilateral partnerships.[170] The Convention includes articles to encourage the establishment of such public and private foundations as well as international fund-raising campaigns (articles 17–18).

McNeely speaks to the lack of funding as a weakness in the World Heritage system:

> I think that the Convention would be taken a lot more seriously, if it had a budget
> of something like 50 million per year to work on World Heritage in Danger
> which to me it is not used as it is supposed to be used. But if it had more meat,
> more money, to be crass, there would be more interest, and especially on the part
> of some of the smaller countries.[171]

In the years leading up to the millennium, some governments recognized the pressing needs at World Heritage Sites and set up mechanisms to provide extra-budgetary support. In 1989, Japan established the Japanese Trust Fund for the Preservation of the World Cultural Heritage; in 1995, the Nordic countries set up the Nordic World Heritage Office to support capacity-building in developing countries; in 1997, France adopted a different model to provide technical and financial support through the France-UNESCO Cooperation Agreement.[172] Lack of funds was a critical issue by 2000. The search for new partnerships and new sources of funding to safeguard the growing number of World Heritage Sites became a priority for the new millennium.

This chapter has documented steady progress in developing mechanisms to understand the state of conservation of World Heritage Sites and to provide assistance to address threats to their outstanding universal value. A unique instrument bringing cultural and natural heritage together, the Convention succeeded in creating a learning platform that merged different approaches and methodologies. The serious challenges of protecting and conserving heritage properties brought together many actors in the World Heritage system, including States Parties, site managers, specialists and UNESCO staff. The rich dialogue that ensued led to new standards and tools for a global approach to conservation. The different reporting processes confirm the link between site values and ongoing management. The

---

170   UNESCO, *Investing in World Heritage,* pp. 47–8. Retrieved from http://whc.unesco.org/documents/publi_wh_papers_02_en.pdf

171   Canada Research Chair, interview McNeely.

172   In the new century, several countries signed formal agreements with UNESCO to support World Heritage activities.

development of reactive monitoring and systematic monitoring enabled the World Heritage Committee to fulfil its oversight role in assessing whether World Heritage Sites maintain their outstanding universal value over time. This positive evolution to recognize the close connection between heritage values and on-site management is one of the great achievements of the World Heritage Convention.

Chapter 5
The Players

Implementation of the World Heritage Convention is entrusted to member governments, technical advisory bodies and UNESCO. Their respective roles and responsibilities are spelled out in the Convention text. Like a three-legged stool, all three groups need to participate actively in order to achieve the full potential of this international treaty. Although it has no formal role in the Convention text, civil society clearly has a contribution to make in the identification and conservation of World Heritage properties. In the period from 1972 to 2000, each group evolved as implementation of the Convention advanced.

States Parties

States Parties are national governments which have ratified or acceded to the World Heritage Convention through formal agreements with UNESCO. In 1973, the United States of America was the first country to ratify the Convention; Switzerland became the twentieth State Party on 19 September 1975, thereby bringing the Convention into force three months later on 19 December 1975. By joining the Convention, States Parties make a commitment to nominate properties on their national territory for inscription on the World Heritage List, to protect the outstanding universal value of such sites both at home and abroad, and to report on their condition. In addition, States Parties collectively are responsible for the governance of the Convention through participation in the General Assembly of States Parties and the World Heritage Committee, including its Bureau and subsidiary bodies.

General Assembly of States Parties

The General Assembly of States Parties meets every two years during the sessions of the General Conference of UNESCO. According to the World Heritage Convention, it has only two formal powers: to elect members to the World Heritage Committee (article 8.1) and to determine a uniform percentage of compulsory contribution to the World Heritage Fund, not to exceed 1 per cent of the contribution to the regular budget of UNESCO (article 16.1). It has no legal power under the Convention to approve general policies and procedures, except its own rules of procedure, nor to review the state of accounts of the World Heritage Fund. However, because it is the whole membership of the Convention, the General Assembly of States Parties has the opportunity to influence what the

World Heritage Committee does, particularly if it is unanimous in its views. In the period up to 2000, most of its energy was spent on Committee elections, an activity that is both time-consuming and competitive. It took many years for the General Assembly to develop an informal policy role.

In 1976 when the General Assembly of States Parties gathered for its first session in Nairobi, 26 countries had joined the Convention. It is interesting to note that one-third of the States Parties came from the European region (including North America) and one-third from the Arab region. Most were represented by diplomats. During the meeting there was a minor skirmish about setting the amount of the compulsory contributions to the World Heritage Fund. Switzerland and Poland proposed a lower amount of .75 per cent, but the majority adopted a higher figure of 1 per cent of a country's contribution to the regular budget of UNESCO.[1] However, there was a major confrontation over the method of electing members of the Committee, foreshadowing a perpetual irritant that has plagued the Convention ever since.

The Convention states that the World Heritage Committee would initially have a membership of 15 States Parties, to be increased to 21 members when the Convention came into force with at least 40 States Parties (article 8.1). Terms were set at six years, with one-third of the Committee being renewed every two years. At the first session of the General Assembly of States Parties, the Syrian Arab Republic proposed that the election process follow the usual practice within UNESCO by distributing seats to five pre-determined electoral groups. The ensuing debate revealed that not all delegations agreed. "This proposal gave rise to a wide exchange of views, in which 19 delegations took part, on the principles to be applied in distributing the seats." Those opposed pointed to the Convention text that says "election of members of the Committee shall ensure an equitable representation of the different regions and cultures of the world" (article 8.2). They argued that it was "necessary to take into account the nature and purpose of the Convention in determining the geographical distribution of the members of the Committee."[2] This disagreement still persists. At the first session, it was side-stepped when Australia proposed a secret vote to elect members to the Committee. The results of the election, with one-third of the elected members from Europe and only one-fifth from the Arab region, set a pattern of perceived and real unfairness.

In 1978, at the second General Assembly, the threshold of 40 States Parties had been reached, thereby increasing Committee membership to the full 21 countries.

---

1   UNESCO, Summary record of the first General Assembly of States Parties to the Convention concerning the protection of the world cultural and natural heritage, Nairobi, 26 November 1976, Paris, 15 February 1977, SHC/76/conf-014/col.9, para. 20. Retrieved from http://whc.unesco.org/archive/1976/shc-76-conf014-col9e.pdf

2   UNESCO, Summary record of the first General Assembly of States Parties to the Convention concerning the protection of the world cultural and natural heritage, Nairobi, 26 November 1976, Paris, 15 February 1977, SHC/76/conf-014/col.9, para. 11. Retrieved from http://whc.unesco.org/archive/1976/shc-76-conf014-col9e.pdf

Once again participants debated the question of how to achieve equitable representation. The proposal to use UNESCO electoral groups as the basis for seat distribution was put forward again, leading to an impasse, a suspension of the session and eventually to a secret vote. Although the Chairperson exhorted States Parties to fulfil their "moral obligation to achieve an equitable distribution" in their voting, countries from Europe were again elected to one-third of the available seats. Only two policy matters were raised, one on the enforcement of the Convention whereby any State Party in arrears with its payment to the World Heritage Fund may not stand for election to the Committee and the other a reminder from IUCN about the need to achieve a balance of natural and cultural sites on the World Heritage List.[3] By 1980 there was no debate in the General Assembly on equitable representation and elections proceeded by secret ballot as the normal course of business.[4]

The principle of fair rotation of membership was not something that came up in the beginning. In 1983, States Parties whose terms were completed (Australia, Egypt and Tunisia) re-submitted their candidatures immediately.[5] It was only in 1987 that concern about equitable representation re-surfaced, now specifically tied to three under-represented regions: "Several delegates ... drew the attention of the Assembly to the under-representation of Latin America, Africa and the Arab states within the Committee."[6] This time there was a call for reform:

> After the elections, several delegates declared that, without putting into question the results of the voting, it was possible to note a lack of balance in the distribution of seats to the different geographical groups. The Chairman was asked to see to it that the procedures for the election of the Committee Members

---

3 UNESCO, Summary record of the second General Assembly of States Parties to the Convention concerning the protection of the world cultural and natural heritage, Paris, 24 November 1978, CC-78/conf-011/6, paras. 11–15, 20. Retrieved from http://whc.unesco.org/archive/1978/cc-78-conf011-6e.pdf

4 UNESCO, Summary record of the third General Assembly of States Parties to the Convention concerning the protection of the world cultural and natural heritage, Belgrade, 7 October 1980, CC-80/conf-018/6. Retrieved from http://whc.unesco.org/archive/1980/cc-80-conf018-6e.pdf

5 UNESCO, Summary record of the fourth General Assembly of States Parties to the Convention concerning the protection of the world cultural and natural heritage, Paris, 28 October 1983, Paris, 28 November 1983, CLT-83/conf-022/6, paras. 13–14. Retrieved from http://whc.unesco.org/archive/1983/clt-83-conf022-6e.pdf

6 UNESCO, Summary record of the sixth General Assembly of States Parties to the Convention concerning the protection of the world cultural and natural heritage, Paris, 30 October 1987, Paris, 31 October 1987, CC-87/conf-013/6, para. 13. Retrieved from http://whc.unesco.org/archive/1987/cc-87-conf013-6e.pdf

be reviewed in order to ensure the universal and cultural representation within the committee foreseen by the Convention.[7]

The 1989 session of the General Assembly of States Parties received a report from the World Heritage Committee on ways of ensuring equitable representation of the different regions and cultures of the world. What followed was "an intensive debate" during which delegates weighed the merits of suggestions to improve rotation, including setting quotas of seats per region, revising the Convention to increase Committee membership to 36 States Parties, creating special observer status for outgoing members and allocating travel money for representatives from developing countries. The only immediate decision was the adoption of a rather weak resolution on rotation: "The General Assembly of States Parties ... invites the States Parties to the World Heritage Convention, whose mandates on the Committee expire, to consider not to stand for re-election during an appropriate period." The other ideas were sent back to the Committee for further consideration. Nonetheless the Chairperson concluded the meeting by noting, somewhat naively, "that the meeting had allowed the drawing up of the main principles for an improved geographical distribution of members on the Committee."[8] It is worth noting that no other policies were discussed and that the election process had become so laborious that the meeting spilled over into a second day in order to complete four rounds of voting.

Despite the animated debate two years earlier, the 1991 General Assembly only took note of working documents prepared by the secretariat, one presenting the periods during which each State Party had been a member of the Committee, the other showing the distribution of Committee membership according to region. With a few exceptions, these charts showed that representation over time had been moderately equitable. Before the 1991 election, the Chairperson of the General Assembly appealed to countries to respect the desire for rotation. Two out-going States Parties chose different paths: Bulgaria withdrew its candidature for a renewed term but Mexico did not and was re-elected. The vote brought little satisfaction. When none of the three African candidates (Kenya, Madagascar and Zimbabwe) was elected, the delegate of Mauritania asked that the minutes record "his regret at the fact that the elections had not in the least improved the geographical distribution, which was recognized by everyone to be necessary. Indeed, no African State south of the Sahara had been elected, and this region was

---

7   UNESCO, Summary record of the sixth General Assembly of States Parties to the Convention concerning the protection of the world cultural and natural heritage, Paris, 30 October 1987, Paris, 31 October 1987, CC-87/conf-013/6, para. 21. Retrieved from http://whc.unesco.org/archive/1987/cc-87-conf013-6e.pdf

8   UNESCO, Summary record of the seventh General Assembly of States Parties to the Convention concerning the protection of the world cultural and natural heritage, Paris, 9 and 13 November 1989, Paris, 13 November 1989, CC-89/conf-013/6, paras. 12, 21. Retrieved from http://whc.unesco.org/archive/1989/cc-89-conf013-6e.pdf

thus represented on the Committee by only one State."[9] This was the State Party of Senegal, previously elected during the 1989 General Assembly.[10]

The 1993 election brought the voting procedure to a head. At the outset, there were 31 States Parties seeking election to seven seats. In addition, five of the seven outgoing members (Brazil, Cuba, France, Italy and the United States of America) were running again, despite the Assembly's appeal for voluntary rotation. In the face of so many candidatures, the Assembly suspended its deliberations in order to allow regional groups to consult among themselves. The Arab and African groups did their work, returning with a reduced slate of two candidates per region. The European group failed to do so and fielded more than half of all candidates. The voting process was long and bitter. There were an astounding nine rounds of voting, two of which failed to produce the required majority for any candidate.

At the close of the 1993 election, the General Assembly called for a review of the rules of procedure since "the requirement of obtaining an absolute majority had necessitated nine ballots, and, on the other hand, the system did not guarantee an equitable representation of the different cultures and regions of the world." While participants suggested revamping the requirement for an absolute majority and creating a screening committee for candidatures, they reserved their strongest criticism for outgoing States Parties who ignored their "obligation ... to abide by the recommendation not to seek immediate re-election."[11] In addition, the inefficiency of the voting process meant that there was little time for debate on other substantive measures. The General Assembly began to think about a larger role in providing leadership for the World Heritage system, leading to the recommendation, for the first time ever, "that its future sessions devote more time to debates of substance aimed at defining general policy directives for the implementation of the Convention."[12]

---

9   UNESCO, Summary record of the eighth General Assembly of States Parties to the Convention concerning the protection of the world cultural and natural heritage, Paris, 2 November 1991, Paris, 6 November 1991, CLT-91/conf-013/6, paras.13, 21. Retrieved from http://whc.unesco.org/archive/1991/clt-91-conf013-6e.pdf

10   UNESCO, Summary record of the seventh General Assembly of States Parties to the Convention concerning the protection of the world cultural and natural heritage, Paris, 9 and 13 November 1989, Paris, 13 November 1989, CC-89/conf-013/6, para. 15. Retrieved from http://whc.unesco.org/archive/1989/cc-89-conf013-6e.pdf

11   UNESCO, Summary record of the ninth General Assembly of States Parties to the Convention concerning the protection of the world cultural and natural heritage, Paris, 29–30 October1993, Paris, 2 November 1993, WHC-93/conf-003/6, paras. 13–29. Retrieved from http://whc.unesco.org/archive/1993/whc-93-conf003-6e.pdf

12   UNESCO, Summary record of the ninth General Assembly of States Parties to the Convention concerning the protection of the world cultural and natural heritage, Paris, 29-30 October1993, Paris, 2 November 1993, WHC-93/conf-003/6, para. 30. Retrieved from http://whc.unesco.org/archive/1993/whc-93-conf003-6e.pdf

The 1995 General Assembly addressed the voting process by adopting changes to its rules of procedure "to avoid an excessive number of ballots." These changes limited the requirement for an absolute majority to four ballots, to be followed by a fifth ballot requiring only a simple majority with a drawing of lots to break a tie vote. It is interesting that delegates, sensitive to State Party sovereignty, rejected a formal "interdiction of an outgoing Committee member to present its candidature for immediate re-election." Nonetheless, unlike the previous session when four outgoing countries were successfully re-elected, the 1995 meeting rejected the candidatures of three outgoing countries seeking immediate re-election (Indonesia, Oman and Thailand), suggesting that delegates had effectively taken on board the idea, if not the rule.[13]

Many countries sought election to the Committee in 1997, necessitating four rounds of voting. The principle of rotation was still not well respected as indicated by the fact that four of the seven outgoing members (China, Egypt, Mexico and Spain) sought immediate re-election although only one (Mexico) was successful. Curiously, there was an error in voting procedure that apparently went unnoticed by all participants. The Chairperson of the General Assembly announced that, according to the rules of procedure, the fourth ballot required only a "simple majority". This rule had been amended two years earlier to state that it was the fifth ballot that required a simple majority, the fourth ballot still requiring an absolute majority. One can only wonder what would have happened if the rules had been properly applied. This error speaks to the chaos of the elections process and the inattention of UNESCO staff and delegates alike.[14]

More significant is the shift in role. To address concerns about monitoring, the 1995 General Assembly held its first substantive policy debate outside the elections process. The issue was the Committee's 1994 approval of new text for its Operational Guidelines to formalize monitoring and reporting procedures on the state of conservation of World Heritage Sites.[15] The policy debate that followed was less about the need for monitoring and more about legalities and state sovereignty. It marked the first open power struggle between the Committee and the General Assembly of States Parties. While one group viewed a voluntary system of on-site monitoring and reporting as a technical process within the purview of the Committee, others saw it as an intrusion on national sovereignty, arguing that

---

13   UNESCO, Summary record of the tenth General Assembly of States Parties to the Convention concerning the protection of the world cultural and natural heritage, Paris, 2-3 November 1995, Paris, 22 November 1995, WHC-95/conf-204/8, paras. 13–14, 34–42. Retrieved from http://whc.unesco.org/archive/1995/whc-95-conf204-8e.pdf

14   UNESCO, Summary record of the eleventh General Assembly of States Parties to the Convention concerning the protection of the world cultural and natural heritage, Paris, 27-28 October 1997, Paris, 18 December 1997, WHC-97/conf-205/7, para. 30. Retrieved from http://whc.unesco.org/archive/1997/whc-97-conf205-7e.pdf

15   See chapter 4 for details on the monitoring and reporting proposal. The new text was approved in 1994 and published in the 1996 version of the Operational Guidelines.

neither the Committee nor the General Assembly could demand reports from individual countries. According to the Convention, only the General Conference of UNESCO could require reports from member states (article 29). After long discussion and no less than eight proposed draft resolutions, the UNESCO legal advisor suggested a way forward. He confirmed that article 29 did indeed require reports to be submitted to the General Conference of UNESCO but suggested that it "could be used in a flexible way and the manner of the reporting could very well be, if the General Conference would so decide, through the General Assembly or the World Heritage Committee."[16]

The proposal was deferred by vote to the next General Assembly in 1997 at which time an agreement was reached to ask the General Conference of UNESCO to activate the procedures of article 29 so that a periodic reporting system could be established under the direction of the World Heritage Committee. The resolution reaffirmed the responsibility of States Parties for managing World Heritage Sites situated in their territories and concluded that monitoring is the responsibility of the State Party concerned. It also highlighted the benefits of a common policy for the protection of cultural and natural heritage as a means of fostering interaction among States Parties and sharing experiences with regard to conservation methods.[17]

The last General Assembly before the turn of the millennium covered several policy issues that were on the upcoming reform agenda of the Cairns Committee meeting in 2000: how to ensure a balanced and representative World Heritage List; how to achieve equitable representation on the World Heritage Committee; how to improve the working methods of the Committee; and how to increase funding for the World Heritage Centre. A proposal to increase the number of seats on the World Heritage Committee led to an explanation from the UNESCO legal advisor on how the Convention might be modified.

The 1999 election took place in this intense atmosphere of reform. While France withdrew its candidature "for the sake of the system of rotation", three of the outgoing members sought immediate re-election (Italy, Japan and Lebanon). One of the high points occurred after two ballots when the United Kingdom withdrew in favour of Egypt. In so doing, it reinforced the principle of rotation, stating that "on this occasion three Western European countries have already been selected to the Committee, but only one representative each of Africa, Asia and Latin America,

---

16   UNESCO, Summary record of the tenth General Assembly of States Parties to the Convention concerning the protection of the world cultural and natural heritage, Paris, 2-3 November 1995, Paris, 22 November 1995, WHC-95/conf-204/8, para. 28. Retrieved from http://whc.unesco.org/archive/1995/whc-95-conf204-8e.pdf

17   UNESCO, Summary record of the eleventh General Assembly of States Parties to the Convention concerning the protection of the world cultural and natural heritage, Paris, 27–28 October 1997, Paris, 18 December 1997, WHC-97/conf-205/7, paras. 23–6. Retrieved from http://whc.unesco.org/archive/1997/whc-97-conf205-7e.pdf

and none from the Arab States. The UK believes strongly in the need for rotation and a proper balance of representation in the work of the Committee."[18]

At the beginning of the period under study, the scant number of States Parties meant that each had many opportunities to participate directly in the work of the Convention. As the number of countries grew and meetings became more formal, States Parties felt disenfranchised unless they were part of the powerful inner circle of the World Heritage Committee. This frustration may explain the shift in the mid-1990s towards substantive policy discussions in the General Assemblies, a forum where all members have an equal voice. Nonetheless, throughout the entire period dissatisfaction continued to fester over the failure of the elections to produce a Committee with "an equitable representation of the different regions and cultures of the world." The issue therefore became part of the reform agenda of the Cairns meeting in 2000.

World Heritage Committee

The World Heritage Committee is responsible for the practical implementation of the Convention. In its start-up phase, it was composed of 15 members; by 1979, it reached its full complement of 21 members with six-year terms. The Committee meets in annual sessions and works through a small Bureau chosen from among its members on a regional basis to carry out preparatory activities between regular sessions. According to the Convention, the Committee's key roles are to identify and list sites with outstanding universal value, to inscribe threatened sites on the Danger List, and to allocate resources from the World Heritage Fund. The duties have been nuanced over time through changes in the Operational Guidelines. In 1977, the first guidelines state that the Committee has four critical functions: to draw up a World Heritage List, to prepare a List of World Heritage in Danger, to determine the best use for the World Heritage Fund, and to assist member states in the protection of their properties of outstanding universal value.[19] This last function disappeared in the 1980 version of the guidelines. In 1994 a new function to monitor the state of conservation of properties was added.[20]

---

18   UNESCO, Summary record of the twelfth General Assembly of States Parties to the Convention concerning the protection of the world cultural and natural heritage, Paris, 28–29 October 1999, Paris, 8 November 1999, WHC-99/conf-206/7, para. 12. Retrieved from http://whc.unesco.org/archive/1999/whc-99-conf206-7e.pdf

19   UNESCO, Operational Guidelines for the implementation of the World Heritage Convention, Final report of the first session of the intergovernmental committee for the protection of the world cultural and natural heritage in Paris, 27 June–1 July 1977, Paris, 20 October 1977, CC-77/conf. 001/8 rev., para. 3. Retrieved from http://whc.unesco.org/archive/1977/cc-77-conf001-8reve.pdf

20   UNESCO, Operational Guidelines for the implementation of the World Heritage Convention, WHC/2/rev. February 1994, para. 3. Retrieved from http://whc.unesco.org/archive/opguide94.pdf. These four functions remained in place until the 2005 Operational Guidelines introduced a more detailed list of duties.

The Convention calls for a Committee composed of professionals working in the heritage field: "States members of the Committee shall choose as their representatives persons qualified in the field of the cultural or natural heritage" (article 9.3). Canadian delegate Peter Bennett reports that at the 1977 Committee meeting "political considerations virtually never intruded."[21] An early participant from UNESCO, Bernd von Droste describes the initial expectations among the creators of the Convention:

> Experts had met in '77 and experts had formulated the criteria. Experts had adopted the Operational Guidelines, rules of procedures. There was the expectation by these experts that the Convention ... would be really something for leading personalities of the cultural and natural world to implement, that they would be supported in their decisions by the expert advice of IUCN, ICCROM and ICOMOS.[22]

Speaking of the 1980s, Jane Robertson Vernhes notes that "definitely there was a trend ... the Bureau meetings were very much the technical meetings, clearing the workload if I can say for the Committee ... Those were people who were experts in cultural heritage or natural heritage."[23] Léon Pressouyre confirms her recollection. "In 1980 we had archaeological colleagues, historians, site managers who really spoke the same language we did."[24] As von Droste says, "This was in the beginning how the Committee saw it should provide for its leadership and how it should deliberate and take the decisions. It really tried to relate all these questions of what is World Heritage, what is World Heritage List value to experts."[25]

François Leblanc, an ICOMOS staff member at the time, provides a fascinating glimpse of an atmosphere of conviviality and collegiality at early Committee meetings. "There was a very positive interaction, not during Committee meetings because we discussed technical matters there, but in the evenings, the meetings, the coffee breaks."[26] He describes his situation as a young ICOMOS representative at the 1980 session of the Committee: "I was facing colleagues who were either

---

21 Peter H. Bennett, "Protecting the World's Cultural and Natural Heritage," Ottawa, 3 May 1978, Parks Canada, World Heritage archives, file 1972–1978.

22 Canada Research Chair on Built Heritage, Université de Montréal, audio interview of Bernd von Droste by Christina Cameron and Mechtild Rössler, Paris, 1 February 2008.

23 Canada Research Chair on Built Heritage, Université de Montréal, audio interview of Jane Robertson Vernhes by Christina Cameron and Mechtild Rössler, Paris, 24 November 2009.

24 Canada Research Chair on Built Heritage, Université de Montréal, audio interview of Léon Pressouyre by Christina Cameron and Mechtild Rössler, Paris, 18 November 2008. "En 80 nous avions des collègues archéologues, historiens, gestionnaires de sites qui vraiment parlaient le même langage que nous."

25 Canada Research Chair, interview von Droste. 2008.

26 Canada Research Chair on Built Heritage, Université de Montréal, audio interview of François Leblanc by Christina Cameron, Ottawa, 7 April 2009. "Il y avait une interaction

doctors of archaeology, specialists in the history of architecture or directors of cultural agencies in their countries. So, I was with professional colleagues."[27] His initial unease about speaking in a convincing way about sites he did not know well soon vanished:

> There was always someone around the table who raised his hand and supported the nomination with professional arguments and so I quickly realized that I was in a situation where I was not opposed to the Committee but I was there to bring them information … colleagues were working with me either to sustain the nomination or to support the fact that it was not recommended.[28]

He observes that meetings during this period "focused more on trying to build capacity in the future generation who had to look after these sites than on carrying out diplomatic exchanges among the different countries."[29]

Complaints about the Committee's lack of expertise in natural heritage surfaced almost immediately. The 1980 Committee reminded States Parties of their obligation to send representatives qualified in natural heritage and proposed that the chairmanship be rotated between natural and cultural heritage experts every two years.[30] Hal Eidsvik confirms that there were few natural heritage experts: "In all those early meetings, and I speak here of the first ten years of the Convention, there was certainly an imbalance of people with knowledge on the natural heritage side compared to the people involved with cultural heritage."[31] Rob Milne adds that "there were probably six to seven cultural heritage people to one natural

---

très positive, pas au niveau du Comité parce qu'on discutait des choses techniques là, mais dans les soirées, les réunions, les pauses café."

27 Canada Research Chair, interview Leblanc. "On faisait face à des pairs, à des collègues qui étaient soit des docteurs en archéologie, des spécialistes en histoire de l'architecture ou des directeurs d'agence culturelle dans leur pays. Donc, on était entre confrères professionnels."

28 Canada Research Chair, interview Leblanc. "Il y en avait toujours un autour de la table qui levait la main et qui soutenait la nomination avec des arguments professionnels et donc, je me suis rapidement senti dans un milieu où je n'étais pas en opposition avec le Comité mais j'étais là pour apporter au Comité des informations … les collègues autour de la table travaillaient avec moi, soit pour maintenir la nomination ou soutenir le fait qu'on ne la recommandait pas."

29 Canada Research Chair, interview Leblanc. "C'était donc une période beaucoup plus de contribution à essayer de former la génération future qui était pour s'occuper de ces sites que de faire des échanges internationaux entre les différents États."

30 UNESCO, Report of the rapporteur on the fourth session of the World Heritage Committee in Paris, 1–5 September 1980, Paris, 29 September 1980, CC-80/conf.016/ WHC/2 rev., paras. 6, 21, 47–9. Retrieved from http://whc.unesco.org/archive/1980/op guide80.pdf

31 Canada Research Chair on Built Heritage, Université de Montréal, audio interview of Hal Eidsvik by Christina Cameron, Ottawa, 3 July 2009.

heritage person within delegations."[32] IUCN's World Heritage coordinator in the 1980s and 1990s, Jim Thorsell observes that "there were never, from the natural side, a lot of natural experts that were attending these meetings. Even when there were, they didn't often have much power to speak or make any interventions. They were always seen as secondary … somebody around who would grill IUCN on … the details." He adds with regret that "We always tried to promote more natural experts coming to the Committee meetings but unfortunately this has never really come to pass except for a couple of years when we did have a number around."[33]

Politicization of the World Heritage Committee
During the period under study, the Committee evolved from a technical body to a more political one. In his 1983 speech as outgoing Chairperson, Australian Ralph Slatyer warned of the difficulties that arise when the nomination of a property is made by a country which is a member of the Committee. He noted the opportunity for lobbying fellow members, stating that such advocacy could be seen "to place considerable pressure on the Committee to make a favourable decision, thereby giving a distinct advantage to nominations from States which were Committee members in comparison to those from States which were not." Calling on the Committee to spell out rules about advocacy in its Operational Guidelines, Slatyer went even further: "I consider objectivity and freedom of bias so important to the quality and interpreting of the World Heritage List that I … ask you to consider the proposition that, whenever a State party is serving on the Committee, none of its nominations should be dealt with."[34] This is the first expression of an oft-repeated idea about inviting countries to withhold presenting nominations during membership on the Committee to avoid a perceived or real conflict of interest.

According to Francesco Francioni, the balance has always been tilted towards the political dimension. "There are of course delegations who come very well equipped with technical expertise both in the environmental and cultural field. But at the end of the day, it is in the nature of the Committee as an organ of States, and not of individuals." Calling it the "original sin of the Committee," he bluntly remarks that "you have diplomats and they call the shots … So there is no way that the epistemic community of the environmentalist, of the naturalist or the people who are involved in cultural heritage are going to have the last word."[35]

---

32   Canada Research Chair on Built Heritage, Université de Montréal, audio interview of Rob Milne by Christina Cameron and Mechtild Rössler, Paris, 2 March 2009.

33   Canada Research Chair on Built Heritage, Université de Montréal, audio interview of Jim Thorsell by Christina Cameron, Banff, 11 August 2010.

34   Ralph Slatyer, Address by the outgoing chairman of the World Heritage Committee, in UNESCO, Report of the rapporteur on the seventh session of the World Heritage Committee in Florence 5–9 December 1983, Paris, January 1984, SC/83/conf.009/8, annex II, pp. 3–5. Retrieved from http://whc.unesco.org/archive/1983/sc-83-conf009-8e.pdf

35   Canada Research Chair on Built Heritage, Université de Montréal, audio interview of Francesco Francioni by Christina Cameron and Mechtild Rössler, Rome, 5 May 2010.

Pressouyre remarks on the acceleration of politicization near the end of the 1980s:

> Little by little we witnessed the arrival of diplomats, ambassadors, ministers, people who were sent by their countries, who transformed the course of the World Heritage Convention and who accentuated, I would say, the political aspect, perhaps not in the best sense of the term, of the Convention ... There was this kind of pressure that was felt through the permanent delegates of UNESCO or political persons starting at the end of the 1980s.[36]

He does not mince his words: "I actually think that in an ideal world, the politicians, if they are to take the floor, should draw much more strongly on the specialists and that they should take into account realities that they do not always know."[37]

Robertson Vernhes agrees that the Committee meetings became more political: "Given the pride, the prestige and everything else ... the heads of the foreign affairs departments or whoever inevitably got into the show. In their delegation they would bring along their expert but I think that more and more the experts were given the back seat and the politicians were given the front seat."[38] Herb Stovel reinforces the point that scientific experts from Committee delegations were rarely heard: "Those voices may be present and those people may be present, but they are sitting in the back row. The person that's taking the microphone is an ambassador, someone with a diplomatic portfolio ... who has trouble dissociating their behaviour in other diplomatic forums from the requirements of the World Heritage Convention."[39] Jim Collinson explains that financial considerations could be a contributing factor: "The cost of somebody sending someone for a week twice a year is prohibitive. The effect then is that the ambassador of that country to UNESCO attends. That ambassador may or may not be qualified in the sense envisioned by the Convention."[40]

---

36 Canada Research Chair, interview Pressouyre. "Je dirais que l'évolution s'est accélérée justement au cours de la période où j'ai été actif entre 80 et 90 ... Petit à petit, nous avons vu arriver des diplomates, des ambassadeurs, des ministres, des personnes qui étaient envoyées, des personnes donc envoyées par le pays qui ont grandement infléchi la marche de la Convention du patrimoine mondial et qui, qui ont accentué, je dirais l'aspect politique, peut-être pas au meilleur sens du terme, de cette Convention ... Il y a eu cette sorte de pression qui s'est fait sentir via des délégués permanents auprès de l'UNESCO ou des personnes politiques à partir de la fin des années 80, je dirais."

37 Canada Research Chair, interview Pressouyre. "Je pense effectivement que dans un monde idéal, il faudrait que les politiques, si c'est à eux de prendre la parole, s'appuient d'une manière beaucoup plus forte sur des spécialistes et qu'ils tiennent compte de réalités qu'ils ne connaissent pas toujours."

38 Canada Research Chair, interview Robertson Vernhes.

39 Canada Research Chair on Built Heritage, Université de Montréal, audio interview of Herb Stovel by Christina Cameron, Ottawa, 3 February 2011.

40 Canada Research Chair on Built Heritage, Université de Montréal, audio interview of Jim Collinson by Christina Cameron, Windsor, 12 July 2010.

There were several key cases that increased the politicization of the Committee. Von Droste claims that initially tension between the scientific and political aspects of the Convention "was triggered by the intention of some State Parties right away to use the World Heritage Convention for a message, a political message and an important message for humanity." When Poland proposed to inscribe Auschwitz Birkenau, von Droste says the secretariat was taken by surprise. "Poland, in particular, thought that the World Heritage List should be opened up by a humanitarian perspective. World Heritage should stand for intercultural understanding, for tolerance and should commemorate the crimes of the Nazi." He confesses to a certain naivety in thinking that Auschwitz was a unique case. The secretariat believed that this would "be the only time where we inscribe such a politically laden property."[41]

The next year, the nomination of the old city of Jerusalem plunged the Committee into a deep political quagmire. The sequence of events is straightforward. Jordan presented the nomination of Jerusalem to the September 1980 session of the World Heritage Committee which "decided to open the established procedure for the examination of this proposal."[42] In May 1981 the Bureau received a positive recommendation from ICOMOS but could not reach consensus on inscription and therefore forwarded the nomination file to the Committee.[43] At the request of 17 States Parties, the Committee held an extraordinary session in September 1981 to consider the nomination. Following a roll-call vote that recorded 14 votes in favour, one against and five abstentions, the Old City of Jerusalem and its Walls was inscribed on the World Heritage List.[44]

The pioneers who participated in the inscription of Jerusalem all recall the meeting as emotional and tense. Von Droste reports that the secretariat did not expect the tabling of the nomination in 1980: "We did not know, we did not anticipate, and a big surprise to the Committee itself."[45] In its professional judgement, ICOMOS was convinced that Jerusalem met the threshold of outstanding universal value, recommending inscription because "Jerusalem is directly and materially associated with the history of the three great monotheistic religions of mankind" and because the major monuments of the city have had "a considerable influence

---

41 Canada Research Chair, interview von Droste, 2008.

42 UNESCO, Report of the rapporteur on the fourth session of the World Heritage Committee in Paris, 1–5 September 1980, Paris, 29 September 1980, CC-80/conf.016/10, para. 16. Retrieved from http://whc.unesco.org/archive/1980/cc-80-conf016-10e.pdf

43 UNESCO, Report of the rapporteur on the fifth session of the Bureau of the World Heritage Committee in Paris, 4–7 May 1981, Paris, 20 July 1981, CC-81/conf.002/4, paras. 7–13.Retrieved from http://whc.unesco.org/archive/1981/cc-81-conf002-4e.pdf

44 UNESCO, Report of the rapporteur on the first extraordinary session of the World Heritage Committee in Paris, 10–11 September 1981, Paris, 30 September 1981, CC-81/conf.008/2, paras. 1, 6–15. Retrieved from http://whc.unesco.org/archive/1981/cc-81-conf008-2reve.pdf

45 Canada Research Chair, interview von Droste, 2008.

Figure 5.1   Old City of Jerusalem and its Walls in 1980

on the development of Christian and Moslem religious architecture."[46] Pressouyre, who presented the case for ICOMOS, says "If there is a site in the world which has universal value, it is Jerusalem. Jerusalem where one finds believers from three religions: Jews, Christians, Muslims. Jerusalem where there is also a considerable accumulation of monuments from the Hellenistic period to the end of the Ottoman period."[47] In its technical analysis, ICOMOS considered Jerusalem as an historic ensemble and argued that "protection must take into account in so far as possible, the whole of the archaeological and monumental heritage." Therefore ICOMOS proposed a list of other monuments and historic buildings that should be added to the protection zone.[48] It was this supplementary list which heightened tension over the status of Jerusalem and raised the issue of sovereignty.

---

46   ICOMOS, The Old City of Jerusalem (Al-Quds) and its Walls, Paris, April 1981, 148 rev. Retrieved from http://whc.unesco.org/archive/advisory_body_evaluation/148.pdf

47   Canada Research Chair, interview Pressouyre. "S'il y a un site au monde qui a une valeur universelle, c'est bien Jérusalem, Jérusalem où se retrouve les croyants de trois religions du Livre : les juifs, les chrétiens, les musulmans et Jérusalem aussi où il y a une accumulation monumentale considérable depuis la période hellénistique jusqu'à la fin de la période ottomane."

48   ICOMOS, The Old City of Jerusalem (Al-Quds) and its walls, Paris, April 1981, 148 rev. Retrieved from http://whc.unesco.org/archive/advisory_body_evaluation/148.pdf

At the first extraordinary session of the Committee in 1981, the political debate was related to jurisdiction over territory. The composition of member delegations also changed. While the 1980 session had had the usual mix of experts and a few diplomats, the extraordinary session in 1981 was attended almost entirely by diplomats and government ministers.[49] Von Droste observes: "This debate on Jerusalem was of course fully politicized and we had all of a sudden a change in the Committee, namely ambassadors, because the experts saw that they could not really handle that question."[50] The discussion unfolded in a highly charged atmosphere. Von Droste describes the meeting as chaotic: "People were crying, had tears in their eyes. It was a highly emotional debate because we are speaking of religions and their symbols, we are speaking about the most emotionally politically-laden property we probably have today on the World Heritage List."[51]

In the months leading up to the extraordinary 1981 session, the two ICOMOS representatives were hounded by all parties. Leblanc reports that "from January 1 until September 11, I can tell you that the ICOMOS Bureau received a huge amount of mail and phone calls and visits, from Israelis, Americans and other western countries and about ten Arab countries, some supporting the nomination and others against."[52] He says that both he and Pressouyre received calls well into the night on the eve of the extraordinary session, to the point that a visibly upset Pressouyre almost refused to attend the meeting. In the end, he did present the ICOMOS evaluation. In his interview, he recalls the difficulty of speaking on behalf of an international professional organization in such a situation:

> I was the one who spoke all the time on behalf of ICOMOS during this session, under the critical eye of the representative of France among others, who did not want me to say things contrary to what France wanted to say. But I said to him, I am not French for the moment.[53]

---

49   UNESCO, Report of the rapporteur on the first extraordinary session of the World Heritage Committee in Paris, 10–11 September 1981, Paris, 30 September 1981, CC-81/conf.008/2, annex 1. Retrieved from http://whc.unesco.org/archive/1981/cc-81-conf008-2reve.pdf

50   Canada Research Chair, interview von Droste, 2008.

51   Canada Research Chair, interview von Droste, 2008.

52   Canada Research Chair, interview Leblanc. "Or du 1er janvier aller jusqu'au 11 septembre, je peux vous dire que le Bureau de l'ICOMOS a reçu énormément de courrier et de coups de téléphone et de visites, soit de représentants israéliens, américains, et de d'autres pays de l'Ouest aussi et d'une dizaine de pays arabes qui les uns soutenaient la nomination et les autres étaient contre."

53   Canada Research Chair, interview Pressouyre. "C'est moi qui ai parlé constamment au nom de l'ICOMOS au cours de cette séance, voilà sous le regard critique du représentant de la France entre autre, qui ne voulait pas que je dise des choses contraires à ce que la France voulait qu'il dise. Mais je lui disais là, je ne suis pas français pour le moment."

Pressouyre considers that the inscription of Jerusalem was a key moment in the implementation of the Convention and that paradoxically it added enormously to the credibility of the World Heritage List. "If we had said, it does not concern us, it's uniquely a political problem, it would mean that no one believed in the universal value of properties on the World Heritage List."[54]

Another political highlight for the Committee was the process leading up to the 1988 inscription of the Wet Tropics of Queensland (Australia). An internal tussle that pitted federal interests against state government rights to resource extraction spilled into the international forum through World Heritage. Thorsell remarks: "Australia at that point in time was kind of leading the pack on World Heritage. They were implementing and using the Convention more than any country ever had before to further the aims of government and conservation in the country."[55] Both levels of government carried out world-wide lobbying efforts in the capital cities of Committee members to persuade them to support their respective positions. Robertson Vernhes confirms that the Queensland case "went to the Ministerial level, it went to the top level and that sort of politicized the whole thing. Obviously there were internal politics admittedly but then it was in the international arena." She goes on to say that: "You can't say that it would always have to remain technical. In such a body, and this is UNESCO in general, you're doing the technical things but there is a political undercurrent and you cannot ignore that."[56]

In 1996, the case of W National Park of Niger drew the Committee into another political situation, not on jurisdictional grounds but on scientific ones. IUCN recommended against inscription arguing that the park did not meet any of the natural criteria under the World Heritage Convention. The Bureau, in June and again in November, referred the site back to the State Party. This meant that, according to the Operational Guidelines, the nomination should not have been examined by the Committee later that same year. Those in favour of inscription lobbied hard to convince the Committee to over-ride IUCN's technical advice and the Bureau decisions. One of the sticking points was the persistent lobbying during the meeting by the concerned State Party which had an added advantage of being a member of the Committee. At the time, the Operational Guidelines frowned on such behaviour, clearly stating that "representatives of a State Party, whether or not a member of the Committee, shall not speak to advocate the inclusion in the List of a property nominated by that State, but only to deal with a point of information in answer to a question."[57] Von Droste describes the session as "one of the most

---

54   Canada Research Chair, interview Pressouyre. "Alors, à ce moment là, à mon sens, c'est un moment clé de la vie de la Convention. Si on avait dit, ça ne nous regarde pas, c'est un problème uniquement politique, ça veut dire que personne ne croyait à la valeur universelle des biens de la Liste du patrimoine mondial."

55   Canada Research Chair, interview Thorsell.

56   Canada Research Chair, interview Robertson Vernhes.

57   UNESCO, Operational Guidelines for the implementation of the World Heritage Convention, WHC/2/rev. February 1996, para. 62. Retrieved from http://whc.unesco.org/

Figure 5.2   W National Park, Niger

negative" and regrets that "the Committee ... closed their eyes before their own Operational Guidelines which could always be changed with two thirds majority but they didn't."[58] After a spirited debate, an open vote was taken with the majority supporting inscription of W National Park on the World Heritage List. Statements from four States Parties annexed to the meeting report reveal the divergence of passionately held views. Under the heading "Statements on the legal significance of the Operational Guidelines," Germany denounced the neglect of several rules relating to inscription criteria and advocacy: "Germany is of the strong opinion that the Operational Guidelines can be overruled by the Committee only by amending them, but not by not applying them in one single case. By not applying the Operational Guidelines, the World Heritage Convention is in Danger to become a mere political instrument." The United States of America also took a strong stand on the Committee's performance:

> We made a sham of our integrity. Why is that important? It is important, because conservation and preservation of the best of this world is a constant battle and an uphill battle at that ... Our most important weapon is our integrity ... We tarnished our integrity by not following our own procedures.

Curiously, the delegate of Italy stated "that he was in agreement with the views expressed by Germany concerning the Guidelines," and then contradicted himself by adding that he "wished to point out that all the decisions taken by the Committee during this session were taken in complete conformity with the

archive/opguide96.pdf

58   Canada Research Chair, interview von Droste, 2008.

existing regulations." To wrap up, the Mexican Chairperson curiously remarked that "whilst respecting the statements of one and all, even although she considered those of the Delegates of Germany and the United States of America unacceptable, the Committee had retained its credibility and competence."[59]

Soon after, another case surfaced concerning the state of conservation of Kakadu National Park in Australia. The government responded at the ministerial level against a proposal to inscribe the site on the Danger List. Kakadu dominated the 1998 Committee meeting in Kyoto and forced an extraordinary meeting in 1999 for this issue alone. In von Droste's opinion, "Kakadu created the most dramatic situation I have seen in World Heritage, even more perhaps than Jerusalem in its way." The situation became so explosive because several components had international implications: the rights of indigenous peoples, the right to extract resources for economic reasons, the commitment to conserve World Heritage Sites, and the growth of a worldwide anti-nuclear movement. There also was the thorny issue of inscribing a site on the Danger List without State Party consent.[60] In the end, the Committee set out conservation requirements but did not put Kakadu on the List of World Heritage in Danger. Von Droste is pragmatic in reflecting on the outcome: "the final purpose of the Convention is to conserve sites and you can only conserve that in a dialogue, in cooperation with the country."[61]

Taken collectively, these five cases explain the impression of several interviewees that the shift towards politicization was complete by the close of the century. Jean-Louis Luxen has the impression that "at the beginning of the 1990s, there was a majority of professionals and experts who attended and by 2000 it was mainly diplomats. Over ten years I observed this change."[62] The growing dominance of a politicized Committee ran counter to its affirmation, during the 1992 strategic review, that it sought balanced participation among those responsible for implementation of World Heritage. The approved guideline stated that "the three pillars on which implementation of the Convention rests, namely, the Committee, the Secretariat and the consultative bodies, should play their role fully and equitably."[63]

---

59   UNESCO, Report of the rapporteur on the twentieth session of the World Heritage Committee in Merida, 2–7 December 1996, Paris, 10 March 1997, WHC-96/conf.201/21,annex IX. Retrieved from http://whc.unesco.org/archive/1996/whc-96-conf201-21e.pdf

60   See chapter 6 for a fuller discussion of inscription on the Danger List without State Party consent.

61   Canada Research Chair, interview von Droste, 2008.

62   Canada Research Chair on Built Heritage, Université de Montréal, audio interview of Jean-Louis Luxen by Christina Cameron, Leuven, 26 March 2009. "J'ai l'impression que jusqu'au début des années quatre-vingt dix, c'était une majorité de professionnels et d'experts qui siégeaient. Et en 2000, c'était une majorité de diplomates. Sur les dix années, j'ai observé cette mutation."

63   UNESCO, Strategic guidelines for the future, sixteenth session of the World Heritage Committee in Santa Fe, 7–14 December 1992, Paris, 16 November 1992, WHC-92/conf.002/4, para. II.4. Retrieved from http://whc.unesco.org/archive/1992/whc-92-conf002-4e.pdf

In responding to a question about balance between the technical and diplomatic aspects of the World Heritage Convention, several interviewees expressed support for a delicate equilibrium. As Luxen puts it, "I do not think the diplomats should necessarily be excluded. I remember some diplomats who participated in the work. They did it in a constructive manner. It is important that it is not just experts ... but it is a question of balance."[64] Matsuura concurs:

> It is an intergovernmental body. Therefore it would be totally illogical to expect those professionals to act independently from their governments. They are appointed by governments as their representatives. One of the important things is governments should appoint somebody who has professional knowledge, professional expertise, knows World Heritage, knows issues, avoid nominating generalists who don't have expertise in this area.

However he nuances this statement by adding, "I nevertheless regret too much politicking is taking place these days. We should avoid it ... We should not be subjected excessively to diplomatic or political pressure. Otherwise we could not maintain the credibility of the World Heritage List."[65]

Others clearly lean towards the scientific dimension. In an interview, former Director-General Amadou-Mahtar M'Bow opts for science: "States should choose professionals to represent them rather than diplomats ... I have nothing against the diplomats, but diplomats do not know everything ... and should choose people who know how to and obviously can safeguard. So, I am very loud and clear."[66] Abdelaziz Touri, former Committee Chairperson also believes that expertise should prevail:

> I think that the technical side, the scientific side should come first. Unfortunately, this is not the case ... The Committee is ... a meeting where countries that have conflicts come to defend themselves, defend their point of view. This is not why there is this Convention. This is not why there is the notion of World Heritage that transcends nationalities, transcends bilateral or multilateral difficulties that countries may have.[67]

---

64   Canada Research Chair, interview Luxen. "Je pense qu'y ne faut pas exclure nécessairement les diplomates. J'ai le souvenir de certains diplomates qui participaient aux travaux, qu'ils le faisaient d'une manière constructive et que c'est important que ce ne soit pas rien que des professionnels, mais c'est une question d'équilibre."

65   Canada Research Chair on Built Heritage, Université de Montréal, audio interview of Koïchiro Matsuura by Christina Cameron and Mechtild Rössler, Paris, 24 November 2009.

66   Canada Research Chair on Built Heritage, Université de Montréal, audio interview of Amadou-Mahtar M'Bow by Mechtild Rössler and Petra Van Den Born, Paris, 22 October 2009. "Les États choisissent les professionnels et non pas des diplomates. Je n'ai rien contre les diplomates, mais les diplomates n'ont pas la science infuse, ... et qu'il faut choisir des gens qui connaissent et qui peuvent évidemment sauvegarder. Alors, là, je suis très clair et net."

67   Canada Research Chair on Built Heritage, Université de Montréal, audio interview of Abdelaziz Touri by Christina Cameron and Petra Van Den Born, Paris,

Milne is blunt in his assessment:

> There has never been an adequate professional representation on delegations.
> That has been widespread, it has been commonplace and it's been unfortunate
> ... I think every nation deserves and must have a political advisor for some of
> these discussions but that's not head of delegation, it's not second in command
> ... For the meaty discussions that should be taking place, they should be taking
> place first and foremost on a professional level.[68]

In looking back over the early decades, there is no doubt that cases like
Auschwitz, Jerusalem, Wet Tropics of Queensland, W National Park and Kakadu
National Park changed the heritage discourse from technical to political. For many
States Parties, the evolution towards a more political Committee was inextricably
linked to the continuing quest to find a fair system for electing Committee
members that would represent "an equitable distribution of the different regions
and cultures of the world." Based on their experience from the past, States Parties
gave high priority to the millennium task force focused on achieving equitable
representation on the World Heritage Committee.

Advisory Bodies

The World Heritage Convention formally names three international professional
organizations from the cultural and natural heritage fields. While several other
organizations were active in the preparatory phase, those identified in the
Convention text are the International Union for Conservation of Nature and
Natural Resources (IUCN), the International Council of Monuments and Sites
(ICOMOS) and the International Centre for the Study of the Preservation and the
Restoration of Cultural Property (ICCROM). The World Heritage Committee
and UNESCO are called on to cooperate with these advisory bodies and to utilize
their services "to the fullest extent possible" in preparing documentation for
Committee meetings and for implementing its decisions (articles 13.7 and 14.2).
According to Anne Raidl, "these three organizations were chosen because they
all had connections to UNESCO, they were powerful in terms of membership,

---

22 June 2011. "Moi je crois qu'il faut que le côté technique, le côté scientifique soit mis
en premier. Ce n'est pas le cas malheureusement ... le Comité devient, si vous voulez,
comment dire, une assemblée où les pays qui ont des conflits viennent, n'est-ce pas, se
défendre, défendre leur point de vue. Ce n'est pas pour ça qu'il y a cette Convention.
Ce n'est pas pour ça qu'il y a la notion de patrimoine mondial. La notion de patrimoine
mondial transcende les nationalités, transcende les difficultés bilatérales ou multilatérales
que des pays peuvent avoir."
68   Canada Research Chair, interview Milne.

representativity and substance, and they covered the three major elements: nature, culture and technical assistance."[69]

IUCN

Establishment

IUCN is the oldest of the three advisory bodies mentioned in the Convention. Founded shortly after the creation of UNESCO and initially called the International Union for the Protection of Nature (IUPN), it is the first truly global organization for nature protection. In 1948, UNESCO issued invitations for a formal constitutive congress of IUPN to be held in Fontainebleau, France. Representatives of 18 governments, seven international organizations and 107 national organizations attended.[70] It changed its name to the International Union for the Conservation of Nature and Natural Resources (IUCN) in 1956. Martin Holdgate, a former Director-General of IUCN, notes in his history of the organization that IUCN benefited greatly from its long-standing collaboration with UNESCO's science sector. He credits Julian Huxley, UNESCO's first Director-General, not only with the creation of IUCN but also for the UNESCO grants "which were IUPN's life-line."[71]

IUCN's had a major hand in drafting the World Heritage Convention through its work on a parallel international agreement in the 1960s and early 1970s.[72] Holdgate recalls that some consideration was even given to situating the World Heritage secretariat function within IUCN: "Gerardo Budowski, Frank Nicholls and Eskandar Firouz wanted IUCN to administer the Secretariat, but Maurice Strong felt that it was more appropriate for UNESCO, which has had the responsibility ever since."[73] Over the years, IUCN's programmes have strongly influenced the development of scientific standards for World Heritage. Its general assemblies and world parks congresses offer a global platform to discuss World Heritage matters, often leading to resolutions on specific sites and general policy recommendations.

---

69   Canada Research Chair on Built Heritage, Université de Montréal, audio interview of Anne Raidl by Christina Cameron and Mechtild Rössler, Vienna, 28 February 2008. "L'ICOMOS, l'ICCROM et l'UICN étaient vraiment trois organismes puissants du point de vue de leurs membres, de leur représentativité et du point de vue de la substance, ils couvraient les trois aspects principaux: la nature, la culture et tout ce qui est l'assistance technique."

70   IUCN, *50 Years of working for protected areas: a brief history of IUCN World Commission on Protected Areas* (Gland, 2010), p. 3. Retrieved from http://cmsdata. iucn.org/downloads/history_wcpa_15july_web_version_1.pdf A slightly different set of numbers appears in Martin Holdgate, *The Green Web: A Union for World Conservation* (London, 1999), pp. 29–31.

71   Holdgate, *Green*, pp. 17, 45. The role of Julian Huxley, the British biologist and first Director-General of UNESCO (1946-1948) should not be underestimated in the creation of the "S" for science in UNESCO and for the establishment of IUCN.

72   Chapter 1 describes the development of IUCN's draft convention.

73   Holdgate, *Green*, p. 114.

IUCN's 1972 General Assembly in Banff in September, sandwiched between the June Stockholm Conference and the November adoption of the World Heritage Convention, endorsed a resolution encouraging global participation in World Heritage to further its conservation goals:

> Recalling proposals by conservationists for the recognition of outstanding natural and cultural areas as constituting the World Heritage and the initiatives taken by UNESCO and IUCN in this connection; Being aware of the draft convention on conservation of the World Heritage that will be considered by the General Conference of UNESCO in Paris in October/November 1972; Noting the endorsement of this draft convention by the UN Conference on the Human Environment (Stockholm 1972); The 11th General Assembly meeting at Banff, Canada, in September 1972: Calls upon all governments to adhere to the convention on the Conservation of the World Heritage; … urges governments to give the widest publicity to the concept of the convention and to take action to enable potential sites to be designated as soon as possible."[74]

Its 1978 General Assembly resolved "that the protection of outstanding natural areas is essential to meeting basic human needs" and urged "all States to nominate natural areas of outstanding universal value with full world-wide representation." It also requested "continuous monitoring of World Heritage Natural Sites, to identify areas in danger and work toward their inclusion in the World Heritage in Danger List."[75] This early reference to monitoring is astonishing since IUCN only presented such reports to the World Heritage Committee several years later.

At the 1984 General Assembly, delegates commended "the World Heritage Committee and UNESCO for the significant successes of the first seven years of full operation of the Convention." Vigilant in the pursuit of its conservation goals, IUCN regretted the continuing imbalance between natural and cultural nominations received for the World Heritage List and between natural and cultural experts on delegations to the World Heritage Committee, calling for "States to include technically proficient natural area specialists in their delegations to the World Heritage Committee." Delegates also encouraged countries with adjacent protected areas to work together to submit trans-boundary nominations in the interests of protecting entire ecosystems.[76] Two years after the adoption of

74 IUCN, Resolutions of the eleventh General Assembly of IUCN, Banff, 16 September 1972, res. 2. Retrieved from http://cmsdata.iucn.org/downloads/resolutions_recommendation_en.pdf

75 IUCN, Resolutions of the fourteenth session of the General Assembly of IUCN, Ashkhabad, 26 September–5 October 1978, res. 17. Retrieved from http://cmsdata.iucn.org/downloads/resolutions_recommendation_en.pdf

76 IUCN, Resolutions of the sixteenth session of the General Assembly of IUCN, Madrid, 5–14 November 1984, res. 16/35. Retrieved from http://cmsdata.iucn.org/downloads/resolutions_recommendation_en.pdf

the World Heritage cultural landscapes category, the 1994 General Assembly welcomed "the development of criteria for the recognition of Cultural Landscapes of Outstanding Universal Value under the World Heritage Convention."[77]

One of IUCN's global networks, the World Commission on Protected Areas (WCPA), played a particularly important role in reflecting on the theory and practice of natural heritage conservation and management. A recent history of the WCPA explains its origins:

> The roots of this Commission can be traced to its earliest days. IUCN established a provisional Committee on National Parks during its 1958 General Assembly at Athens and … its purpose was to "strengthen international cooperation in matters relating to national parks and equivalent reserves in all countries throughout the world." Supporters for this work on protected areas included UNESCO and the Food and Agriculture Organization of the United Nations (FAO). In 1960, IUCN raised the status of the Committee to that of a permanent Commission, with the creation of the Commission on National Parks. [78]

The world parks congresses, organized every ten years by IUCN and WCPA, provide opportunities for in-depth reflection on the conservation of natural sites that ultimately contributes to World Heritage work. The Second World Conference on National Parks, organized in collaboration with the United States National Park Service, UNESCO and FAO, was held at Yellowstone National Park in 1972, the year of the Stockholm summit and the adoption of the World Heritage Convention. The WCPA history notes that "the 1972 conference was a landmark: it consolidated world wide experience in park policies and management approaches, and marked a shift towards a more professional form of management."[79] The links between these congresses and World Heritage gradually grew closer. The proceedings from the 1982 Third World Congress on National Parks held in Bali include a special chapter on World Heritage.[80] At the 1992 Fourth World Congress in Caracas, Thorsell, an executive member of WCPA and World Heritage coordinator for IUCN, organized a workshop to review the overall implementation of the Convention. As part of this review, the workshop recommended revisions to World Heritage natural criteria which were adopted by the Committee later that year.[81] Subsequent

---

77   IUCN, Resolutions of the nineteenth session of the General Assembly of IUCN, Buenos Aires, 17–26 January 1994, res. 19/40. Retrieved from http://cmsdata.iucn.org/downloads/resolutions_recommendation_en.pdf

78   IUCN, *50 Years*, p. 3.

79   IUCN, *50 Years*, p. 5.

80   IUCN, *50 Years*, p. 9.

81   Hemanta Mishra and N. Ishwaran, "Summary and conclusions of the workshop on the World Heritage Convention held during the IV World Congress on national parks and protected areas Caracas, Venezuela, February, 1992," *World Heritage: twenty years later. Based on papers presented at the World Heritage and other workshops held during the IVth*

world congresses have discussed World Heritage issues and featured IUCN's contribution to the implementation of the Convention through its membership, ranging from site managers to governmental and non-governmental organizations. The congresses also have raised key issues of World Heritage concern including sustainable use of sites, mining and protected areas, and the use of Danger listing as a tool in conservation.

IUCN at World Heritage

IUCN has used its extensive network and technical expertise to play an active role at the World Heritage Committee meetings. Its representatives have crafted criteria, evaluated nominations, monitored properties, developed training initiatives and contributed to the Operational Guidelines and other framework tools. Between 1977 and 2000, three people sequentially represented IUCN at Committee sessions: Hal Eidsvik (1977–1980), Jeff McNeely (1981–1983) and Jim Thorsell (1983–2002).

Eidsvik, a Canadian, was involved with World Heritage from the beginning, first as an IUCN staff member in the role of executive officer for the Commission for National Parks and Protected Areas, then as its volunteer chair. In an interview, he explains that World Heritage was a small part of IUCN's work. With its strong focus at that time on endangered species, IUCN opted to delegate responsibility for World Heritage to the group that worked on protecting and managing natural areas. Eidsvik recalls that "it eventually came down to, not IUCN, not the Commission, but it came down to an individual. Everybody else was busy, they did their own stuff. They said 'okay, your responsibility is the World Heritage Convention'."[82]

The 1978 evaluations of natural site nominations to World Heritage were brief one-pagers with comments on values and integrity but no field missions. Eidsvik's covering letter says that "in all cases except one there was either direct field knowledge of the site or our files contacts were sufficiently comprehensive to arrive at a judgement." His letter also reveals that at that time IUCN had set up a panel to consider nominations, a practice that faded away until re-introduced by Thorsell a few years later.[83] In response to a Committee request for comparisons of properties of a similar type, IUCN began in 1980 to prepare a global inventory of natural sites through worldwide distribution of questionnaires and the organization

---

World Congress on National Parks and Protected Areas, Caracas, Venezuela, February 1992 (Gland and Cambridge, 1992), p. 14; UNESCO, Revision of the Operational Guidelines for the implementation of the World Heritage Convention, Paris, 11 October 1992, WHC-92/conf.002/10, paras. 6–7. Retrieved from http://whc.unesco.org/archive/1992/whc-92-conf002-10e.pdf

82   Canada Research Chair, interview Eidsvik.

83   Letter from Harold K. Eidsvik to Bernd von Droste concerning World Heritage Sites screening process, Morges, 31 May 1978. Retrieved from http://whc.unesco.org/archive/advisory_body_evaluation/024.pdf

of a series of expert meetings over two years. The work was intended to give guidance about what kind of natural properties might be appropriate for World Heritage listing.[84]

McNeely, an American, took over as executive officer for the Commission in 1981. He says that he only worked part-time on World Heritage: "That was just one of many things that I was doing. I didn't give it the attention that we were able to give it later ... I couldn't visit the World Heritage sites. I didn't have time and so I depended more on written comments from people who knew the areas."[85] McNeely produced nomination evaluations which continued to be brief. He also completed the global inventory of natural sites entitled *The World's Greatest Natural Areas: an indicative inventory of natural sites of World Heritage quality*.[86] He remarks on its seminal importance: "As I wrote it, I could say whatever I wanted. I look back at that and a lot of these places are now on the World Heritage List."[87]

IUCN's presence at World Heritage during the period of this study is inextricably linked with Thorsell, a Canadian scientist who has worked with the WCPA from 1983 to the present. He held various titles, beginning as executive officer responsible for the Commission as well as for World Heritage, then as head of the natural heritage programme and finally as senior advisor for natural heritage. At the beginning, World Heritage was a minor part of his workload but by 1989 it became his fulltime job.[88] Under his leadership the evaluation of nominations grew more robust, introducing greater rigour in the comparative analysis of sites and emphasizing requirements for integrity, conservation and management.

Thorsell began by following the approach of his predecessors but soon introduced field missions and an expert panel to improve the quality of the work. He explains: "So in 84 I did the evaluations just by writing people I knew in the region, getting their opinions and then going to the literature." But in 1985 he concluded that missions to nominated sites were essential, saying, "If we are going to take this seriously we are going to have to start doing field evaluations." In addition to achieving a better understanding of the properties, he notes that field missions also played a big role in increasing awareness about World Heritage. Referring to newspaper articles and television coverage of a field trip to Hunan Province in China, he recalls that "the media interest in

---

84 UNESCO, Report of the rapporteur on the fourth session of the World Heritage Committee in Paris, 1–5 September 1980, Paris, 29 September 1980, CC-80/conf.016/10, paras. 20, 47. Retrieved from http://whc.unesco.org/archive/1980/cc-80-conf016-10e.pdf

85 Canada Research Chair on Built Heritage, Université de Montréal, audio interview of Jeff McNeely by Mechtild Rössler, Gland, 17 September 2010.

86 IUCN Commission on National Parks and Protected Areas, *The World's Greatest Natural Areas: an indicative inventory of natural sites of World Heritage quality* (Gland, 1982), pp. 1–69.

87 Canada Research Chair, interview McNeely.

88 Canada Research Chair, interview Thorsell.

Figure 5.3   Jim Thorsell on mission in Canaima National Park, Venezuela

visits to evaluate a site and just the political interest it would generate and the meetings and the profile we would get ... we were reaching hundreds of millions of people."[89]

While operating in his words like a "one-man band", Thorsell regularly consulted an informal network of IUCN experts who worked at a global level in specific areas like botany, species and forestry. In the early 1990s, an IUCN World Heritage panel was formally constituted in tandem with the creation of the position of vice-chair in World Heritage. Panel membership slowly expanded beyond IUCN headquarters to include outside experts from different regions.[90]

In a recent institutional history, IUCN refers to World Heritage work as "among the most politically exposed" of its activities and notes its "reputation for adopting consistently high and rigorous standards in its evaluations of nominations and frank, even fearless, assessments of threats to World Heritage."[91] This reputation can be attributed in large measure to Thorsell who was renowned for holding the highest standards. His predecessor McNeely says "Jim Thorsell was very strict. He was more strict than I was frankly."[92] As Thorsell succinctly explains it, IUCN's

---

89   Canada Research Chair, interview Thorsell.
90   Canada Research Chair, interview Thorsell.
91   IUCN, *50 Years*, pp. 9–10.
92   Canada Research Chair, interview McNeely.

approach was exclusive: "Only the best meet the test."[93] Former Chairperson Francioni expresses his admiration for the integrity and reliability of IUCN's "very strong stand ... speaking even against a certain preference shown by States in favour of adopting a positive decision on a nomination." He contrasts this approach with that of ICOMOS and surmises that it may be easier "to rely on hard scientific data of biodiversity with regard to natural sites than to rely on hard scientific evidence of outstanding universal cultural value because ... there is no such thing as universal because every culture is to be appreciated for its own singularity and specificity."[94]

From the beginning, IUCN used comparative analysis, drawing on its internal scientific capacity and especially on its volunteer expert network in the WCPA to prepare global frameworks and thematic studies to help countries identify properties of World Heritage significance. Based on this research, IUCN soon understood the need to update its 1982 global inventory on *The World's Greatest Natural Areas: an indicative inventory of natural sites of World Heritage quality* "to incorporate sites for which further information had been gathered or to include new sites which had been recently discovered."[95]

The 1990s saw further reflection on the nature of eligible natural heritage sites with the emergence of cultural landscape theory and a cultural global strategy to attain a representative World Heritage List. In terms of humanized landscapes, IUCN was constrained by the legal definition of natural heritage in the Convention (article 2) which made no reference to human interaction with the environment. It must also be acknowledged that Thorsell had a personal preference for pristine natural sites. Pressouyre, his ICOMOS counterpart, disagreed with this view of nature and recounts an incident when he lost his temper over it. For the evaluation of the nomination of Meteora (Greece) in 1988, Pressouyre asked IUCN for advice, only to be told that there were no exceptional natural values because the mountains were not as high as other mountains and the species were nothing special. Insisting on the strong association of culture and nature, Pressouyre claims he told Thorsell that "the monks set themselves up on the high mountains. When they are up there, they have to bring provisions up using pulleys ... They did that to be closer to heaven ... If you say that there is no link between cultural and natural value, you are making an enormous mistake."[96]

---

93   Canada Research Chair, interview Thorsell.

94   Canada Research Chair, interview Francioni.

95   IUCN Commission on National Parks and Protected Areas, *The World's Greatest Natural Areas: an indicative inventory of natural sites of World Heritage quality* (Gland, 1982), pp. 1-69. UNESCO, Report of the rapporteur on the eighth session of the World Heritage Committee in Buenos Aires, 29 October-2 November 1984, Buenos Aires, 2 November 1984, SC/84/conf.004/9, para. 20. Retrieved from http://whc.unesco.org/archive/1984/sc-84-conf004-9e.pdf

96   Canada Research Chair, interview Pressouyre. "Les moines sont allés se jucher sur des hautes montagnes. Quand ils sont là-haut, il faut les alimenter à la poulie et descendre leur débris aussi à la poulie. Ils ont fait ça pourquoi, pour être près du ciel ... Si vous dites qu'il n'y a pas un lien entre la valeur culturelle et la valeur naturelle, vous faites une énorme bêtise."

Figure 5.4   Meteora, Greece

While this incident reveals differing views on natural heritage, IUCN nonetheless participated regularly in World Heritage initiatives to elaborate cultural landscape theory beginning with rural landscapes in 1984 through the early 1990s, eventually taking on a consultative role to ICOMOS for the evaluation of cultural landscape nominations.[97] The 1996 expert meeting at Parc national de la Vanoise in France made an important proposal to recognize human activity in World Heritage natural sites as a potential complement to natural heritage values. The meeting also recommended a comprehensive global strategy that would reinforce the unifying concept of World Heritage by embracing both cultural and natural heritage.[98]

---

97   UNESCO, Report of the rapporteur on the ninth session of the World Heritage Committee in Paris, 2–6 December 1985, Paris, December 1985, SC-85/conf.008/9, paras. 25-8. Retrieved from http://whc.unesco.org/archive/1985/sc-85-conf008-9e.pdf; UNESCO, Report of the rapporteur on the twelfth session of the World Heritage Committee in Brasilia, 5-9 December 1988, Paris, 23 December 1988. SC-88/conf.001/13, paras. 9, 30–31. Retrieved from http://whc.unesco.org/archive/1988/sc-88-conf001-13e.pdf

98   UNESCO, Report of the expert meeting on evaluation of general principles and criteria for nominations of natural World Heritage Sites, Parc national de la Vanoise, France,

By the year 2000, IUCN had an impressive record of thematic studies to help States Parties determine which natural sites might best fill the gaps in the World Heritage List. Drawing on their networks and databases, IUCN published global research on fossil sites (1996), wetland and marine ecosystems (1997), forests (1997), human use in natural properties (1998), geological features (1998), sites of exceptional biodiversity (1999) as well regional thematic studies carried out in collaboration with the WCMC.[99]

In addition to its work on nominations, IUCN was instrumental in moving conservation and management effectiveness to the agenda of the Committee. It led the way in developing monitoring mechanisms to assess the state of conservation of natural World Heritage Sites. Thorsell's own field experience made him particularly well suited to keeping track of the condition of properties after inscription. Having been encouraged by the 1983 session of the Committee "to collect information through their contacts and to inform the Committee on the state of conservation of World Heritage properties," IUCN made a formal report on four natural World Heritage Sites in 1984.[100] The following year, it reported on the state of conservation of 12 properties. In addition it continued its campaign to improve monitoring procedures, describing to the Committee its impressive internal capacity in no way matched by the other two advisory bodies:

> The IUCN system is based at the Conservation Monitoring Centre at Cambridge (United Kingdom) and has close links with the Global Environmental Monitoring System of the United Nations Environment Programme. IUCN is assisted by 4000 voluntary correspondents located in 126 countries who report regularly to the Conservation Monitoring Centre. Thus, IUCN is in a position to obtain reliable and up-to-date information on almost all natural World Heritage properties.[101]

22–24 March 1996, Paris, 15 April 1996, WHC-96/conf.202/inf. 9, para. 2c. Retrieved from http://whc.unesco.org/archive/1996/whc-96-conf202-inf9e.pdf

99   UNESCO, Progress report, synthesis and action plan on the global strategy for a representative and credible World Heritage List, Paris, 27 October 1998, WHC-98/conf.203/12, pp. 8–11. Retrieved from http://whc.unesco.org/archive/1998/whc-98-conf203-12e.pdf; UNESCO, Progress report on the implementation of the regional actions described in the global strategy action plan adopted by the Committee at its twenty-second session, Paris, 21 October 1999, WHC-99/conf.209/8, pp. 31–4. Retrieved from http://whc.unesco.org/archive/1999/whc-99-conf209-8e.pdf All studies are available at http://www.unep-wcmc.org/world-heritage-thematic-studies_519.html

100   UNESCO, Report of the rapporteur on the eighth session of the World Heritage Committee in Buenos Aires, 29 October–2 November 1984, Buenos Aires, 2 November 1984, SC/84/conf.004/9, para. 40. Retrieved from http://whc.unesco.org/archive/1984/sc-84-conf004-9e.pdf

101   UNESCO, Report of the rapporteur on the ninth session of the World Heritage Committee in Paris, 2–6 December 1985, Paris, December 1985, SC-85/conf.008/9, para. 16. Retrieved from http://whc.unesco.org/archive/1985/sc-85-conf008-9e.pdf

From that point on, IUCN reported regularly on the state of conservation of natural World Heritage properties and periodically produced directories with up-to-date information prepared by the Protected Areas Data Unit (PADU) of the WCMC.[102]

On the twentieth anniversary of the Convention, Thorsell documented the steady improvement in monitoring systems and urged further strengthening as a means of ensuring the survival of the outstanding universal value of natural sites.[103] Five years later, he presented for the first time the ten most important threats to World Heritage Sites, cautioning that "the natural places of the world are gradually disappearing and are under increasing threat. World Heritage is not the only answer to addressing the problem but it is an important part of the tool-kit that can be used effectively for those special places that humankind can not afford to lose."[104] Through its publications and field missions, IUCN raised public awareness about World Heritage and built capacity among site managers and others involved in protected areas management.

In the 1990s, pressure grew within the Committee to hear from diverse voices within IUCN, a situation eerily reminiscent of Pressouyre's situation as ICOMOS coordinator a decade earlier. Thorsell's influence in the 1980s and 1990s was immense. As Ishwaran puts it, "he was a one man show and he went to almost every place himself so he had a first-hand view of things."[105] Some worried about such concentration of power. Observing that "Jim came to really dominate IUCN's role in the Convention," Eidsvik recalls his concern in the 1980s that "there is too much value, too much danger, too much risk in having one person responsible for the total Convention."[106] Milne also expresses his unease at IUCN's dependence on "one person's perspective, no matter how professional or judicious":

> Enrichment, appreciation and understanding properties would have been enhanced earlier on by greater, far greater utilisation of people closer to the site and the scale of economy and the scale of capability, that could have pointed out earlier issues that might have been overlooked by one or two people coming by on a one-time swing.[107]

102   UNESCO, Report of the rapporteur on the thirteenth session of the World Heritage Committee in Paris, 11–15 December 1989, Paris, 22 December 1989, SC-89/conf.004/12, para. 15. Retrieved from http://whc.unesco.org/archive/1989/sc-89-conf004-12e.pdf

103   Jim Thorsell, "From Strength to Strength: World Heritage in its 20th Year," *World Heritage: twenty years later. Based on papers presented at the World Heritage and other workshops held during the IVth World Congress on National Parks and Protected Areas, Caracas, Venezuela, February 1992* (Gland and Cambridge, 1992), pp. 19–25.

104   Jim Thorsell, "Nature's Hall of Fame: IUCN and the World Heritage Convention," *Parks*, 7/2 (1997), p. 7.

105   Canada Research Chair on Built Heritage, Université de Montréal, audio interview of Natarajan Ishwaran by Christina Cameron and Mechtild Rössler, Paris, 24 November 2009.

106   Canada Research Chair, interview Eidsvik.

107   Canada Research Chair, interview Milne.

From 1992 to 1997, New Zealander Bing Lucas joined Thorsell at World Heritage sessions in order to add landscape expertise and a view from a different region. In 1998, Thorsell handed over the lead World Heritage role to David Sheppard and remained as senior advisor. In addition, the team also added IUCN's director for global programmes and other staff to deal with the troublesome issue of Kakadu in Australia. The era of the single evaluator was over.

During this period, IUCN was a key global player in creating frameworks and tools to evaluate outstanding universal value and to assess the state of conservation of natural heritage sites. It used iconic natural sites like the Galapagos Islands (Ecuador) or Ngorongoro (Tanzania) to showcase the tools and mechanisms that the World Heritage Convention could use to achieve conservation gains. Stovel considers IUCN the most impressive of the three advisory bodies, combining as it does the qualities of a non-governmental organization with the strength of a strong institutional organization.[108] As the years unfolded, IUCN influenced the evolution of natural heritage discourse particularly through its global lists, thematic studies and reflections on landscapes. Its insistence on effective conservation and management practices may be the organization's greatest contribution to the implementation of the Convention.

ICOMOS

Establishment
ICOMOS emerged in the context of UNESCO's initiatives in the 1960s to develop international cooperation in the cultural field. Hiroshi Daifuku, who worked in the Museums and Monuments Division of UNESCO at the time, explains that:

> ICOMOS came into being as two initiatives converged. UNESCO's advisory committee recognized that demand for specialized services related to monuments preservation was growing. It was beyond our capacity to meet the need. The committee recommended that UNESCO explore the creation of a new non-governmental organization (NGO). I participated in the initial conversations to explore this idea with the French Ministry of Culture. Soon thereafter in 1964 the II International Congress of Architects and Technicians of Historic Monuments met in Venice, Italy ... and recognized the need for this new NGO system. I was given the assignment to organize steps to bring it into being.[109]

---

108 Canada Research Chair on Built Heritage, Université de Montréal, audio interview of Herb Stovel by Christina Cameron, Ottawa, 16 March 2011.

109 Russell V. Keune, "An interview with Hiroshi Daifuku," *CRM: The Journal of heritage stewardship*, 8/1 and 2 (2011) p. 41. Retrieved from http://crmjournal.cr.nps. gov/03_spotlight_sub.cfm?issue=Volume%208%20Numbers%201%20and%202%20 Winter%2FSummer%202011&page=4&seq=1

Figure 5.5   Piero Gazzola

The driving force for its creation came from Italian architect Piero Gazzola, "the true founder of ICOMOS" according to its first Secretary General Raymond Lemaire.[110] Gazzola and Daifuku had worked together in earlier international protection efforts for cultural heritage including the 1954 Hague Convention for the protection of cultural property in the event of armed conflict. In a tribute to Gazzola after his death, Lemaire credits his understanding of the need to create such an organization to his friendship with Georges-Henri Rivière, the founder of the International Council of Museums (ICOM). Gazzola was inspired by the effectiveness of ICOM in fostering an international professional dialogue. At the First Congress of Architects and Technicians of Historic Buildings held in Paris in 1957, a motion was passed in favour of the creation of an international assembly of architects and specialists of historic buildings in order to ensure a better exchange of knowledge and experience.[111] It was Gazzola who invited the group to a second meeting in Venice in 1964.

Held in collaboration with UNESCO, the Second Congress of Architects and Technicians of Historic Buildings adopted an international code for conservation

---

110   Raymond Lemaire, "Report of the President of ICOMOS Piero Gazzola 1965-1975: a tribute to Piero Gazzola," *Thirty Years of ICOMOS* (Paris, 1995), p. 87. Retrieved from http://openarchive.icomos.org/254

111   Lemaire, "Report of the President," p. 87.

practice, the International Charter for the Conservation and Restoration of Monuments and Sites (Venice Charter) and passed a resolution put forward by UNESCO for the creation of the International Council of Monuments and Sites (ICOMOS). Gazzola calls the results of the Venice meeting momentous. He describes the newly-created ICOMOS as "the institution which constitutes the court of highest appeal in the area of the restoration of monuments and of the conservation of ancient historical centres, of the landscape and in general of places of artistic and historical important." He was especially proud of the Venice Charter which he calls "the official code in the field of the conservation of cultural properties."[112] Conceived in Venice, ICOMOS was established officially in Warsaw at the first ICOMOS General Assembly in June 1965.

ICOMOS was both a professional organization and a non-governmental body advising on heritage matters and promoting the conservation, protection, use and enhancement of monuments, buildings and sites. In addition to advising UNESCO on cultural heritage, ICOMOS developed its own charters often related to the themes of its general assemblies. During the period of this study, ICOMOS charters cover historic gardens (1981), historic towns and urban areas (1987), archaeological heritage (1990), underwater cultural heritage (1996), tourism (1999) and built vernacular heritage (1999).

ICOMOS at World Heritage
ICOMOS actively contributed to the intellectual development of the World Heritage Convention. According to Leblanc, it was a minor part of ICOMOS's activities at the beginning: "At that time, work for the World Heritage Convention, the contribution of ICOMOS, was something very small ... the focus of ICOMOS at that time was to serve its members, to arrange international conferences, to advance the field of conservation."[113] In the early years, ICOMOS contributed to the concepts and wording of the cultural heritage criteria for inscribing properties on the World Heritage List. Reflecting a challenge that has persistently dogged the organization, ICOMOS President Lemaire said at the first World Heritage Committee meeting that ICOMOS "recognized the difficulty of drafting criteria to be applied to cultural property throughout the world and of translating concepts into words that were meaningful on a universal scale."[114]

---

112   *The monument for the man: records of the II International Congress of Restoration* (Venice, 1964), foreword. Retrieved from http://www.icomos.org/en/about-icomos/mission-and-vision/history?id=411:the-monument-for-the-man-records-of-the-ii-international-congress-of-restoration&catid=157:publications

113   Canada Research Chair, interview Leblanc. "À cette époque, le travail pour la Convention du patrimoine mondial, la contribution de l'ICOMOS, c'était quelque chose de très mineur ... le focus de l'ICOMOS à cette époque était de servir ses membres, de faire des conférences internationales, de faire avance le domaine de la conservation."

114   UNESCO, Final report of the first session of the intergovernmental committee for the protection of the world cultural and natural heritage in Paris, 27 June–1 July 1977,

ICOMOS played a key role in providing technical advice on cultural site nominations. Surprisingly, before 1980 there were no written site evaluations from ICOMOS. In 1978, the first year of inscriptions, all cultural nominations were covered by a single letter from Ernest Connally, Secretary General of ICOMOS indicating which ones met the cultural criteria.[115] In 1979, there were brief comments on a few nominations signed by three ICOMOS members, French art historian André Chastel, American architectural historian Henry Millon and French inspector general of historical monuments Jean Taralon. According to two witnesses, the early evaluations were simple and delivered orally. Pressouyre cites Lemaire as saying in 1980: "Until now, we did not have any written evaluation. We went before the Committee and said 'this one is very good, you must inscribe it. That one is less good, you should reflect on it, perhaps it will come back'."[116] Leblanc confirms this story, adding that "sometimes at the last minute, a few days before the Committee meeting, someone arrived with a file asking us to evaluate it."[117]

In 1980, Connally acted on two fronts to formalize the ICOMOS process for nominations. First, on the advice of Chastel and Lemaire, he hired French archaeologist Professor Léon Pressouyre to improve the quality of the evaluations; secondly, he successfully proposed an annual cycle for receiving and evaluating nominations. Described by his colleague Leblanc as diplomatic and academic, Pressouyre brought a more sophisticated and systematic approach to the work. He conducted desk research and consulted widely, although he did not visit the sites. Leblanc describes him in action:

> Léon Pressouyre applied his personal knowledge to sites that he knew. He knew them either because he had deeply studied them or because he had visited them personally. He knew the experts who worked or had worked on the sites ... Léon used the telephone and the fax to communicate with his colleagues to verify a certain number of facts ... For sites that he did not know or for which he did not feel sufficiently expert to make a recommendation, he used the ICOMOS network.[118]

Paris, 20 October 1977, CC-77/conf.001/9, para. 19. Retrieved from http://whc.unesco.org/archive/1977/cc-77-conf001-9e.pdf

115   ICOMOS, letter from Secretary General Ernest Allen Connally to Firouz Bagerzadeh, Chair of World Heritage Committee, Paris, 7 June 1978. Retrieved from http://whc.unesco.org/archive/advisory_body_evaluation/004.pdf

116   Canada Research Chair, interview Pressouyre. "Jusqu'à présent on a roulé comme ça, on n'a pas d'évaluation écrite, on allait devant le Comité, on disait: ça c'est très bien, il faut le prendre, vous l'inscrivez, ça c'est moins bien, il faut réfléchir, peut-être que ça reviendra."

117   Canada Research Chair, interview Leblanc. "À la dernière minute des fois, à quelques jours avant la réunion du Comité, on vous arrivait avec un dossier en vous demandant: pouvez-vous nous évaluer ça."

118   Canada Research Chair, interview Leblanc. "Léon Pressouyre se servait de ses connaissances personnelles pour des sites que lui-même connaissait. Il les connaissait, soit

**Figure 5.6   Léon Pressouyre on mission in Mostar, Bosnia and Herzegovina
in the late 1990s**

Pressouyre confirms this statement but comments on the weakness of ICOMOS at that time: "This is essentially a network of architects and conservation specialists of monuments and we would like to see geographers, ethnologists, well, people you never used to see in ICOMOS."[119] Pressouyre submitted his recommendations for approval to the ICOMOS Bureau consisting of the president, secretary general, head of finance and five vice presidents.

pour les avoir beaucoup étudiés ou soit pour y être allé personnellement et il connaissait les professionnels qui travaillaient ou qui avaient travaillé sur les sites ... Léon utilisait beaucoup le téléphone et le fax pour communiquer avec ses collègues pour revérifier un certain nombre de faits. Pour les sites qu'il ne connaissait pas ou pour lesquels il ne se sentait pas suffisamment expert pour faire une recommandation, on utilisait le réseau de l'ICOMOS."

119   Canada Research Chair, interview Pressouyre. "Le réseau de l'ICOMOS a la faiblesse de l'ICOMOS elle-même, c'est-à-dire que c'est un réseau essentiellement composé d'architectes et de spécialistes de la conservation des monuments et on aimerait y voir des géographes, des ethnologues, bon, des gens qu'on ne verra pas à l'ICOMOS avant, avant belle lurette."

A curious question arises about the early ICOMOS evaluations that were presented orally to the Committee. How is it that written ICOMOS evaluations now appear in the official UNESCO documents for properties inscribed in 1978 and 1979? Pressouyre provides the explanation. In an interview, he reports that he was asked by UNESCO and ICOMOS to write evaluations after the fact in order to have a complete record on file. He describes his incredulous reaction: "You are making fun of me. There were some sessions. I was not there. You want me to say what went on during these sessions? Both UNESCO and ICOMOS said, yes, that is exactly what we want." While sceptical, Pressouyre undertook what he calls "this memorable exercise" by interviewing people who had attended the sessions: "Through searching their memories they often recalled the arguments, the elements of discussion, the criteria. And I transcribed them as faithfully as possible."[120]

By 1981, ICOMOS and IUCN were asked by the Committee "to be as strict as possible" in their evaluations and that "the manner of the professional evaluation carried out by ICOMOS and IUCN should be fully described when each nomination is presented."[121] Slatyer repeated this message in 1983, appealing to both organizations to "raise their standards even higher" and "act with the highest integrity and objectivity, avoiding favouritism or prejudice."[122] Pressouyre complains that the Committee's request for robust evaluations was thwarted by UNESCO and ICOMOS who asked for reports on a single page:

> We know that you are an archaeologist, a scientific person. We are not asking you to make a detailed report on each site. We are asking you to say if it's good or if it's bad and why and what criteria. So for years, I found it a little bit unfortunate but in the end that was how it worked. We did extremely brief papers, extremely brief, that were rather, I would say, position statements.[123]

It was only at the end of the 1980s that more developed reports were allowed. In hindsight Pressouyre muses that the evaluations were tailored to meet the

---

120   Canada Research Chair, interview Pressouyre. "Vous vous fichez de moi. Il y a eu des sessions, je n'étais pas là, vous voulez que je dise ce qui s'est passé pendant ces sessions. Et on m'a dit: à l'UNESCO et à l'ICOMOS, des deux côtés, oui, c'est exactement ce qu'on veut. En fouillant dans la mémoire ils ont souvent retrouvé les arguments, les éléments de discussion, les critères. Et je les ai transcrits le plus fidèlement possible."

121   UNESCO, Report of the rapporteur on the fifth session of the World Heritage Committee in Sydney, 26–30 October 1981, Paris, 5 January 1982, CC-81/conf/003/6, paras. 26-27. Retrieved from http://whc.unesco.org/archive/1981/cc-81-conf003-6e.pdf

122   Slatyer, Address, p. 5.

123   Canada Research Chair, interview Pressouyre. "On sait que vous êtes un archéologue, un scientifique, etc., on ne vous demande pas de faire un rapport détaillé sur chaque site. On vous demande de dire si c'est bien ou si c'est mal et pourquoi et quels critères. Bien donc, pendant des années, j'ai trouvé un petit peu dommage mais enfin, c'était comme ça qu'on fonctionnait, on a fait des papiers extrêmement brefs, extrêmement brefs qui étaient, je dirais plutôt des prises de position."

expectations of the Committee at any particular moment and that the advisory bodies merely complied.

Pressouyre resigned as World Heritage coordinator after the 1990 session in Banff, following a dispute with Stovel, at that time Secretary General of ICOMOS. In his interview, Pressouyre barely mentions this incident but Stovel is more fulsome. He recalls that several States Parties raised concerns at the 1990 session about "whether or not ICOMOS was actually preparing the evaluations or whether they were coming from someone called Léon Pressouyre all by himself without any ICOMOS input." Stovel reports raising the issue with Pressouyre:

> He didn't take kindly to the conversation. And by the time I put it down on paper and sent him a letter saying that from now on it might be a good idea to be a little more collegial, can we not work toward a kind of sharing of information and a more open system of decision-making, he took offense with my letter or he took offense with the proposition or he took offense with both, I am not sure, but he quickly resigned and he resigned retroactively.[124]

ICOMOS president Roland Silva confirms that Pressouyre sent a letter in March 1991 vacating the post, effective 31 December 1990. He says he made several appeals to him to change his mind but to no avail.[125]

According to Stovel, Pressouyre's resignation plunged ICOMOS into a fundamental debate about whether or not to continue participating in World Heritage work at all. Stovel argued in favour of continuing, on condition that ICOMOS would "invent a system which brings in all the ICOMOS expertise and doesn't just leave it to the coordinator."[126] Eventually a decision was made to stay the course.

The 1991 World Heritage meeting sorely tested the capacity of ICOMOS. Left unprepared, the organization cobbled together recommendations for cultural nominations by seeking advice on an emergency basis from what Stovel calls "a kind of brains trust of ICOMOS experts at the highest level who would stand in and help us go through the nomination documents." Among the group were former ICOMOS officers including Lemaire, Cevat Erder, Helmut Steltzer, Abdelaziz Daoulatli and Jean Barthélemy. "We linked them to the executive committee, not all the time but from time to time, so we expanded the circle up to another twenty people and by the end of the process we were pretty sure that we'd made reasonable decisions."[127] Despite this effort, the ICOMOS evaluations were much criticized at the 1991 World Heritage meeting in Carthage in part because Stovel took ill en route to the meeting and was unable to present the nominations himself.

---

124   Canada Research Chair, interview Stovel, February 2011.

125   Canada Research Chair on Built Heritage, Université de Montréal, audio interview of Roland Silva by Christina Cameron, Victoria, 12 October 2011.

126   Canada Research Chair, interview Stovel, February 2011.

127   Canada Research Chair, interview Stovel, February 2011; Canada Research Chair, interview Silva.

Figure 5.7   Henry Cleere on mission in China

In 1992, Stovel asked British archaeologist Henry Cleere to take over as ICOMOS World Heritage coordinator, a role that he carried out for a decade. Luxen praises Cleere for his scholarship and his knowledge of World Heritage processes, calling him a "work horse who knew many things, a scholar, a scientist."[128] In addition, he spoke English and French with some capacity in other languages "so there was some sense that he could communicate in international fora."[129]

Cleere recalls the situation at that time:

> The successive presidents of that time rather jealously kept all this work to themselves. But when Roland Silva from Sri Lanka became President and Herb Stovel as Secretary General, the whole thing opened up. So I was invited to take over from Léon Pressouyre, a very distinguished French historian, archaeologist, a great friend of mine who had his own idiosyncratic, shall we say, autocratic almost approach to this work. He was not very happy about the new regime. They tried to do without him for a year then they brought me on board. Whether that was the right thing or not, I leave it to you to judge.[130]

---

128   Canada Research Chair, interview Luxen. "une grande force de travail, qui connaissait beaucoup de choses, enfin c'était un scholar, un savant."

129   Canada Research Chair, interview Stovel, February 2011.

130   Canada Research Chair on Built Heritage, Université de Montréal, audio interview of Henry Cleere by Christina Cameron, London, 24 January 2008.

Cleere made significant changes to the way ICOMOS did its evaluations of cultural nominations, strengthening consideration of values and authenticity, and introducing conservation and management aspects. Von Droste sees Stovel and Cleere as part of an anglophone tradition: "They were not so strong probably on the art historian perspective but they were very strong on the management issues, a bit like IUCN. They came closer to IUCN and how they looked at the properties. So they really brought into the evaluation the management issue."[131] It was during this period that ICOMOS began to make field missions to each nominated site, a practice that IUCN had adopted seven years earlier. Stovel credits Cleere with improving the overall quality of the evaluations:

> If you go through the evaluations of ICOMOS from 92 on, you can see that each year they got bigger and bigger and bigger, and they begin to cover more ... One of the significant advances was to represent in the ICOMOS evaluation "what did the State Party claim?" to reproduce their view on the criteria of choice, their view on the justification, their view on management, and then to, in parallel, talk about the ICOMOS view. So it became more clear, more open, more transparent what the starting point was and what ICOMOS went through.[132]

Cleere was well-known for his animated style of delivery. Wanner calls him "gutsy ... a prickly guy who would spit back at people." He specifically admires Cleere's resistance to American pressure to defer the nomination of Hiroshima: "But here's where tough old Henry Cleere, he was on the scene ... he would have none of our politicking and told us, 'these are the facts, and this is ICOMOS's view and these are our reasons for it, and we're going to stick with it'."[133]

In addition to inputs from the ICOMOS Bureau and eventually the entire executive committee, evaluations also drew on the ICOMOS scientific committees. The 1990s saw the number grow from 11 to 24 under the encouragement of Silva: "To me the scientific committees of ICOMOS were the front line in terms of policy, principles and preservation of World Heritage listed monuments ... We were sharpening the edge and the principles covering the need to be more specific in our understanding of the World Heritage monuments."[134]

Despite this collective approach, concerns emerged again about a single coordinator. The sheer volume of cultural nominations after 1995 created an impossible workload for one person. It is a tribute to Cleere's stamina and work ethic that he was able to cover them all. Stovel explains that the questions raised concerning Pressouyre's role began to come back:

---

131 Canada Research Chair, interview von Droste, 2008.

132 Canada Research Chair, interview Stovel, February 2011.

133 Canada Research Chair on Built Heritage, Université de Montréal, audio interview of Ray Wanner by Christina Cameron, Springfield, 18 May 2011.

134 Canada Research Chair, interview Silva.

To what extent is Henry Cleere actually sharing? Is there an ICOMOS process that actually brings Henry's viewpoints early enough in the stage of developing the thinking, to the attention of the larger ICOMOS circle and, you know, is there any way to build in that expertise into where Henry ends up or does Henry pre-decide everything and that's it?[135]

Luxen insists that Cleere was not acting alone: "He was accused of being a bit personal because, unfortunately, it is human that, in making the presentations, he did it as if it was his. But in reality, I made sure that the work he presented was the fruit of a significant collective effort."[136] Milne is more philosophical: "For ICOMOS [it] was quite similar to IUCN with primary responsibilities lying in the hands of one person. And that one person's perspective, no matter how professional or judicious, is still one person and may not have reflected the full depth and significance and particularly issues facing sites."[137]

In carrying out its evaluation, ICOMOS needed but did not have many comparative studies. Von Droste criticizes ICOMOS for failing to make them a priority:

These comparative studies were completely neglected in the first phase of the Convention. This is probably the most, the biggest flaw of ICOMOS, that there were not enough comparative studies, there was not enough leaning back and saying, we cannot yet evaluate, we need first to assess more and to do our inventory, we need to consult a network, or build up a network. ICOMOS was too weak to handle this agenda of comparative studies. It was too weak even to build up a more coherent and in-depth network.[138]

ICOMOS was at first reluctant to go beyond a comparative analysis of specific nominations.[139] In 1982, it began to compile a global inventory of cultural heritage sites and prepared its first comparative study on Jesuit missions in North and South America.[140] The next year the Committee asked ICOMOS to prepare a preliminary typological study based on all the properties listed or proposed for

---

135   Canada Research Chair, interview Stovel February 2011.

136   Canada Research Chair, interview, Luxen. "On l'a accusé un petit peu d'être personnel parce que malheureusement, c'est humain dans la présentation, il le faisait un peu comme si c'était lui, mais dans la réalité et ça je me suis battu pour que le travail qu'il présentait, était le fruit quand même d'un travail collectif très important."

137   Canada Research Chair, interview Milne.

138   Canada Research Chair, interview von Droste, 2008.

139   UNESCO, Report of the rapporteur on the fourth session of the World Heritage Committee in Paris, 1–5 September 1980, Paris, 29 September 1980, CC-80/conf.016/10, paras. 49–50. Retrieved from http://whc.unesco.org/archive/1980/cc-80-conf016-10e.pdf

140   UNESCO, Report of the rapporteur on the sixth session of the World Heritage Committee in Paris, 13–17 December 1982, Paris, 17 January 1983, CLT-82/conf.015/8, paras. 17, 56. Retrieved from http://whc.unesco.org/archive/1982/clt-82-conf015-8e.pdf

nomination as well as suggestions on how to better interpret the cultural criteria for historic towns and for properties representing events, ideas or beliefs.[141] In 1986, ICOMOS presented a study on contemporary architectural structures noting the challenges in dealing with this type.[142] Responding to Committee dissatisfaction with this ad hoc approach, ICOMOS proposed in 1988 to carry out "a retrospective and prospective reflection on the Convention" to identify "entities according to different parameters of coherence – chronological, geographical, ecological, functional, social, religious, etc."[143] This idea was subsumed into the global study and subsequently the global strategy. Although ICOMOS agreed that its comparative studies were mainly reactive, a 1996 list reveals that some were prospective, focusing on regions whose heritage was still under-represented on the World Heritage List.[144]

In addition to its primordial role in evaluating cultural nominations, ICOMOS also provided advice on working tools and doctrinal issues. The organization participated in the many revisions to the Committee's Operational Guidelines and collaborated with IUCN in preparing a draft framework for inscribing properties on the Danger List. In 1984, it worked on a handbook for managing World Heritage properties.[145] Drawing on the Venice Charter and other texts, ICOMOS worked on various doctrinal issues including refinements to the cultural criteria for nominations. With regard to monitoring, the Committee rejected a formal reporting system in 1983 and turned instead to ICOMOS to collect information through its experts on the state of conservation of World Heritage properties.[146] Three years later, ICOMOS introduced an ambitious monitoring system for cultural properties as requested by the Committee. This proposal was turned down

---

141   UNESCO, Report of the rapporteur on the seventh session of the World Heritage Committee in Florence 5–9 December 1983, Paris, January 1984, SC/83/conf.009/8, paras. 23–4. Retrieved from http://whc.unesco.org/archive/1983/sc-83-conf009-8e.pdf

142   UNESCO, Report of the rapporteur on the tenth session of the Bureau of the World Heritage Committee in Paris, 16-19 June 1986, Paris, 15 September 1986, CC-86/conf.001/11, para. 12. Retrieved from http://whc.unesco.org/archive/1986/cc-86-conf001-11e.pdf

143   UNESCO, Report of the rapporteur on the twelfth session of the World Heritage Committee in Brasilia, 5–9 December 1988, Paris, 23 December 1988, SC-88/conf.001/13, para. 14. Retrieved from http://whc.unesco.org/archive/1988/sc-88-conf001-13e.pdf

144   UNESCO, Comparative and related studies carried out by ICOMOS, 1992-1996, twentieth session of the World Heritage Committee in Merida, 2-7 December 1996, Paris, 18 September 1996, WHC-96/conf.201/inf. 11, pp. 1–8. Retrieved from http://whc.unesco.org/archive/1996/whc-96-conf201-infl1e.pdf

145   UNESCO, Report of the rapporteur on the eighth session of the World Heritage Committee in Buenos Aires, 29 October–2 November 1984, Buenos Aires, 2 November 1984, SC/84/conf.004/9, paras. 12, 37. Retrieved from http://whc.unesco.org/archive/1984/sc-84-conf004-9e.pdf. See chapter 4 for comments from Feilden on this project.

146   UNESCO, Report of the rapporteur on the seventh session of the World Heritage Committee in Florence 5–9 December 1983, Paris, January 1984, SC/83/conf.009/8, para. 41. Retrieved from http://whc.unesco.org/archive/1983/sc-83-conf009-8e.pdf

and a version modified by the secretariat was approved only on an experimental basis. By 1989, the Committee expressed its dissatisfaction with the monitoring of cultural sites, especially in comparison to IUCN's system for monitoring natural sites, and suggested that the role given to ICOMOS be reviewed and that other expertise be sought.[147]

In assessing the overall performance of ICOMOS, several pioneers begin by emphasizing that it was a small non-governmental body with insufficient financial resources to carry out its World Heritage work. Stovel notes that ICOMOS "has worked very hard over time to try to establish a kind of consistent line in responding to the expectations of the Committee," but agrees that, with elections every three years, ICOMOS has an inherent institutional limitation due to lack of continuity.[148] Touri picks up on this lack of consistency. While praising ICOMOS for its general technical competence and quality of expertise, he says: "I say in all honesty, they do not have the same rigor ... and then as they do not have the same rigorous approach, it puts them in situations of difficulty."[149]

Others contextualize ICOMOS's performance by pointing to the inherent difficulty of assessing cultural sites. Francioni believes: "when you are dealing with culture that there is no such thing as universal because every culture is to be appreciated for its own singularity and specificity. And so it is much less reliable and much more difficult to deny the merits of a certain site on cultural grounds."[150] Writing on the twentieth anniversary of the Convention, Batisse expressed a similar view: "Because the significance of different cultural attributes cannot usually be compared, the situation is much more difficult for cultural nominations. Thus, ICOMOS, with its limited means, is often placed in a very delicate position between technical considerations and national or local sensitivities."[151] Several pioneers make the point that ICOMOS's difficulties grew as participation in World Heritage became truly global. In Dawson Munjeri's words: "It's a very delicate assignment that ... ICOMOS has to handle because of the peculiarities of each cultural heritage ... I know all those components make it very difficult. Of course the issues relate also to personal feelings of identity, national pride. And the sheer volume and multiplicity and variety make it very difficult." He emphasizes the need for continuous learning. "The problem is again of expertise, the expertise

---

147   UNESCO, Report of the rapporteur on the thirteenth session of the World Heritage Committee in Paris, 11-15 December 1989, Paris, 22 December 1989, SC-89/conf.004/12, para. 18. Retrieved from http://whc.unesco.org/archive/1989/sc-89-conf004-12e.pdf

148   Canada Research Chair on Built Heritage, Université de Montréal, audio interview of Herb Stovel by Christina Cameron, Ottawa, 5 April 2011.

149   Canada Research Chair, interview Touri. "Je le dis en toute honnêteté, ils n'ont pas la même rigueur ... et donc comme ils n'ont pas la même rigueur d'approche, ça les met dans des situations de difficulté."

150   Canada Research Chair, interview Francioni.

151   Michel Batisse, "The struggle to save our world heritage," *Environment*, 34/10 (1992), p. 17.

in the various areas. It's a constantly evolving changing landscape. What was cultural heritage in the past, what it is turning into, particularly following the global strategy, makes it very difficult to say you have expertise in this area."[152] Von Droste echoes this view: "It was obvious that the expertise was there for European sites and it was not in place for other continents because simply ICOMOS did not have the network ... It's not the fault of ICOMOS. It's the fault of the breadth and scope of World Heritage."[153] An important partner in the 1980s, ICOMOS saw its influence wane in the second half of the 1990s with the emergence of a robust and competitive World Heritage Centre and mobilized States Parties.

ICCROM

Establishment

ICCROM is the third advisory body specifically mentioned in the Convention. Known under various names, ICCROM was established in 1959 by UNESCO in collaboration with the Italian government. It grew out of a proposal from the UNESCO General Conference "to create an international centre for the study of the preservation and restoration of cultural property ... to be located in Rome, where it will be able to profit from the assistance of the Istituto Centrale des [sic] Restauro and other specialized scientific institutes."[154] It was known as the UNESCO Rome Centre through the 1960s until director Paul Philippot changed the name in 1971 to the International Centre for Conservation. In 1977, director Bernard Feilden changed it to ICCROM. Former staff member Jukka Jokilehto notes that "ICCROM was from its foundation always a very close collaborator with UNESCO. ICCROM was, in a way, almost like a department of UNESCO at the beginning for practical issues. Then with time it started taking more independence."[155]

While maintaining close links with UNESCO, ICCROM is nonetheless an autonomous intergovernmental organization funded by member countries. In his recent history of the organization, Jokilehto documents ICCROM's research and training activities over its first half century. He cites a 1983 speech by UNESCO's Director-General M'Bow which captures the essence of the organization's focus at that time: "ICCROM's work is probably of the most decisive importance when it comes to training. It is obviously a most important task to make available to experts in all branches of restoration work, from craftsmen to scientists, the widest possible range of new knowledge and techniques needed for the protection

---

152   Canada Research Chair on Built Heritage, Université de Montréal, audio interview of Dawson Munjeri by Christina Cameron, Brasilia, 3 August 2010.

153   Canada Research Chair, interview von Droste, 2008.

154   UNESCO, Records of the General Conference ninth session, Paris, 1957, p. 24. Retrieved from http://unesdoc.unesco.org/images/0011/001145/114585e.pdf

155   Canada Research Chair on Built Heritage, Université de Montréal, audio interview of Jukka Jokilehto by Christina Cameron and Mechtild Rössler, Rome, 5 May 2010.

and preservation of cultural property."[156] ICCROM remains a highly respected international organization for training cultural heritage specialists.

ICCROM at World Heritage

According to Stovel, ICCROM worked hard to be included as an advisory body in the World Heritage Convention. "It was not envisioned initially that there would be a need for a second cultural heritage body but ICCROM pushed in the early going." ICOMOS had already been identified as an independent cultural organization able to provide neutral assessment of nominations without undue influence from governments. Stovel identifies ICCROM staff member Gael de Guichen as the person who convinced the drafters of the Convention to include ICCROM among the advisory bodies on the grounds that there would be post-inscription needs for training and technical assistance, services that his organization could offer.[157] Its role differs from the other two advisory bodies in that it does not evaluate nominations.

At UNESCO's invitation, ICCROM participated in the World Heritage preparatory meetings in Morges (1976) and Paris (1977), contributing proposals for evaluation criteria for cultural nominations and other aspects.[158] Yet its involvement with World Heritage developed slowly; indeed in the 1980s there were three years when no representative of ICCROM attended Committee meetings.[159] Jokilehto attributes the organization's low profile to the attitude of ICCROM's director at the time, Philippot, who "was very much concerned about the fact that this Convention would identify an exclusive list of major monuments and forget about the rest of heritage." Jokilehto recalls agreeing with this view: "Yes, we will do this World Heritage because we are part of it, but we had our own agenda to take care of heritage more generally speaking, and not just these major significant monuments."[160] The degree of ICCROM's involvement with World Heritage, as Raidl observes, "depended also a bit on their director … some took it very seriously and others attached less importance to it."[161]

Jokilehto confirms however that the debates at Committee meetings concerning threats to cultural properties influenced the design of ICCROM's training courses, research strategies and manuals during those years. In an effort to strengthen the

156   Jukka Jokilehto, *ICCROM and the Conservation of Cultural Heritage. A History of the Organization's First 50 Years 1959–2009* (Rome, 2011), p. 80.

157   Canada Research Chair, interview Stovel, March and April 2011.

158   UNESCO, Final report of informal consultation of intergovernmental and non-governmental organizations in the implementation of the Convention concerning the protection of the world cultural and natural heritage, Morges, 19–20 May 1976, CC-76/WS/25, annex II. Retrieved from http://unesdoc.unesco.org/images/0002/000213/021374eb.pdf

159   ICCROM was not present at Committee meetings in 1980, 1984 and 1987.

160   Canada Research Chair, interview Jokilehto.

161   Canada Research Chair, interview Raidl. "Ça se dépendait aussi un peu de leur directeur … il y en avait qui prenait ça très au sérieux et qui participaient bien et puis d'autres qui y attachaient peut-être moins d'importance."

Figure 5.8   1994 World Heritage Committee meeting in Phuket, Thailand from left to right: Jukka Jokilehto (ICCROM) and Henry Cleere, Jean-Louis Luxen and Carmen Añón Feliu (ICOMOS)

skills of those responsible for World Heritage properties, ICCROM prepared its first manual on the management of cultural sites. UNESCO's Anne Raidl made the suggestion at a 1983 meeting with the advisory bodies. As part of the rotating secretariat for World Heritage, she was well positioned to see the emerging need for guidance in preparing management plans and effective conservation strategies. Written by Feilden and later revised by Jokilehto, *Management Guidelines for World Cultural Heritage Sites* was eventually published in 1993.[162] In the 1990s, the issue of risk preparedness arose in Committee discussions thereby inspiring an ICCROM publication by Stovel on *Risk Preparedness: a Management Manual for World Cultural Heritage*. Part of a set of management guidelines for site officials, Stovel's manual focused on general principles and specific strategies for addressing major risks such as earthquakes, flooding, fire, armed conflict and civil unrest.[163]

While ICCROM regularly gave advice on specific requests for international assistance, its most important contribution to World Heritage is arguably the development of a comprehensive training strategy for cultural properties.

162   Bernard Feilden and Jukka Jokilehto, *Management Guidelines for World Cultural Heritage Sites* (Rome, 1993), pp. 1–122; Jokilehto, *ICCROM*, p. 107.

163   Herb Stovel, *Risk Preparedness: a Management Manual for World Cultural Heritage* (Rome, 1998), pp. 1–145.

Acknowledging that a lack of management and conservation skills contributed to the loss of heritage values, the Committee decided in the mid-1990s to give priority to training and capacity building for site managers and others involved in stewardship activities. At its 1995 session the Committee considered two proposals for training, one from IUCN for natural sites and one from ICCROM for cultural sites. The previous Bureau had asked ICCROM to develop a conceptual and methodological framework for cultural heritage training in consultation with regional institutions and partners.[164] The two separate proposals were adopted as an integrated global training strategy for managers of both cultural and natural heritage sites.[165]

Training activities gained momentum at the end of the century when ICCROM succeeded in securing resources for the first time at the 1999 session in Marrakesh. Stovel, then an ICCROM staff member, recounts how early opposition to the ICCROM funding request turned to support through the intervention of the delegate of Benin who spoke in support of ICCROM's African capacity development for museums and its new Africa 2009 programme for built heritage and community relations. "He was then followed by eight or nine other speakers who in equally enthusiastic terms supported ICCROM and so the pendulum swung."[166]

Stovel notes that ICCROM began to devote significant amounts of its own money to World Heritage: "We were trying very hard to identify emerging problems, emerging threats, emerging issues, which would benefit from a concentrated focus of expenditure or activity ... sometimes on a regional basis, sometimes on a thematic basis, sometimes putting the two together."[167] Two projects from this initiative are particularly noteworthy. The first one, Africa 2009, launched in 1998, was intended to improve conditions for cultural heritage in those countries through training and capacity building using World Heritage as a key framework for enlisting the interest of governments and for influencing heritage conservation practices. The second was ICCROM's program for Integrated Territorial and Urban Conservation (ITUC) which was first presented by Jokilehto to the 1996 World Heritage Committee. The ITUC program was meant to assist decision-makers in historic cities. Stovel claims that much of the testing ground was in World Heritage Sites. "This gave us an opportunity again in the developing world to think of things we can do to strengthen management capacity for historic cities."[168]

---

    164   UNESCO, Strategy for training in the field of natural heritage, Paris, 1 October 1995, WHC-95/conf.203/inf. 11A; Training strategy in the conservation of cultural heritage sites, Paris, 11 October 1995, WHC-95/conf.203/inf. 11B. Retrieved from http://whc.unesco.org/archive/1995/whc-95-conf203-inf11be.pdf

    165   UNESCO, Report of the rapporteur on the nineteenth session of the World Heritage Committee in Berlin, 4–9 December 1995, Paris, 31 January 1996, WHC-95/conf.203/16, para. XII.12. Retrieved from http://whc.unesco.org/archive/1995/whc-95-conf203-16e.pdf

    166   Canada Research Chair, interview Stovel, February 2011.

    167   Canada Research Chair, interview Stovel, April 2011.

    168   Canada Research Chair, interview Stovel, February 2011.

By the year 2000, ICCROM had become an effective partner in the World Heritage system. It responded to training needs in different regions through well-designed courses, capacity building programs and guidance manuals for cultural heritage property managers and other stakeholders.

UNESCO

Rotating Secretariat

In the period leading up to the adoption of the World Heritage Convention in 1972, UNESCO's support came from two distinct sectors, one focused on natural sciences and the other on culture and communications. The Convention states that the World Heritage Committee shall be assisted by a secretariat appointed by the Director-General of UNESCO (article 14.1). Although World Heritage brought cultural and natural sites together, UNESCO was slow to adjust its administrative structure to align with this vision. Therefore, the secretariat rotated between these two UNESCO sectors from the beginning until 1991.

Both collaborated in the preparatory meetings prior to the first session of the World Heritage Committee in 1977. From that point forward, two staff members served in rotation as secretary to the World Heritage Committee: Anne Raidl from the Cultural Heritage Division and Bernd von Droste from the Ecological Sciences Division. They and their staff took turns supporting the activities of the General Assembly of States Parties, the World Heritage Committee and its Bureau. In particular, they prepared all documents for statutory meetings, including information on the status of State Party ratifications, World Heritage Sites, budget, regulations, Operational Guidelines and so forth. In addition, they reviewed nominations from State Parties to ensure completeness before forwarding them for technical evaluation to ICOMOS or IUCN. They also worked in an ill-defined collaboration with the advisory bodies to prepare reports on the state of conservation of World Heritage Sites.

The rotation of the secretariat between the two divisions was an imperfect solution. The working styles were different and the two groups did not communicate well with each other. Indeed, there was often outright competition between them. Pressouyre, who lived through this early period, describes the discordant voices: "There was no way to make these two sections of the Convention work together since these two groups ... were physically separated by the UNESCO premises and were also separated as a result of the presence of staff who did not work together."[169] Stovel confirms this impression, recalling that "there was competition

---

169   Canada Research Chair, interview Pressouyre. "On n'avait aucune chance de faire marcher ensemble ces deux volets de la Convention puisque ces deux volets de la Convention étaient séparés physiquement par les locaux de l'UNESCO et étaient séparés aussi du fait de l'existence de personnel qui ne travaillait pas ensemble."

Figure 5.9   1983 World Heritage Committee meeting in Florence, Italy from left to right: Jürgen Hillig (Science Sector, UNESCO), Makaminan Makagiansar (Assistant Director-General Culture, UNESCO), Jorge Gazaneo (Argentina, acting Chairperson), Bernd von Droste (Ecological Sciences, UNESCO) and Anne Raidl (Cultural Heritage, UNESCO)

between the cultural heritage unit and the natural heritage unit inside UNESCO and this was not very healthy … there were sometimes very unpleasant exchanges behind the scenes, sometimes even in front of the Committee." He recalls that he found it hard to find a consistent way to work with States Parties because "there were so many things going on inside UNESCO that you could never really see clearly what that one way should be."[170]

A chronic irritant for the secretariat was lack of funding. Despite additional responsibilities, no new resources were allocated by UNESCO. Raidl confirms that "there were no supplementary posts from UNESCO for culture" and von Droste reiterates that "UNESCO made no financial provisions whatsoever." He says people were asked to serve on a voluntary basis in addition to their regular duties.[171] This meant that the secretariat asked the Committee each year for resources from the World Heritage Fund to hire "temporary" staff. The Committee expressed its reluctance, citing the Convention that reserves the Fund for international assistance for World Heritage Sites (article 20). The Convention explicitly defines international

170   Canada Research Chair, interview Stovel, April 2011.
171   Canada Research Chair, interview Raidl. "Aucun poste supplémentaire de l'UNESCO … à la culture;" Canada Research Chair, interview von Droste, 2007.

assistance: studies, provision of experts, training, equipment and loans (article 22). Nowhere does it indicate that the Fund may be used for salaries of UNESCO staff. As a point of principle, the Committee considered that the secretariat should be resourced from the regular UNESCO budget.

At the close of the first Committee session, rapporteur Michel Parent made an appeal to UNESCO for additional resources for the secretariat. The report of the meeting cites his remarks:

> in view of the volume and complexity of the administrative work involved both in the preparation of documentation for the sessions of the Committee and in implementing its decisions, which would be particularly heavy as from 1979, he suggested that UNESCO should carefully examine the situation and provide the additional staff support necessary for the work related to the World Heritage Convention.[172]

Starting in 1978 and continuing through 1991, the Committee approved "temporary assistance" to the secretariat in amounts ranging from $70,000 to $260,000 a year. At the same time it repeatedly asked UNESCO to assume financial responsibility for the secretariat, warning the Director-General in 1978 that temporary assistance would be "for a one-year period and drawing his attention to the need for additional permanent staff support financed by the Regular Programme and Budget of the organization."[173] In the next couple of years, the Committee kept asking for permanent staff but continued to renew temporary assistance for yet another year.[174]

In 1982 UNESCO pushed back. The representative of the Director-General reminded the Committee that the Convention states that management of the World Heritage Fund should be carried out according to UNESCO's financial regulations for trust funds (article 15.2) and indicated that UNESCO's practice was to take 14 per cent of these funds for general management costs. "In the case of the Convention, the funds for assistance to the secretariat to cover management costs which have thus far been requested are considerably less than those which the

---

172 UNESCO, Final report of the first session of the intergovernmental committee for the protection of the world cultural and natural heritage in Paris, 27 June–1 July 1977, Paris, 20 October 1977, CC-77/conf.001/9, para.68. Retrieved from http://whc.unesco.org/archive/1977/cc-77-conf001-9e.pdf

173 UNESCO, Report of the rapporteur on the second session of the World Heritage Committee in Washington, 5-8 September 1978, Paris, 9 October 1978, CC-78/conf.010/10 rev., para. 62. Retrieved from http://whc.unesco.org/archive/1978/cc-78-conf010-10reve.pdf

174 UNESCO, Report of the Rapporteur on the third session of the World Heritage Committee in Cairo and Luxor, 22–26 October 1979, Paris, 30 November 1979, CC-79/conf.003/13, para. 53. Retrieved from http://whc.unesco.org/archive/1979/cc-79-conf003-13e.pdf ; UNESCO, Report of the rapporteur on the fifth session of the World Heritage Committee in Sydney, 26-30 October 1981, Paris, 5 January 1982, CC-81/conf/003/6, para. 34. Retrieved from http://whc.unesco.org/archive/1981/cc-81-conf003-6e.pdf

organization could legitimately claim."[175] The Committee backed off for a couple of years although the secretariat continued to report that "the number of staff working for the implementation of the Convention had remained the same since the Convention became operational."[176]

In 1988, in response to the Committee's request for six additional posts from UNESCO, the Director-General refused in the short term but held out some hope: "There were indications that this situation could be resolved progressively in the near future, thereby releasing the funds allocated for temporary assistance." The Committee snapped back that an allocation for temporary assistance to the secretariat would probably not be granted by the Committee in future years.[177]

The system finally broke down at the 1991 Committee session in Carthage under the burden of an impossible workload and general disorganization. Von Droste describes the Carthage meeting as "completely dysfunctional", attributing the chaos to a lack of staff, a lack of money and the bipolar nature of the Convention: "the natural and the cultural part were dissociated and people did not know to whom to address."[178]

World Heritage Centre

Exhausted and disenchanted, Raidl retired from UNESCO after the Carthage meeting. This vacancy was von Droste's cue to revamp the secretariat in order to bring the two streams together. Inspired by the North American model followed by Parks Canada and the United States National Park Service that combines cultural sites and national parks under one administration, he pitched his concept to Federico Mayor, then Director-General of UNESCO. Von Droste began with a basic premise:

> There would be no separation between culture and nature. It would be just a matter of degree of involvement of the natural or cultural components depending on what kind of conservation issue we have to deal with, and that UNESCO would be in a unique position to overcome the traditional dichotomy between culture and nature because it has responsibilities in both fields but it needs to unite these within one secretariat so State Parties would also know to whom to address in the future.[179]

---

175   UNESCO, Report of the rapporteur on the sixth session of the World Heritage Committee in Paris, 13–17 December 1982, Paris, 17 January 1983, CLT-82/conf.015/8, para. 39. Retrieved from http://whc.unesco.org/archive/1982/clt-82-conf015-8e.pdf

176   UNESCO, Report of the rapporteur on the ninth session of the World Heritage Committee in Paris, 2–6 December 1985, Paris, December 1985, SC-85/conf.008/9, para. 10. Retrieved from http://whc.unesco.org/archive/1985/sc-85-conf008-9e.pdf

177   UNESCO, Report of the rapporteur on the twelfth session of the World Heritage Committee in Brasilia, 5–9 December 1988, Paris, 23 December 1988, SC-88/conf.001/13, para. 53. Retrieved from http://whc.unesco.org/archive/1988/sc-88-conf001-13e.pdf

178   Canada Research Chair, interview von Droste, 2007.

179   Canada Research Chair, interview von Droste, 2007.

Beyond this premise, von Droste harboured a grandiose vision for the World Heritage Centre. In an interview, he says that Mayor knew that this new institutional model would be a sensitive issue with member states and advised him to develop it in secret. "I had proposed to establish a World Heritage Institute" based on the model of the International Educational Institute in Paris which at that time had a "good reputation, functioned well and had autonomy." It would be "a semi-autonomous body within UNESCO since it has a different set of States belonging to it than the General Conference of UNESCO and therefore it needs to be more or less seen as an autonomous matter." Von Droste recalls his tripartite plan. First an administrative centre would be established to manage the Convention. Secondly, he proposed the creation of "a World Heritage Academy which would be the hundred leading people of more scientific standing to look at the ethical dimension and at the same time ...to have a publication outlet." For the academy, he envisaged acquiring "the world's largest collection of books, particularly on art history but also natural history ... so we would have a very strong scientific basis for World Heritage." Thirdly, a media committee would be created to raise awareness about the conservation goals of the Convention.[180]

In 1992, Mayor issued a blue note unilaterally announcing the creation of a new World Heritage Centre and appointing von Droste as its first director to report directly to him. A UNESCO executive member, Mounir Bouchenaki recounts the events leading up to this announcement. As the new Director of the Cultural Heritage Division in early 1992, Bouchenaki along with von Droste attended a meeting with Mayor who said that, in light of the difficult Carthage meeting, it was impossible to continue in the same vein. He announced his intention to create a World Heritage Centre and "asked each of us to take some colleagues who followed the files in the Cultural Heritage Division and the Ecological Sciences Division and this is what created the nucleus of the World Heritage Centre because he did not create any positions."[181]

Mayor came to the 1992 Committee session in Santa Fe to explain his decision:

> I have recently ... established a World Heritage Centre, bringing together a secretariat previously divided along cultural and natural lines ... This new unified World Heritage Centre, working in close co-operation with the other sectors of the secretariat, should be better equipped to assist the Committee in its various objectives, such as the building up of sound monitoring systems, the launching of appeals and fund-raising activities for the World Heritage Fund, and action to promote greater public awareness of the Convention. It should

---

180   Canada Research Chair, interview von Droste, 2007.

181   Canada Research Chair on Built Heritage, Université de Montréal, audio interview of Mounir Bouchenaki, Paris, 29 February 2009. "Il nous a demandé à chacun d'entre nous de prendre quelques collègues qui suivaient les dossiers au secteur à la Division du patrimoine culturel et à la Division des sciences écologiques et c'est ce qui a créé le noyau du Centre du patrimoine mondial parce qu'il n'a pas créé des postes."

Figure 5.10   World Heritage Centre staff in the mid-1990s

also make for closer and easier co-operation with the technical advisory bodies – notably ICOMOS, ICCROM and IUCN -- which have consistently provided UNESCO with such excellent service in the implementation of the World Heritage Convention.[182]

In his interview Mayor says this action was a way to overcome UNESCO's compartmentalization: "UNESCO for me is the world and this is why I thought it should rather be a Centre that receives cross-cutting knowledge not only from the cultural, historical, philosophical institutions of UNESCO" but from elsewhere:

> I thought it was better to have a Centre that receives all these contributions and which could give the Committee not only purely technical support but also strategic and policy advice. That was the reason that I made this structural change ... to be able to gather information from everyone and not just from a few units that have become too specialised, sometimes too arrogant.[183]

---

182   UNESCO, Report of the rapporteur for the sixteenth session of the World Heritage Committee in Santa Fe, 7–14 December 1992, 14 December 1992, WHC-92/conf.002/12, annex [IX], p. 4. Retrieved from http://whc.unesco.org/archive/1992/whc-92-conf002-12e.pdf

183   Canada Research Chair on Built Heritage, Université de Montréal, audio interview of Federico Mayor by Christina Cameron, Madrid, 18 June 2009. "L'UNESCO était trop sectorialisé ... l'UNESCO pour moi c'est le monde et j'ai pensé ça doit être plutôt un Centre qui reçoit cette contribution transversale pas seulement de les institutions culturelles historiques, philosophiques de l'UNESCO ... Donc, c'est pour cette raison

Von Droste picks up the story: "So I became the first Director of World Heritage, of a unified secretariat with an autonomy which was not formally approved but *de facto*, because Mr. Mayor was directly my boss. I was not subject to any sector and I considered my work as inter-sectoral."[184] While not precisely the grand semi-autonomous institute envisaged by von Droste, the Centre's mission included many of the same features.

The idea of putting culture and nature together proved to be acceptable to most States Parties since it addressed an inherent weakness in the original system. But they tried to pin down the details in order to understand exactly what had been created. At its 1993 meeting, the Committee deliberated long and hard, eventually adopting a text addressed to the Director-General seeking confirmation of the Committee's understanding of the role and function of the new Centre. In summary, the text states that the Centre would work under Committee direction and in accordance with UNESCO policies. It identified the following functions: to serve as the secretariat to the organs of the 1972 Convention; to act as a clearing house for information on conservation of cultural and natural heritage; to oversee implementation of training, monitoring and technical assistance by States Parties, ICOMOS, IUCN and ICCROM and others, to cooperate with other sectors in UNESCO including field offices; and to serve as the primary instrument for facilitating decisions of Committee and primary contact with States Parties on technical matters related to the Convention.[185]

As views about the World Heritage Centre flew back and forth between States Parties and the Director-General, four negative issues emerged: the need for more resources, perceived duplication between the advisory bodies and the secretariat, decentralization and especially the Centre's autonomy.

In terms of resources for the World Heritage Centre, von Droste claims that Mayor initially promised him 25 staff positions.[186] At the beginning, the Centre was staffed by the transfer of nine people from the culture and science sectors. In 1993, the Americans called for the allocation of ten posts and operational funding from UNESCO.[187] In 1994 the Director-General added three new posts

---

que j'ai pensé que c'était mieux d'avoir un centre qui reçoit toutes ces contributions et qui peut les demander et que finalement peut donner au Comité un appui pas seulement exclusivement technique mais aussi stratégique et politique ... il y a cette capacité de pouvoir se fournir des données de tout le monde et pas seulement de quelques unités qui deviennent trop spécialistes parfois, trop arrogantes aussi."

184   Canada Research Chair, interview von Droste, 2007.

185   UNESCO, Report of the rapporteur for the seventeenth session of the World Heritage Committee in Cartagena, 6–11 December 1993, Paris, 4 February 1994, WHC-93/conf.002/14, para. VII.10.1 Retrieved from http://whc.unesco.org/archive/1993/whc-93-conf002-14e.pdf

186   Canada Research Chair, interview von Droste, 2007.

187   UNESCO, Report of the rapporteur for the seventeenth session of the World Heritage Committee in Cartagena, 6–11 December 1993, Paris, 4 February 1994, WHC-

and beginning in 1997 an additional eight for a grand total of 20.[188] At the same time, States Parties began in earnest to finance national experts to augment the staff, thereby generating an unprecedented level of activity.

A source of tension came from a lack of clarity between the roles of the World Heritage Centre and the advisory bodies. With regard to perceived overlaps and duplication of work, the confusion was real as UNESCO staff and associate experts expanded their scientific activities. The Convention calls for UNESCO to utilize "to the fullest extent possible" the services of IUCN, ICOMOS and ICCROM in preparing documentation and implementing decisions (article 14.2). The 1992 strategic orientations confirmed that "the three pillars on which implementation of the Convention rests, namely, the Committee, the Secretariat, and the consultative bodies, should play their role fully and equitably."[189] But in this same year, the Director's vision for the World Heritage Centre appeared to differ.

Some of the confusion stemmed from a shared responsibility within UNESCO for implementing parts of the Convention. Bouchenaki explains that the Cultural Heritage Division retained responsibility for operational follow-up: "There were communications difficulties because the Centre had responsibility for the preparation, for receiving the files, for looking after the nomination files, but in parallel the Heritage Division also undertook operational work, sent missions and followed the work at the site."[190]

But the bigger issue was tension between the advisory bodies and the World Heritage Centre. While taking care to recognize the excellence and helpfulness of individual staff members, those pioneers who worked for the advisory bodies at that time recall the ambitions of the Centre. Within weeks of starting its operations, Stovel remembers being astounded when "Right off the bat, Bernd von Droste came over to ICOMOS and ... simply stated as a matter of almost fact, 'I've pretty well got Henry Cleere's office ready for him'."[191] Cleere, then ICOMOS World

---

93/conf.002/14, para. VII.1. Retrieved from http://whc.unesco.org/archive/1993/whc-93-conf002-14e.pdf

188   UNESCO, Report of the rapporteur on the eighteenth session of the World Heritage Committee in Phuket, 12–17 December 1994, Paris, 31 January 1995, WHC-94/conf.003/16, para. I.8. Retrieved from http://whc.unesco.org/archive/1994/whc-94-conf003-16e.pdf; UNESCO, Report of the rapporteur on the twentieth session of the World Heritage Committee in Merida, 2-7 December 1996, Paris, 10 March 1997, WHC-96/conf.201/21, para. I.10. Retrieved from http://whc.unesco.org/archive/1996/whc-96-conf201-21e.pdf

189   UNESCO, Report of the rapporteur for the sixteenth session of the World Heritage Committee in Santa Fe, 7–14 December 1992, 14 December 1992, WHC-92/conf.002/12, annex II.4. Retrieved from http://whc.unesco.org/archive/1992/whc-92-conf002-12e.pdf

190   Canada Research Chair, interview Bouchenaki. "Il y a eu des difficultés de communication parce que le Centre avait la responsabilité de la préparation, de recevoir les dossiers, d'instruire les dossiers d'inscription mais parallèlement la Division du patrimoine faisait aussi un travail opérationnel, envoyait des missions, suivait le travail sur le terrain."

191   Canada Research Chair, interview Stovel, April 2011.

Heritage coordinator, says that the Director told him that "he was intending to build up a centre of excellence and gradually the roles of the advisory bodies would disappear."[192] Thorsell, then IUCN coordinator, tells a similar tale, observing that it was problematic when the Centre "started to expand beyond its original mandate to take on the technical roles ... there was a pretty major empire-building effort going on there and I think it went too far."[193] This view is echoed by Adrian Phillips, former chair of the World Commission on National Parks and Protected Areas: "The one issue that I didn't really feel comfortable with was the extent to which the secretariat would duplicate the expertise which IUCN had or could secure ... The temptation to build an empire and to duplicate is there all the time."[194]

Stovel claims he was at first speechless at what he saw as attempts to absorb the advisory body functions within the Centre. He made the case to von Droste that "it's important that ICOMOS maintain its independence and we need to not just maintain the independence of the production of the documents but the appearance of independence. We can't be within your grasp, because otherwise, we will lose our credibility before States Parties." The area that he considers at the heart of the "muddy relationship" is the post-inscription question of monitoring: "The Convention simply didn't go far enough to talk about how that process should be managed over time."[195] Guo from the Chinese delegation notes the confusion between the advisory bodies and the Centre staff and concludes "We have to distinguish the advisory bodies and the World Heritage Centre ... I don't think they [staff] have the time nor do they have an advantage to make some professional studies."[196]

To the Committee, UNESCO went out of its way to emphasize the important contribution of ICOMOS, ICCROM and IUCN in implementing the Convention. Yet the perception of overlap and duplication led the World Heritage Centre in 1996 to prepare Memoranda of Understanding with the three advisory bodies as a way of clarifying the roles and strengthening collaboration. This measure did not really resolve the issue which continued to fester and became an item for the reform agenda of 2000.[197]

Another impact of the creation of the World Heritage Centre came as a result of UNESCO's intention to decentralize some implementation to its own field offices.[198] Although many countries favoured decentralization of some UNESCO functions,

---

192   Canada Research Chair, interview Cleere.

193   Canada Research Chair, interview Thorsell.

194   Canada Research Chair on Built Heritage, Université de Montréal, audio interview of Adrian Phillips by Christina Cameron, London, 24 January 2008.

195   Canada Research Chair, interview Stovel, April 2011.

196   Canada Research Chair on Built Heritage, Université de Montréal, audio interview of Guo Zhan by Christina Cameron, Brasilia, 3 August 2010.

197   UNESCO, Report of the rapporteur on the twentieth session of the World Heritage Committee in Merida, 2–7 December 1996, Paris, 10 March 1997, WHC-96/conf.201/21, para. IV.9. Retrieved from http://whc.unesco.org/archive/1996/whc-96-conf201-21e.pdf

198   UNESCO, Report of the rapporteur for the seventeenth session of the World Heritage Committee in Cartagena, 6–11 December 1993, Paris, 4 February 1994, WHC-

certain States Parties, particularly France and Italy, were opposed to regional offices
for World Heritage. The issue blew up at the 1995 Berlin meeting when it became
apparent that the Director-General had signed a three-year agreement with the
government of Norway to create a Nordic World Heritage Office without consulting
the World Heritage Committee. In the face of denials from the Centre, the French
delegation proved the contrary by circulating copies of the actual agreement that
UNESCO had signed. It is noteworthy that the official report of the Berlin meeting
incorrectly states that the Centre had informed the Committee about the agreement
at the beginning of the meeting.[199] Some Committee members "expressed strong
concern about the creation of a network of World Heritage offices."[200] The Norwegians
explained the purpose and funding model as well as the evaluation requirements
for the new office. Some members were still not satisfied with the process and the
session ended in chaos, although the minutes are somewhat understated:

> At the conclusion of the discussions a copy of the agreement was made available
> to the members of the Committee; but they did not have the opportunity to
> express their views on this text. The Director of the World Heritage Centre
> agreed to prepare a report for the next session of the Bureau on the subject of
> decentralization as it relates to World Heritage.[201]

Two years later, other States Parties began to show interest in further decentralization
of World Heritage activities. Referring to the annual report from the Nordic World
Heritage Office, Japan asked about the policy for establishing similar regional
offices to strengthen implementation of the Convention, confirming its interest
in playing such a role in Asia. The Republic of Korea also expressed its wish to
contribute to regional programs for World Heritage Sites.[202]

93/conf.002/14, para. VII.5. Retrieved from http://whc.unesco.org/archive/1993/whc-93-
conf002-14e.pdf

199   Several witnesses at this session confirm this sequence of events although
the official report incorrectly states that the Centre had informed the Committee at the
beginning of the meeting. UNESCO, Report of the rapporteur on the nineteenth session
of the World Heritage Committee in Berlin, 4-9 December 1995, Paris, 31 January 1996,
WHC-95/conf.203/16, para. IV.3. Retrieved from http://whc.unesco.org/archive/1995/whc-
95-conf203-16e.pdf

200   UNESCO, Report of the rapporteur on the nineteenth session of the World
Heritage Committee in Berlin, 4–9 December 1995, Paris, 31 January 1996, WHC-95/
conf.203/16, para. XV.2. Retrieved from http://whc.unesco.org/archive/1995/whc-95-
conf203-16e.pdf

201   UNESCO, Report of the rapporteur on the nineteenth session of the World
Heritage Committee in Berlin, 4–9 December 1995, Paris, 31 January 1996, WHC-95/
conf.203/16, para. XV.2. Retrieved from http://whc.unesco.org/archive/1995/whc-95-
conf203-16e.pdf

202   UNESCO, Report of the rapporteur on the twenty-first session of the World
Heritage Committee in Naples, 1–6 December 1997, Paris, 27 February 1998, WHC-97/

The thorniest issue was the proposed autonomy for the World Heritage Centre. Francioni became involved in World Heritage as a result of it. As an international lawyer, he was asked by Italy and France to prepare a background paper on the administrative structure of the Centre: "There was some concern that this sort of privatization of the UNESCO World Heritage Centre would diminish the sovereign control over the World Heritage.... I was asked as a lawyer to provide legal arguments for not having the Centre become a creature of its own with too much autonomy."[203] He describes the 1993 meeting in Cartagena as very controversial because there were tensions over what the World Heritage Centre should be. The United States of America, supported by Germany and Thailand, advocated for delegated staffing authority, streamlined administrative procedures and Committee responsibility for the Centre's work plan.[204] France on the other hand feared the consequences of greater autonomy and a potential "divorce between the Centre's policy and that of UNESCO."[205] By the end of the 1993 session, the issue appeared to be resolved: "The Committee agreed on the importance of the World Heritage Centre as a unified body within the secretariat of UNESCO."[206] Francioni concurs: "What came out of the meeting was a fairly balanced solution that I think strengthened the World Heritage Centre and at the same time tranquilized a little bit the countries such as Italy and France, major contributors to UNESCO who wanted to continue to have a governmental oversight."[207]

But UNESCO came back for another round in 1994, suggesting that the Committee might endorse "an effective functional autonomy in regard to administrative and financial aspects" modelled on the International Institute for Educational Planning (IIEP) and the International Bureau of Education (IBE).[208]

---

conf.208/17, paras. III.12-13. Retrieved from http://whc.unesco.org/archive/1997/whc-97-conf208-17e.pdf

203   Canada Research Chair, interview Francioni.

204   UNESCO, Report of the rapporteur for the seventeenth session of the World Heritage Committee in Cartagena, 6–11 December 1993, Paris, 4 February 1994, WHC-93/conf.002/14, para. VII.1. Retrieved from http://whc.unesco.org/archive/1993/whc-93-conf002-14e.pdf

205   UNESCO, Report of the rapporteur for the seventeenth session of the World Heritage Committee in Cartagena, 6–11 December 1993, Paris, 4 February 1994, WHC-93/conf.002/14, para. VII.4. Retrieved from http://whc.unesco.org/archive/1993/whc-93-conf002-14e.pdf

206   UNESCO, Report of the rapporteur for the seventeenth session of the World Heritage Committee in Cartagena, 6–11 December 1993, Paris, 4 February 1994, WHC-93/conf.002/14, para. VII.9. Retrieved from http://whc.unesco.org/archive/1993/whc-93-conf002-14e.pdf

207   Canada Research Chair, interview Francioni.

208   UNESCO, Report of the rapporteur on the eighteenth session of the World Heritage Committee in Phuket, 12–17 December 1994, Paris, 31 January 1995, WHC-94/conf.003/16, para. I.8. Retrieved from http://whc.unesco.org/archive/1994/whc-94-conf003-16e.pdf

Italy and France challenged this model of functional autonomy. France opposed "creating a unit which might lead to a separation from UNESCO ... The evolution of the Centre should be administrative and structural within the Organization," concluding that "a private foundation cannot be created in the shadow of a Convention between States Parties, which is what appeared to be envisaged."[209] Italy found it unacceptable, arguing that "the Centre was meant to be simply a secretariat for coordination, monitoring of the Convention's implementation, information and cooperation with the States Parties in order to assure follow-up actions." According to Italy, UNESCO's proposal "seems to lead on the contrary to a full autonomy of the Centre by giving it functional and administrative autonomy." The delegation believed that all actions of UNESCO should be united to achieve a major impact and judged that the institutional set-up of the models presented was inappropriate:

> These have been established within the General Conference of UNESCO, which means that all Member States of UNESCO are included, and not just some, as is the case with the Centre. Moreover, the internal structure is quite different: the IIEP and IBE have each an administration council which, however, does not exist in the case of the Centre, as this is directly under the Director-General of UNESCO and is, as such, a simple Secretariat.[210]

In the face of this opposition, the Committee asked for further studies to explain precisely what functional autonomy meant.

This dispute was exacerbated at the 1994 Committee meeting by a presentation of a professional marketing and branding scheme for World Heritage prepared at the request of the Director-General.[211] Justified by the need to raise funds in the private sector, the scheme focused on re-branding the World Heritage concept and upgrading or replacing the existing World Heritage logo to make it more "emotive" and "commercially marketable". A new logo design was even presented at the meeting. The Committee reacted angrily at the possible dissociation of World Heritage from UNESCO. In addition, in the words of the delegate from Brazil, "the World Heritage logo should not be seen as a trademark but rather

---

209   UNESCO, Report of the rapporteur on the eighteenth session of the World Heritage Committee in Phuket, 12–17 December 1994, Paris, 31 January 1995, WHC-94/conf.003/16, para. VIII.4. Retrieved from http://whc.unesco.org/archive/1994/whc-94-conf003-16e.pdf

210   UNESCO, Report of the rapporteur on the eighteenth session of the World Heritage Committee in Phuket, 12–17 December 1994, Paris, 31 January 1995, WHC-94/conf.003/16, paras. VIII.2, VIII.10. Retrieved from http://whc.unesco.org/archive/1994/whc-94-conf003-16e.pdf

211   Charles de Haes and David Mitchell, Strategic recommendations for promoting and fundraising for World Heritage, October, 1994, Paris, 18 November 1994, WHC-94/conf.003/11 add. Retrieved from http://whc.unesco.org/archive/1994/whc-94-conf003-11adde.pdf

as a symbolic representation of the philosophy and high values consecrated in the World Heritage Convention." This debacle resulted in the cancellation of new funding for marketing and increased tension between the Director of the World Heritage Centre and the Committee.[212]

A continuing atmosphere of mistrust was palpable in 1996 when the Director proposed a celebratory review of World Heritage on its twenty-fifth anniversary and instead received a request for an external audit of the World Heritage Fund and a management review of the secretariat's operations. He saw the twenty-fifth anniversary as "a historic occasion to strengthen international cooperation for the implementation of the World Heritage Convention: it is a time to critically review achievements and failures and to chart the course of actions for the future."[213] But the Committee thought otherwise. It did not support the Director and refused to allocate the funding. Instead it proposed, in the framework of the commemoration of the twenty-fifth anniversary, an external audit of the financial procedures and management practices of the World Heritage Centre after its five years of operation "in order to disperse all ambiguity and seek a satisfactory solution for the preparation of the statement of accounts and provisional budgets."[214] The results were presented to the 1997 Committee which set up groups to study the auditor's work. The Director commented that the recommendations "would be useful for planning the work of the Centre in the future and would also ensure greater effectiveness and visibility of the Convention."[215]

Still searching for clarity, the following Bureau asked the Director-General for a report on the tasks and functions of the World Heritage Centre specifically with regard to cooperation with other UNESCO sectors and to the use of funds. Instead of providing a report, the Director-General took action in light of the impending retirement of von Droste. Motivated by the need to develop greater synergies with the Cultural Heritage and Ecological Sciences Divisions, he issued an internal green note appointing Bouchenaki, then Director of the Cultural Heritage Division,

212   UNESCO, Report of the rapporteur on the eighteenth session of the World Heritage Committee in Phuket, 12–17 December 1994, Paris, 31 January 1995, WHC-94/conf.003/16, paras. XV.1–XV.19. Retrieved from http://whc.unesco.org/archive/1994/whc-94-conf003-16e.pdf

213   UNESCO, Report of the rapporteur on the twentieth session of the World Heritage Committee in Merida, 2–7 December 1996, Paris, 10 March 1997, WHC-96/conf.201/21, para. IV.15. Retrieved from http://whc.unesco.org/archive/1996/whc-96-conf201-21e.pdf

214   UNESCO, Report of the rapporteur on the twentieth session of the World Heritage Committee in Merida, 2–7 December 1996, Paris, 10 March 1997, WHC-96/conf.201/21, paras. XIII.12, XIV.5. Retrieved from http://whc.unesco.org/archive/1996/whc-96-conf201-21e.pdf

215   UNESCO, Report of the rapporteur on the twenty-first session of the World Heritage Committee in Naples, 1–6 December 1997, Paris, 27 February 1998, WHC-97/conf.208/17, para. III.2. Retrieved from http://whc.unesco.org/archive/1997/whc-97-conf208-17e.pdf

to the additional role of Director of the World Heritage Centre.[216] The Committee reacted negatively, envisaging a takeover by the cultural sector and fearing that the distinct identity of the Centre would be diluted or absorbed into another part of UNESCO. It also worried about the potential loss of the important balance between cultural and natural heritage. The Committee asked that "the Centre remain a unit specifically dedicated to provide secretariat services to the World Heritage Convention under the direct authority of the Director-General." The representative of Director-General Mayor remarked pointedly that the Committee had "reaffirmed year after year that the World Heritage Centre ... should be considered as a unit of the secretariat. Consequently, it remains the prerogative of the Director-General to take ... the measures he deems necessary for the organization and the functioning of the World Heritage Centre." He reassured the Committee that the Centre would remain a distinct unit within UNESCO specifically assigned to work as the secretariat of the World Heritage Convention.[217]

In 1999, the new Director-General, Koïchiro Matsuura moved quickly to place the World Heritage Centre under the Assistant Director-General for Culture as part of a general restructuring of the cultural sector. The Centre retained some administrative autonomy including management of the World Heritage Fund. He explains that his previous experience with World Heritage led him to make the change:

> When I chaired the Kyoto committee, DG didn't come, DDG didn't come either, and ADG Culture didn't come, ADG Natural Sciences didn't come. And people argued, it's important [to be] attached to the DG directly. But the DG did not pay attention to it at all. So one of the first things [when] I arrived here was to put the Centre under the responsibility of the ADG Culture, of course being aware that 20 per cent of the World Heritage Sites are natural sites, therefore ADG Natural Sciences must be involved. The next step ... was to put the Cultural Heritage unit under the Centre so the Centre would cover not only the World Heritage Sites but also non-sites ... From then on, ADG Culture was always at the Committee meeting from the beginning to the end. [218]

While the initial idea of combining culture and nature in one unit was conceptually sound, the World Heritage Centre got off to a rocky start. Von Droste acknowledges these difficulties highlighting in particular the lack of space for the

---

216   UNESCO, Preservation and presentation of cultural and natural heritage, DG note 98/53, Paris, 23 November 1998, Kyoto, 30 November 1998, WHC-98/conf.203/11 add, annex 4. Retrieved from http://whc.unesco.org/archive/1998/whc-98-conf203-11adde.pdf

217   UNESCO, Report of the rapporteur on the twenty-second session of the World Heritage Committee in Kyoto, 30 November–5 December 1998, Paris, 29 January 1999, WHC-98/conf.203/18, paras. IX.25, 29, 31. Retrieved from http://whc.unesco.org/archive/1998/whc-98-conf203-18e.pdf

218   Canada Research Chair, interview Matsuura.

new unit: "My staff for World Heritage was sitting in the different sectors in the different buildings." Robertson Vernhes corroborates this fact:

> Even though physically we were supposed to be in the same organisation, they [culture] went to Fontenoy, and the science part, the natural heritage part that we were dealing with was left back in Bonvin. And therefore there was a split literally in the people and that meant there was no day to day contact, no intellectual exchange of ideas between people. So there was this huge disadvantage, for example, of separation, disadvantage of separation, and I think this led to an awful lot of confusion, personally I think a dilution of the World Heritage idea.[219]

Von Droste also expresses his disappointment that the promised administrator never materialized. He points to the complexity of managing the World Heritage Fund, extra-budgetary funds as well as income from film rights and books. "The Centre was left without any administrative office and I had ... to do that myself. The Committee very soon found the weakness of this director is that he cannot reply to all administrative questions. It was true. I was simply over burdened with all the different things to handle."[220] Milne concurs: "The Director of the Centre was not fully supported by his appropriate management structure, immediate structure. I think that was unfortunate."[221]

Although the grand vision for a World Heritage Institute never materialized and there were some start-up pains, the secretariat at the turn of the millennium had an impressive record of achievement. It had become the unique focal point at UNESCO for World Heritage matters. The integration of the World Heritage system under one director led to more coherent promotion of the Convention, information management, educational programmes and partnerships. The World Heritage Newsletter was published regularly and the glossy World Heritage Review was available in English, French, Spanish, Japanese and other languages. By 1995, World Heritage documents were available on the UNESCO website, an innovative step at the time. In terms of promotion, the Centre facilitated partnerships with the private sector, leading to the production of hundreds of films and publications in several languages. Japan and Germany were both outstanding media partners.[222] With regard to education, an inter-regional project for young people's participation in World Heritage preservation and promotion, initiated in collaboration with the Education sector of UNESCO, mobilized schools and local communities. The project spawned the very popular educational kit *World Heritage in Young Hands*,

---

219   Canada Research Chair, interview Robertson Vernhes.
220   Canada Research Chair, interview von Droste, 2007.
221   Canada Research Chair, interview Milne.
222   For an overview of communications projects from 1980 to the present, see the doctoral thesis of Juan Shen, *Patrimoine et audiovisuel, l'enjeu de la médiation scientifique dans la communication des connaissances* (Paris, 2010).

**Figure 5.11   1998 launch of *World Heritage in Young Hands* in Paris from left
to right: Sarah Titchen, Taro Komatsu, Elisabeth Khawaykie,
Bernd von Droste and Breda Pavlic**

a manual produced in partnership with the Nordic World Heritage Office and the
Rhône-Poulenc Foundation. By the turn of the century, the World Heritage Centre
was and remains the global focal point for the Convention.

Civil Society

The significant role that civil society plays in protecting and conserving heritage
sites is undeniable, yet the World Heritage Convention does not assign any official
role to outside organizations and groups. States Parties, the three advisory bodies
and the UNESCO secretariat all enjoy a statutory legitimacy that outsiders do
not. This chapter focuses on the involvement of civil society in World Heritage
Committee processes. What is not captured is the involvement and dedication of
local authorities, site managers or owners of World Heritage properties who in
many cases do not have any role within the national official bodies but yet are
responsible ultimately for putting into effect many of the decisions taken by the
Committee and for discharging the responsibilities of the State Party for care of
World Heritage Sites. While not the focus of this book, it is acknowledged that
local involvement is essential to the successful implementation of the Convention.

Voices from community groups and non-governmental organizations have no official place in World Heritage processes and may only be heard at the discretion of the chairperson of the Committee. As Chair of the 1987 session, Collinson recalls dealing with this issue in the case of the Tasmanian wilderness nomination when state government officials as well as mining and forestry companies wanted their views to be heard: "I had to explain to them that unless they were a recognized member of the Australian delegation we weren't in a position to hear what they had to say. But I said that doesn't prevent you from talking to people in the hall. And so they talked to a lot of people."[223]

Environmental groups had been active at the 1972 Stockholm summit although their interest extended well beyond discussions about World Heritage to a broad range of environmental issues. In the early years of World Heritage, attendance by outside groups at Committee sessions was intermittent. At the first session in 1977, only one group, the International Organization for the Protection of the Arts, participated as an observer.[224] The following year in Washington, ten international governmental and non-governmental organizations and "a wider public audience" were present, a mark of increased interest in the Convention in that region.[225]

During the 1980s, three groups regularly attended World Heritage Committee meetings: the International Council of Museums (ICOM), the International Union of Architects (UIA) and the International Federation of Landscape Architects (IFLA). The latter had a specific interest in discussions on rural landscapes that began in 1984 and culminated in the adoption of the cultural landscape category in 1992. Other organizations appeared occasionally including the Arab Educational, Cultural and Scientific Organization (ALECSO) in 1980, the Organization for Museums and Sites of Africa (OMMSA) in 1982, the World Wildlife Fund (WWF) in 1983 and the International Fund for the Promotion of Culture (IFPC) in 1984 and 1985. The 1988 session in Brazil drew representatives from United Nations organizations like UNICEF, UNHCR and UNDP that traditionally work in developing countries.[226]

In the 1990s, participation from civil society grew in tandem with the increase in States Parties and designated sites. One significant development was the creation of an organization specifically centred on World Heritage cities. An international

---

223 Canada Research Chair, interview Collinson.

224 UNESCO, Final report of the first session of the intergovernmental committee for the protection of the world cultural and natural heritage in Paris, 27 June–1 July 1977, Paris, 20 October 1977, CC-77/conf.001/9, annex 1, p. 5. Retrieved from http://whc.unesco.org/archive/1977/cc-77-conf001-9_en.pdf

225 UNESCO, Report of the rapporteur on the second session of the World Heritage Committee in Washington, 5-8 September 1978, Paris, 9 October 1978, CC-78/conf.010/10 rev., para. 3. Retrieved from http://whc.unesco.org/archive/1978/cc-78-conf010-10reve.pdf

226 UNESCO, Report of the rapporteur on the twelfth session of the World Heritage Committee in Brasilia, 5–9 December 1988, Paris, 23 December 1988, SC-88/conf.001/13, para. 5. Retrieved from http://whc.unesco.org/archive/1988/sc-88-conf001-13e.pdf

symposium in Quebec City in 1991 resulted in the creation of the Organization of World Heritage Cities (OWHC) the following year.[227] Although the Committee refused its request for official status, the organization participated regularly at World Heritage sessions in order to follow debates and recruit new members to its network following the inscription of urban properties. Near the end of the millennium, two cases drew the attention of unprecedented numbers of individuals and organizations. The Kakadu case in Australia attracted representatives from anti-nuclear groups, indigenous peoples and non-governmental organizations like Friends of the Earth. The proposed salt production plant at the whaling sanctuary at El Vizcaino, Mexico attracted the attention of global organizations like the Natural Resources Defence Council, the International Fund for Animal Welfare, the Environmental Diplomacy Institute and Pro Esteros Mexico. It provoked a flood of over twenty thousand individual protest letters addressed to the World Heritage Centre and the chairperson of the Committee.[228]

In 1999, attendees ranged from well-known international non-governmental organizations with broad interests in protected areas to smaller single-interest groups. In addition to the usual participants, the list includes les Amis du patrimoine du Maroc, the Arch Foundation, l'Association pour la sauvegarde de la Casbah d'Alger, the Fondation patrimoine historique international, High Tech Visual Promotion Centre, the International Council on Metals and the Environment, the International Federation of Shingon Buddhism, the International Fund for Animal Welfare, the Natural Resources Defence Council, the OWHC, and Pro Esteros Mexico.[229]

At the turn of the millennium, the increase in the number and diversity of organizations attending World Heritage sessions illustrates the commitment of civil society to the protection of places of outstanding universal value. Participants came to Committee meetings in force and sought, not always successfully, to have their voices heard. Campaigns to lobby for the protection of specific properties were well-organized and global, often taking advantage of modern communications technology to build momentum. At the local level, groups put pressure on governments to nominate properties for inscriptions and raised alarms about threats to World Heritage Sites. The lack of a formal role for civil society in the World Heritage system is, in Matsuura's view, "something the 1972

---

227   An innovative management guide prepared as part of the symposium's documentation is Herb Stovel, *Safeguarding historic urban ensembles in a time of change: a Management Guide* (Hull, 1991), pp. 1–71.

228   Second letter from the chairperson of the World Heritage Committee to the environmental groups concerning the whale sanctuary of El Vizcaino (Mexico), Paris, 25 May 1999. Retrieved from http://whc.unesco.org/uploads/news/news-151-1.pdf

229   UNESCO, Report of the rapporteur on the twenty-third session of the World Heritage Committee in Marrakesh, 29 November to 4 December 1999, Paris, 2 March 2000, WHC-99/conf.209/22, annex 2, pp. 70-2. Retrieved from http://whc.unesco.org/archive/1999/whc-99-conf209-22e.pdf

Convention should have more carefully looked into." Emphasizing its importance, he continues "Letters from local community members are always very helpful … They draw attention of the World Heritage people to what is officially not reported with regard to the risks. There are risks. So local community members have enormous responsibility in that sense."[230]

This overview of the various players involved in the implementation of the World Heritage Convention reveals the dedication and enthusiasm that each brings to the cause. From relatively modest beginnings, States Parties, technical advisory bodies and UNESCO expanded their activities and resources to deal with the complexity and diversity of World Heritage issues. This growth within the organizations of all official interveners is matched by an increased interest in civil society and its diverse communities. This evolution bears witness to the attractive and promising vision of World Heritage to safeguard and conserve the world's most significant cultural and natural heritage places.

---

230   Canada Research Chair, interview Matsuura.

Chapter 6

Assessment of the World Heritage System: 1972–2000

From the outset to the year 2000, the World Heritage Convention gathered strength as it evolved through years of implementation. Its rising popularity confirmed the interest and need for an international conservation instrument to look after those places of greatest value to the world. By the year 2000, the number of States Parties reached an impressive 161 countries, representing more than three-quarters of the nations of the world. Through key policy and site-specific decisions, the World Heritage system affected the way that heritage values were perceived and conservation strategies were formulated. Part of its success stems from its role as a platform for discussing innovative concepts in national conservation policies and promulgating them with the authority of World Heritage behind them.[1] Its influence on global practice in cultural and natural heritage conservation is undeniable.

As the decades unfolded, the strengths and weaknesses of the World Heritage Convention became more obvious. On the positive side, it contributed to an extraordinary international dialogue on heritage matters, fostering a new understanding of heritage theory and practice. On the other hand, flaws in the listing process, insufficient funds for a robust programme of international cooperation, threats from urbanization and mass tourism as well as incidents of blatant politicization were sobering reminders that reform was necessary. As the new millennium dawned, the high-minded ideals that inspired the initial vision for World Heritage were under pressure and the need for renewed commitment was evident.

---

1   The 1979 Australia ICOMOS guidelines for the conservation of places of cultural significance (Burra Charter, revised in 1981, 1988 and 1999) is a national charter that proposes an enlarged understanding of heritage value. It influenced international practice by dealing with the broad notion of place rather than specific monuments and sites, and by introducing a new concept of cultural significance that includes aesthetic, historic, scientific or social value. The 1999 revision moves towards recognition of intangible and associative values. Retrieved from http://australia.icomos.org/wp-content/uploads/Burra-Charter_1979.pdf

Key Policy and Site-specific Decisions

From 1972 to 2000, the World Heritage governing bodies developed key policies
that affected the implementation of the Convention itself. Their impact was felt
globally because such thinking reverberated around the world and was then
adapted to suit local heritage situations. In Herb Stovel's view, "it was really the
place where most of the important thinking about scientific developments in the
world of conservation took place or started."[2] Among the many policies emerging
in these early decades, the most significant relate to the evolving concept of
outstanding universal value and the development of systematic monitoring.[3]

In terms of determining heritage value, the careful wording for the ten
selection criteria and other requirements shaped the way the concept of
outstanding universal value is interpreted.[4] Subsequent amendments to the
criteria over time are of critical importance since they reflect the evolving notion
of the concept of heritage. As the number of States Parties to the Convention
grew, diverse ideas confronted each other, forging for the first time a global
exchange and understanding of what constitutes heritage. This international
dialogue led to other important decisions affecting the processes for identifying
eligible World Heritage properties, including the introduction of the cultural
landscapes category that bridged the gap between cultural and natural heritage,
the expanded interpretation of authenticity for cultural sites emerging from the
Nara conference, and the creation of an open-ended global strategy to encourage
nominations of different kinds of heritage sites. The concept of heritage value
gradually expanded beyond a focus on the physical places themselves to include
socio-cultural processes. In Catherine Dumesnil's view, openness to new
categories of heritage was a decisive step in the evolution of the Convention.[5]
Jukka Jokilehto shares this view: "I think that one of the issues that has come

---

2   Canada Research Chair on Built Heritage, Université de Montréal, audio interview
of Herb Stovel by Christina Cameron, Ottawa, 5 April 2011.

3   Important evaluations of the World Heritage system and its policies took place in the
run-up to the twentieth anniversary in 1992 and later. See Jim Thorsell and Jacqueline Sawyer
(eds), *World Heritage Twenty Years Later* (Gland, 1992), pp. 1–191; Azedine Beschaouch,
Towards an evaluation of the implementation of the Convention, Paris, December 1991,
WHC-92/conf.002/3, annex IV. Retrieved from http://whc.unesco.org/archive/1992/whc-
92-conf002-3e.pdf; Jim Thorsell, *World Heritage Convention: effectiveness 1992–2002
and lessons for governance* (Gland, 2003), pp. 1–22.

4   Jukka Jokilehto et al., *The World Heritage List: What is OUV? Defining
the Outstanding Universal Value of Cultural World Heritage Properties* (Berlin, 2008),
pp. 1–111.

5   Canada Research Chair on Built Heritage, Université de Montréal, audio interview
of Catherine Dumesnil by Christina Cameron and Mechtild Rössler, Paris, 25 November
2009. "Je vois l'ouverture plus large de la Liste à des catégories nouvelles. Ça a influencé
la mise en œuvre de la Convention. Je crois que c'est plutôt l'ouverture à de nouvelles
catégories qui a été une étape décisive."

out, especially in the 1990s, was this increasing attention to vernacular ... cultural landscapes and so on... and the increasing interest in intangible heritage associated with physical heritage."[6]

On the conservation front, although countries were slow to transform into reality the management requirements that began to appear in the Operational Guidelines in the early 1980s, important policy achievements were made. From a policy perspective, the establishment of a formal monitoring model is a major advance. After two decades of exploration and endless discussions at World Heritage Committee sessions, the decision at the 1997 General Assembly of States Parties to set in place the periodic reporting process established a global benchmark for monitoring activities. This successful outcome is the result of international collaboration by all those involved in World Heritage to build a systematic approach to understanding and improving the state of conservation of internationally significant sites.[7]

Beyond formulating broad policies on outstanding universal value and conservation, the Committee took decisions on individual sites or groups of sites that set important markers for the subsequent implementation of the Convention. In terms of inscriptions, three precedents stand out as particularly significant. They concern sites of conscience, sites associated with persons and sites inscribed for their human creativity. In terms of collective responsibility for protecting World Heritage Sites, events at the Old City of Dubrovnik (Croatia) and Kakadu National Park (Australia) serve to illustrate the scope and limits of the authority of the World Heritage Committee and the Convention.

Sites of Conscience

Sites of conscience commemorate the darker side of human behaviour and usually arouse strong passions. At the outset, some delegations took the position that "the intent of the Convention on the cultural side has always been to commemorate man's great creative activities and not his negative accomplishments."[8] Others disagreed. The Committee set an early precedent with the 1978 inscription of the slave site at the Island of Gorée (Senegal) and the 1979 listing of the concentration camp at Auschwitz-Birkenau (Poland).[9] Both these sites were listed under criterion

---

6   Canada Research Chair on Built Heritage, Université de Montréal, audio interview of Jukka Jokilehto by Christina Cameron and Mechtild Rössler, Rome, 5 May 2010.

7   UNESCO, Summary record of the eleventh General Assembly of States Parties to the Convention concerning the protection of the world cultural and natural heritage, Paris, 27-28 October 1997, Paris, 18 December 1997, WHC-97/conf. 205/7, paras. 22–4. Retrieved from http://whc.unesco.org/archive/1997/whc-97-conf205-7e.pdf

8   Letter from Peter H. Bennett to M. Raletich-Rajicic, Ottawa, 24 November 1978, United States National Park Service, World Heritage archives, file 1976–1978, p. 1.

9   The site was renamed in 2007 as Auschwitz Birkenau German Nazi Concentration and Extermination Camp (1940–1945).

(vi) alone for values associated with negative events in the world's history.[10] This paved the way for consideration in the 1990s of two other sites of conscience: the Hiroshima Peace Memorial (Genbaku Dome) (Japan) and Nelson Mandela's prison at Robben Island (South Africa).

In 1996, the Hiroshima Peace Memorial was inscribed at the Merida session in a tense political atmosphere exacerbated by numerous television crews waiting in the corridor for the Committee's decision. The site was listed as a significant witness to the most destructive force ever created by humankind and as a symbol of hope for world peace and the elimination of nuclear weapons. Two Committee members openly disagreed with the inscription. China expressed its reservations and the United States of America dissociated itself from the decision on the basis of lack of historical perspective. The United States further stated its general view that war sites were outside the scope of the Convention.[11] Ray Wanner, a desk officer at the State Department at the time, explains that the American position was influenced by the controversial Enola Gay exhibition at the Smithsonian earlier that year. Although Wanner was tasked with developing arguments to oppose the inscription, he expresses his admiration for John Reynolds, then working at the United States National Park Service, who supported it on the basis that it was "the right thing to do" and for ICOMOS coordinator Henry Cleere who refused to budge from a favourable technical evaluation.[12] The 1999 inscription of Robben Island occurred in calmer circumstances. There was unanimous Committee support for listing this site of conscience as a symbol of the triumph of the human spirit, of freedom and of democracy over oppression. Touri, Chairperson of the Committee which inscribed Robben Island, relishes the successful outcome: "This finally marked the end of apartheid in a concrete way at an international level and of segregation that might exist elsewhere."[13]

Von Droste attributes the inscription of sites of conscience to a transformation of the Convention from a system for listing exceptional properties to one of protecting symbols of global ethics. He suggests that the Committee is an "invisible

---

10   For further reflection on the intangible dimensions of this criterion, see UNESCO, Report of the international World Heritage expert meeting on criterion (vi) and associative values held in Warsaw, 28–30 March 2012. Retrieved from http://whc.unesco.org/uploads/events/documents/event-827-15.pdf

11   UNESCO, Report of the rapporteur on the twentieth session of the World Heritage Committee in Merida, 2–7 December 1996, Paris, 10 March 1997, WHC-96/conf.201/21, para. C.1, annex 5. Retrieved from http://whc.unesco.org/archive/1996/whc-96-conf201-21e.pdf

12   Canada Research Chair on Built Heritage, Université de Montréal, audio interview of Ray Wanner by Christina Cameron, Springfield, 18 May 2011.

13   Canada Research Chair on Built Heritage, Université de Montréal, audio interview of Abdelaziz Touri by Christina Cameron and Petra Van Den Born, Paris, 22 June 2011. "Robben Island est une autre réussite parce que elle a fait l'unanimité autour de ce site là où était prisonnier Mandela etc. et donc qui a mis fin finalement d'une manière concrète au niveau international à l'apartheid et donc à la ségrégation qui pouvait exister ailleurs, n'est-ce pas?"

stakeholder on behalf of the interests of future generations." Citing as symbols of global ethics the sites at Auschwitz, Robben Island and the Island of Gorée (the need to respect basic human rights) and Hiroshima (the need to deal with the threats of misuse of modern technology) he continues:

> The misconception from the beginning is that people thought we [were] protect[ing] the wonders of the world. This is not true at all. We are not protecting in World Heritage the wonders of the world ... We are dealing here with a tremendous phenomenon that something which is locally important is recognized by UNESCO as important for humanity and transported through the prestige of the UN system to ... part of globalization.[14]

Sites and Individuals

On the question of recognizing individuals under the World Heritage Convention, the Committee began with a clear policy stance. In the preparatory phase in the early 1970s, ICOMOS had recommended including "properties associated with and essential to the understanding of globally significant persons," giving as an example the laboratory of Thomas Edison in the United States.[15] While the original wording for criterion (vi) included properties associated with persons of outstanding historical importance or significance, the Committee heeded Michel Parent's advice against creating "a sort of competitive Honours Board for the famous men of different countries"[16] that could foster a nationalistic approach in contradiction to the universal perspective of the Convention. In 1980, the Committee removed "persons" from criterion (vi).

Notwithstanding, the Committee subsequently ignored its own rules and set precedent in 1984 by listing a group of buildings called the Works of Antoni Gaudi (Spain) to mark his outstanding architectural creativity in the late nineteenth and early twentieth centuries. A decade later, the Committee made another such inscription, marking the exceptional contribution of Renaissance architect Andrea Palladio by inscribing twenty-six buildings in a site known as the City of Vicenza and the Palladian Villas (Italy). Other States Parties with ambitions to recognize

---

14   Canada Research Chair on Built Heritage, Université de Montréal, audio interview of Bernd von Droste by Christina Cameron and Mechtild Rössler, Paris, 1 February 2008.

15   UNESCO, Final report of informal consultation of intergovernmental and non-governmental organizations in the implementation of the Convention concerning the protection of the world cultural and natural heritage, Morges, 19–20 May 1976, CC-76/WS/25, annex III, p 5. Retrieved from http://unesdoc.unesco.org/images/0002/000213/021374eb.pdf

16   Michel Parent, Comparative study of nominations and criteria for world cultural heritage, principles and criteria for inclusion of properties on the World Heritage List, Paris, 11 October 1979, CC-79/conf.003/11 annex, p. 22. Retrieved from http://whc.unesco.org/archive/1979/cc-79-conf003-11e.pdf

architects and builders continue to cite these early examples as benchmarks, in spite of the Committee's clear policy decision in 1980 to exclude individuals from consideration.

Sites and Human Creativity

In the first decades, the concept of creativity found in cultural criterion (i) essentially referred to aesthetics. Properties were required to "represent a unique artistic achievement, a masterpiece of the creative genius."[17] Following the adoption of the global strategy, the wording was changed in 1996 to diminish the aesthetic emphasis, so that properties thereafter were required to "represent a masterpiece of human creative genius."[18] A test case at the next session in 1997 demonstrates that not all States Parties agreed with the new orientation. In considering Las Médulas (Spain), the argument was made that criterion (i) could be applied to recognize innovative Roman knowledge of hydraulic technology that was used to extract gold from this mountainous landscape in the first and second centuries AD. During the debate, several delegates disagreed on two grounds: the absence of beauty and environmental degradation at the site. Despite these objections, the Committee accepted the argument that the site met the requirement for human creativity and validated its inscription under criterion (i). Three States Parties formally dissociated themselves from the decision. Germany and Finland supported the delegate of Thailand who is reported to have said that "in applying criterion (i), among others, to signify human creativity, he could only consider this site as a result of human destructive activities as well as harmful to the noble cause of environmental promotion and protection."[19] The decision stands as a benchmark for a broad interpretation of human creative genius.

Sites and the Limitations of the Convention

Two threatened World Heritage Sites tested the powers and limitations of the Convention. In one instance, the Committee acted on behalf of the site in the absence of a functioning national government; in the other example, it held back and balked at acting against the wishes of a concerned State Party.

---

17   UNESCO, Operational Guidelines for the implementation of the World Heritage Convention, February 1994, WHC/2/revised, para. 24 a (i). Retrieved from http://whc.unesco.org/archive/opguide94.pdf

18   UNESCO, Operational Guidelines for the implementation of the World Heritage Convention, February 1996, WHC/2/revised, para. 24 a (i). Retrieved from http://whc.unesco.org/archive/opguide96.pdf

19   UNESCO, Report of the rapporteur on the twenty-first session of the World Heritage Committee in Naples, 1–6 December 1997, Paris, 27 February 1998, WHC-97/conf.208/17, para. VIII.9. Retrieved from http://whc.unesco.org/archive/1997/whc-97-conf208-17e.pdf

Figure 6.1   Old City of Dubrovnik, Croatia

The first case involves the Old City of Dubrovnik (Croatia), a mediaeval walled city on the Dalmatian coast. It was nominated by the government of Yugoslavia in 1979 and listed for its exceptionally well-preserved fortifications and architectural monuments in Gothic, Renaissance and Baroque styles. In 1991, when the dissolution of the Socialist Federal Republic of Yugoslavia launched the Balkan Wars, the historic centre of Dubrovnik was damaged by bombardments, thereby challenging the determination of the international community to make good on its commitment to protect World Heritage Sites. Dubrovnik was not on the formal agenda of the Committee's 1991 session in Carthage but Henri Lopes, Assistant Director-General for Culture, put it there. In his opening remarks, Lopes "expressed UNESCO'S dismay about the destruction already caused by this conflict in the old town of Dubrovnik ... which has to be safeguarded in conformity with the stipulations of the World Heritage Convention."[20] Hans Caspary, a delegate from Germany who attended the 1991 session recalls: "We started on Friday and in the newspapers, on the first page, there was the report on Dubrovnik, the shelling of Dubrovnik ... During the bombing, an observer from UNESCO was on-site. And after the bombing ended, money was immediately sent for the damaged roofs."[21]

---

20   UNESCO, Report of the rapporteur on the fifteenth session of the World Heritage Committee in Carthage, 9–13 December 1991, Carthage, 12 December 1991, SC-9/conf.002/15, para. 6. Retrieved from http://whc.unesco.org/archive/1991/sc-91-conf002-15e.pdf

21   Canada Research Chair on Built Heritage, Université de Montréal, audio interview of Hans Caspary by Christina Cameron and Mechtild Rössler, Mainz, 1 July 2011. "On

Committee members were torn between the evident need to protect a threatened World Heritage Site and a strong challenge from an observer who identified himself as a counsellor from the Yugoslav embassy in Tunisia. He argued that the Committee did not have the right to act unilaterally against the wishes of his government nor should it single out this site alone in the Balkan war zone. Placed in this uncomfortable political situation when governance of the territory was in flux, the Committee rose to the occasion and chose to act on behalf of the site. It decided to inscribe Dubrovnik on the List of World Heritage in Danger using article 11.4 of the Convention.[22] The decision is carefully crafted to take into account the concerns raised by all parties:

> Aware of the fact that it represents 123 [sic] States, including Yugoslavia, which are signatories of the Convention, the Committee expressed deep concern about the armed conflict devastating a region that comprises several sites inscribed on the World Heritage List, in particular the Old City of Dubrovnik. It decided to urge the parties in conflict to do their best so that a ceasefire which allows as soon as possible for the repair of the damage already caused in the fighting area, in particular in Dubrovnik, in response to the appeal by the Director-General of UNESCO for international solidarity.[23]

The Committee justified its decision on the grounds of the "state of exceptional emergency," noting that this action should be seen as "the affirmation that all States Parties to the Convention are involved in this situation where a World Heritage city was seriously damaged by an armed conflict."[24]

With regard to what he calls the *ex officio* placing of Dubrovnik on the Danger List, Francesco Francioni observes that "the Convention is an extremely important tool in times of crisis, in times when there is grave resonance of an attack on a natural or cultural site." He calls particular attention to the fact that the decision was taken by the Committee and not by countries from the "dissolving Yugoslavia":

---

a commencé le vendredi et dans les journaux, sur la premier page, il y avait le rapport sur Dubrovnik, le bombardement de Dubrovnik ... on a envoyé un observateur, durant le bombardement, un observateur de l'UNESCO était sur place. Et après, à la fin des bombardements, on a tout de suite envoyé de l'argent pour les toits endommagés."

22   UNESCO, Convention concerning the protection of the world cultural and natural heritage, Paris, 1972, article 11.4. Retrieved from http://whc.unesco.org/archive/convention-en.pdf

23   UNESCO, Report of the rapporteur on the fifteenth session of the World Heritage Committee in Carthage, 9–13 December 1991, Carthage, 12 December 1991, SC-9/conf.002/15, para. 28. Retrieved from http://whc.unesco.org/archive/1991/sc-91-conf002-15e.pdf

24   UNESCO, Report of the rapporteur on the fifteenth session of the World Heritage Committee in Carthage, 9–13 December 1991, Carthage, 12 December 1991, SC-9/conf.002/15, paras. 29-30. Retrieved from http://whc.unesco.org/archive/1991/sc-91-conf002-15e.pdf

Figure 6.2   Jabiluka mine site at Kakadu National Park, Australia

The great strength of the World Heritage is in establishing the idea that a site that has been placed in the List somehow creates a sort of *ergo omnis* obligation by states. We have an obligation that is integral, is not bilateral between one state and UNESCO but is a sort of public law obligation that one takes toward the international community as a whole. Also in that sense there is also a reciprocal sense of obligation to intervene, to help, to respond. This has been always true in situations of crisis.[25]

For Dubrovnik, the Committee used its authority to protect a threatened site in the absence of a viable State Party. The case of Kakadu National Park (Australia) tested the Committee's authority to put a site on the World Heritage List in Danger with the active disagreement of the concerned State Party. The issue emerged in the context of a proposed uranium mine within Kakadu National Park. The property had been listed as World Heritage under both natural and cultural criteria. In the State Party's nomination file, three mining leases were excised from the property as enclaves, a proposal that was accepted by the technical advisors and the Committee at the time. When news of the possible opening of a new mine in one of the enclaves was reported to the Committee in 1996, it expressed concern for potential negative impacts on the site's ecosystem and on the cultural heritage of the Mirrar indigenous community. The intense debate that ensued tested but did not resolve the issue of the Committee's powers.

---

25   Canada Research Chair on Built Heritage, Université de Montréal, audio interview of Francesco Francioni by Christina Cameron and Mechtild Rössler, Rome, 5 May 2010.

Thailand held the view, shared by some Committee members, that "the World Heritage Committee has the authority under the Convention (article 11.4) to place any World Heritage property threatened by serious and specific dangers on the List of World Heritage in Danger at any time in case of urgent need." Citing article 11.4, the advisory bodies joined forces to urge the Committee to use its powers to inscribe the site on the Danger List regardless of the State Party's resistance. On the other hand, Australia used the same article to show that the Committee did not have the authority to do so without a request for assistance from the State Party, a request that Australia did not intend to make.[26] Early on, Ralph Slatyer, Chairperson of the 1983 Committee, had analysed article 11.4 and anticipated this sort of stand-off, observing that "this wording creates a problem since States parties may not need to request assistance or may not wish to do so. The property could then not be placed on the World Heritage List in Danger though it may be under threat." Acknowledging that "it is only the State party which can guarantee the protection of property, so listing without its consent may not assist protection," he nonetheless saw Danger listing as key to mobilizing international assistance to save a property.[27]

The battle lines were drawn. An extraordinary meeting of the World Heritage Committee was scheduled to further discuss the matter. The Committee sent its chairperson to Australia to see the situation on the ground; Australia launched a diplomatic offensive in the foreign capitals of Committee members to gain support for its position; non-governmental organizations moved into high gear targeting those same Committee members; and the Mirrar indigenous group came to Paris to plead its case. At the 1999 extraordinary session of the Committee, States Parties arrived with binders full of technical analysis and confidential government legal advice tucked in their briefcases.[28] When the session finally concluded, the Committee decided not to put Kakadu National Park on the Danger List, asking instead that Australia provide a progress report the next year on the scientific issues, the cultural mapping of the mining site and the social benefits proposed for the Aboriginal communities of Kakadu.[29] Francioni expresses his disappointment that obligations under the Convention were not clarified:

---

26   UNESCO, Report of the rapporteur on the twenty-second session of the World Heritage Committee in Kyoto, 30 November–5 December 1998, Paris, 29 January 1999, WHC-98/conf.203/18, para. VII 28. Retrieved from http://whc.unesco.org/archive/1998/whc-98-conf203-18e.pdf

27   Ralph Slatyer, Address by the outgoing chairman of the World Heritage Committee, in UNESCO, Report of the rapporteur on the seventh session of the World Heritage Committee in Florence 5–9 December 1983, Paris, January 1984, SC/83/conf.009/8, annex II, pp. 6–7. Retrieved from http://whc.unesco.org/archive/1983/sc-83-conf009-8e.pdf

28   UNESCO, Report of the Rapporteur on the third extraordinary session of the World Heritage Committee in Paris, 12 July 1999, Paris, 19 November 1999, WHC-99/conf.205/5rev. Retrieved from http://whc.unesco.org/archive/1999/whc-99-conf205-5reve.pdf

29   UNESCO, Report of the rapporteur of the third extraordinary session of the World Heritage Committee, Paris, 12 July 1999, Paris, 19 November 1999, WHC-99/conf.205/5rev,

I would have liked to see more courage, a bolder Committee ... Kakadu was a very important case because of the ... natural value but also because of the local communities ... That was a decision I would have liked to see on the part of a treaty body like the World Heritage Committee that unfortunately was not made.[30]

So the question raised by the proposed mine at Kakadu National Park never fully tested the Committee's power to act against the wishes of a concerned State Party in situations of crisis. The ambiguity remains.[31]

Strengths and Weaknesses

In terms of overall assessment of the World Heritage Convention from its inception until the 2000 reforms, strengths and shortcomings became more apparent as implementation unfolded. Although interviewees point with regret to some of the weaknesses of the system, they celebrate the many successes with admiration and enthusiasm. Some troubling issues include the structural design of the listing process, inadequate resources for international assistance, inappropriate mass tourism and creeping politicization. On the other hand, the success in fostering a global conversation and the development of new dimensions to heritage theory and practice are achievements on a grand scale.

Listing Process

Despite the lofty ambition to identify cultural and natural heritage of outstanding universal value and efforts to achieve a representative, balanced and credible World Heritage List, several interviewees believe that the listing process is flawed. Léon Pressouyre is pragmatic about the fact that the Convention requires sites to be presented by national governments, but regrets that certain sites will therefore probably never be brought forward for political reasons. He suggests that this approach limits the achievement of a truly global list:

---

pp. 24–5. Retrieved from http://whc.unesco.org/archive/1999/whc-99-conf205-5reve.pdf

30   Canada Research Chair, interview Francioni.

31   During major revisions to the Operational Guidelines at the turn of the millennium, the issue of State Party consent for Danger listing came up again. UNESCO's legal advice argues that case law (decisions of the Committee) demonstrates that it is not necessary to have State Party consent. On the other hand, many States Parties believed the contrary. See the debate in UNESCO, Report of the rapporteur on the sixth extraordinary session of the World Heritage Committee in Paris, 17-22 March 2003, summary record, Paris, 1 June 2004, WHC-03/6 ext.com/inf.8, paras. 4.1–4.91. Retrieved from http://whc.unesco.org/archive/2003/whc03-6extcom-inf08e.pdf and the decision in UNESCO, Decisions adopted by the World Heritage Committee at its sixth extraordinary session, Paris, 17–22 March 2003, Paris, 27 May 2003, WHC-03/ext.com/8, para. 6 ext. com. 4. Retrieved from http://whc.unesco.org/archive/6extcom.htm#6extcom4

The World Heritage Convention will not achieve its potential as long as properties that should be listed are not there. So it is true that if the process adopted had been a process of selection by experts and not one of national selection, one would perhaps have arrived more rapidly at a balanced result. [32]

He admits that he naively believed for many years that countries could be convinced to present nominations of sites of obvious global value and is saddened by their absence from the list. Hans Caspary points to a bias in favour of majority cultures as an element of this phenomenon: "Countries control the decision on whether or not to nominate something from their state ... It is always the national culture that dominates tentative lists ... It is a political decision to ignore minority cultures. The Convention has no tools to force these countries not to forget the minority."[33]

Jim Collinson is also critical of a process that impedes the achievement of a truly global list:

In the case of a World Heritage List, over time you'd like to think that, if you put all of the sites together, you would have a good description of the evolution of the world and the place of human population in it. I'm not sure you get that by designating sites that are specific to a certain thing at a certain period of time. I'm not saying that's wrong. I just don't think that you can put them together and expect to find what you might like to find in terms of a complete perspective of the world and its evolution ... So we end up with a series of sites without the web that holds them together.[34]

32 Canada Research Chair on Built Heritage, Université de Montréal, audio interview of Léon Pressouyre by Christina Cameron and Mechtild Rössler, Paris, 18 November 2008. "Je pense que la Convention du patrimoine mondial n'aura pas atteint son potentiel tant que des biens qui devraient impérativement figurer sur la Liste n'y seront pas. Alors, c'est vrai que si la démarche adoptée avait été une démarche de sélection par experts et pas une démarche de sélection nationale, on serait arrivé peut-être plus vite à un résultat équilibré ... Alors, pendant très longtemps on s'est dit qu'on arriverait à convaincre un certain nombre de pays de présenter des choses évidentes ... Je pense qu'il y a encore des choses très importantes qui devraient entrer sur, d'une manière ou d'une autre, sur la Liste du patrimoine mondial qui n'y entrent pas pour des raisons politiques et qui doivent entrer."
33 Canada Research Chair, interview Caspary. "Que qu'on laisse la décision de nominer ou de ne pas nominer quelque chose au pays ... C'est toujours la culture nationale qui domine dans les listes tentatives ... C'est une décision politique de ne pas penser aux cultures minoritaires. Il ya des pays collaborent avec leurs cultures minorités, d'autres ne les font pas. La Convention n'a aucun instrument pour obliger ces pays de ne pas oublier la minorité."
34 Canada Research Chair on Built Heritage, Université de Montréal, audio interview of Jim Collinson by Christina Cameron, Windsor, 12 July 2010.

Funding for International Assistance

The pioneers reserve their harshest criticism for the failure of the World Heritage system to live up to its goal of international cooperation. Jane Robertson Vernhes bluntly states that the World Heritage Fund is completely inadequate:

> I don't think anybody did their sums, quite frankly. When they were talking about money for the Fund to help save the world's natural and cultural heritage, if only somebody had done their sums on the back of an envelope ... and added it all up, what does that come to? It [the World Heritage Fund] comes to a silly little sum when you look at the World Heritage and what the needs are."

The implication, as she points out, is partnering with foundations, benefactors, the private sector and so forth in order to get "the competent people in the right places to do the right things."[35]

Carmen Añón Feliu regrets that international solidarity is only words on paper. She finds it shameful that money is wasted on international meetings and expert missions when the real needs are at the community level. "The Committee gives aid to under-developed countries and sometimes it only goes for experts' travel to the site."[36] Luxen concurs with her that the little aid that is available is often not appropriately allocated, explaining that "solidarity is not only about repairing a church ... Real solidarity is to see how people ... could improve their living conditions through adapting their heritage ... The goal of international cooperation is to ensure that people's living conditions are dignified."[37]

Rob Milne reproaches the Committee and the secretariat for failing to harness external resources effectively:

---

35   Canada Research Chair on Built Heritage, Université de Montréal, audio interview of Jane Robertson Vernhes by Christina Cameron and Mechtild Rössler, Paris, 24 November 2009.

36   Canada Research Chair on Built Heritage, Université de Montréal, audio interview of Carmen Añón Feliu by Christina Cameron, Madrid, 18 June 2009. "Il n'existe pas, la solidarité internationale. Ce sont des mots sur le papier. Mais vraiment quand on voit ce qu'on dépense. On dépense pour la guerre, on dépense pour n'importe quoi, une exhibition, une réunion, des congrès. On fait des réunions qui coûtent des milliards de ... pour ne dire rien et n'arriver à rien ... Et ça c'est l'aide que le Comité donne à des pays sous-développés et que parfois, ils vont seulement dans les voyages que font les experts pour un bien. Vergogne nationale et internationale."

37   Canada Research Chair on Built Heritage, Université de Montréal, audio interview of Jean-Louis Luxen by Christina Cameron, Leuven, 26 March 2009. "La solidarité ne doit pas fonctionner uniquement pour aller réparer une église ... La vraie solidarité, c'est de voir comment des gens ... peuvent améliorer leur condition de vie en adaptant leur patrimoine ... c'est un problème de coopération internationale pour veiller à ce que les conditions de vie des gens soient dignes."

I think it is almost there for the asking but somebody has not reached out adequately ... Whether people recognize them or admit to them, there is probably an issue eating away and corroding the values in almost all properties. Whether it is a question of time before the values are lost or whether for some of these corrosive elements, no matter what anybody were to do, the values still will be lost. I think for every site lost, it is a failure of the Convention.[38]

Agreeing that the World Heritage Fund is "less than enough" and that efforts to reach out are inadequate, Natarajan Ishwaran observes that "the reality doesn't match the noble aspirations ... The level of resourcing to support World Heritage Sites, particularly in less developed countries, is way behind what is expected." He goes on to criticize the lack of connection between threatened sites and important international funding agencies: "It's nice to say, GEF [Global Environment Facility] and all these big financial instruments give priority to World Heritage Sites in their own decision-making, but the problem is, the Member States don't know that the GEF is prioritizing World Heritage Sites ... There is no connection between GEF and the World Heritage Committee."[39]

Eidsvik agrees. "The amount of money that the World Heritage Convention had in the World Heritage Fund is really negligible ... So that was not very effective at that time." He goes on to make a similar point about opportunities for partnerships: "But when the Convention on Biodiversity was written, and suddenly you have an involvement by the World Bank, you've got an involvement by another Convention, which says within it to protect global biodiversity." Eidsvik makes the link between sites of outstanding value from a biodiversity point of view and the World Bank fund: "So more money has flowed in recent years, but not because of the World Heritage Convention directly, but because the opportunity has been there to use it in accompaniment with other conventions."[40]

Mass Tourism
During the period of this study, tourism gradually came to be seen as a negative force for the conservation of World Heritage Sites. In the early years of implementation, it was not an important consideration and management issues related to tourism were rare. In a review of Committee records from 1977 to 1986, tourism is only mentioned six times. A participant from the outset, Vlad Borrelli confirms this impression: "Tourism was not the disaster that it later became. There was no mass tourism then. At the beginning, it was not even

---

38   Canada Research Chair on Built Heritage, Université de Montréal, audio interview of Rob Milne by Christina Cameron and Mechtild Rössler, Paris, 2 March 2009.

39   Canada Research Chair on Built Heritage, Université de Montréal, audio interview of Natarajan Ishwaran by Christina Cameron and Mechtild Rössler, Paris, 24 November 2009.

40   Canada Research Chair on Built Heritage, Université de Montréal, audio interview of Hal Eidsvik by Christina Cameron, Ottawa, 3 July 2009.

a question ... On the contrary, one wanted to make the heritage sites better known."[41] But by the 1990s, the phenomenon of mass tourism had become a prime motivator to seek World Heritage status. In 1993, the World Heritage Centre along with UNEP carried out a survey on tourism at natural World Heritage Sites. The results showed that site managers regarded tourism as a key management issue.[42] Despite the perceived economic benefits, mass tourism can have serious negative impacts on the physical condition of sites and quality of visitor experience. Guo sees the economic motivation for inscribing World Heritage Sites as a failure of the Convention, noting that sometimes local communities "don't care what the capacity of the site is, what should be the proper limitation of that. They never care. They prefer more and more tourists. And I think this is the wrong trend and practice of the Convention."[43] Luxen sums it up succinctly:

> For me, tourism is not about solidarity. As I see it, tourism endangers heritage. Some argue that it brings resources, etc. but it is dangerous for heritage, physically and at the intangible level. The quality of the site can be killed by visitors ... I know lots of cases where that was the major reason that people fought for listing. For me, it is an aberration. It is commercialization and exploitation of heritage which sometimes ends, if the Committee is not vigilant enough, in the Disneyland effect, a falsification where one kills authenticity because the goal is not conservation but has simply become a tool for tourism promotion.[44]

---

41   Canada Research Chair on Built Heritage, Université de Montréal, audio interview of Licia Vlad-Borrelli by Christina Cameron and Mechtild Rössler, Rome, 6 May 2010. "Quand on a commencé, probablement le tourisme n'était pas ce désastre qu'il est devenu après en fait. C'était pas ce grand tourisme de masse et alors au début je crois qu'on s'est pas posé la question ... Au début, on essayait au contraire de faire connaître les lieux du patrimoine."

42   Arthur Pederson, *Managing Tourism at World Heritage Sites: a Practical Manual for World Heritage Site Managers* (Paris, 2002), p. 16.

43   Canada Research Chair on Built Heritage, Université de Montréal, audio interview of Guo Zhan by Christina Cameron, Brasilia, 3 August 2010.

44   Canada Research Chair, interview Luxen. "Pour moi le tourisme, ce n'est pas la solidarité. Comme je connais, comme je vois le tourisme, le tourisme est un danger pour le patrimoine. ... on explique, ça apporte des ressources, etc. . mais c'est un danger pour le patrimoine, physiquement et au plan immatériel. La qualité du site peut être tuée par la fréquentation touristique. ... je connais beaucoup de cas dans lesquels ça a été la raison majeure pour laquelle les gens se battaient pour l'inscription. Pour moi, ça c'est une dérive, c'est la commercialisation, l'instrumentalisation du patrimoine qui aboutit parfois à, si le Comité n'est pas assez vigilant, à ce qu'on appelle la Disneyland effet, du faux, on tue l'authenticité parce que l'objectif n'est pas la conservation mais est devenu simplement un outil de promotion touristique."

Luxen distinguishes between tourism exploitation and appropriate economic development. He confirms that heritage practitioners have been reluctant to openly promote the use of heritage for economic development, illustrating his point by describing the situation in 1985 when the European ministers of culture "found it a bit low and vulgar to deal with the economy of heritage. It is quite an extraordinary change now. Everyone recognizes that heritage has a value for development." He concludes that "one recognizes today that, independent of tourism, well-conserved heritage is an asset for balanced human development."[45]

Politicization

As implementation proceeded, several cases led to situations where national interests trumped a global perspective.[46] Most observers choose the mid-1990s as the period when political considerations increased significantly. Harald Plachter from the German delegation recalls positively that "in 1992 there was equilibrium between specialists and diplomats and administrators."[47] At about the same time, however, Pressouyre observes that the system "clearly subordinates 'scientific' choices to 'political' choices" and suggests that, as a counterbalance, UNESCO might consider the creation of an advisory scientific council of international experts from academic and scientific institutions.[48]

Francioni explains the reasons behind what he calls a creative tension between sovereignty and general interest:

> The fact that the Convention has this very interesting institutional framework combining the governmental representatives with the scientific components of IUCN and ICOMOS, I think that is the most important substantive expression of this idea of the general interest of humanity. Obviously then, the story of the Convention is a constant tension between this interest and the national claim, the sovereignty claims, that States are always reluctant to relinquish. But that is the interesting part of the whole genesis and evolution of the Convention.[49]

---

45   Canada Research Chair, interview Luxen. "Les ministres de la Culture, en 85, trouvaient que c'était un peu subalterne et vulgaire de s'occuper de l'économie du patrimoine. C'est quand même un changement extraordinaire maintenant. Tout le monde reconnaît qu'un patrimoine est une valeur de développement … On reconnaît aujourd'hui que, indépendamment du tourisme, un patrimoine bien conservé est un atout pour un développement humain équilibré."

46   See chapter 5 for details on the cases of Auschwitz-Birkenau, the Old City of Jerusalem, the Wet Tropics of Queensland, W National Park and Kakadu National Park.

47   Canada Research Chair on Built Heritage, Université de Montréal, audio interview of Harald Plachter by Christina Cameron and Mechtild Rössler, Mainz, 1 July 2011.

48   Léon Pressouyre, *The World Heritage Convention, Twenty Years Later* (Paris, 1996), p. 46.

49   Canada Research Chair, interview Francioni.

Dawson Munjeri believes that politicization undermines the original purpose of the Convention:

> Increasingly this common approach seems to be dissipating. We now have, I'm sorry to say, … but I have no doubt, I don't mind it being said, the original intentions seem to be giving way more to the political dimensions. That which pushed us, most of us, to really look at the Convention as a solution to many of our situations is increasingly moving away." [50]

Robertson Vernhes agrees: "It's failing has been its politicization … I am being very candid about this because I have seen it … That's a shame because I think that its value has been its prestige, its selectivity."[51] Munjeri regrets the destructive force of nationalism, when "State Parties are so concerned about themselves, don't care what is happening elsewhere and they want to push their case to the detriment of the international community," and calls for renewed commitment to the Convention. "Let's collectively come together and that way we will have the necessary critical mass. If we go back to that spirit, then it will work."[52]

Pressouyre looks at the impact of national interests from the perspective of stewardship. Noting that the intergovernmental structure of the Convention means that individual states are responsible for properties within their territories, he regrets that "the concept of oversight of these sites by the international community, which is in theory admissible, is not put into practice." He observes that the international community is limited to drawing attention to threats to properties but has no real power:

> This means that, on the one hand, there is a generous idea that belongs perhaps to the twenty-first century to the effect that it is [the responsibility] of everyone, and a less generous idea which belongs to history which says that these are nations, these are State Parties who look after the sites that they have accepted to propose for the World Heritage List.[53]

---

50   Canada Research Chair on Built Heritage, Université de Montréal, audio interview of Dawson Munjeri by Christina Cameron, Brasilia, 3 August 2010.

51   Canada Research Chair, interview Robertson Vernhes.

52   Canada Research Chair, interview Munjeri.

53   Canada Research Chair, interview Pressouyre. "Il s'agit d'une Convention intergouvernementale, que les biens sont proposés par des États, que ce sont les États qui sont chargés de leur conservation et que le concept d'un contrôle de la communauté internationale sur ces biens, qui est théoriquement admis, n'est pas mis en pratique … Ce qui fait que il y a, d'une part, une idée très généreuse qui appartient peut-être au XXIe siècle, c'est-à-dire, c'est à tout le monde et, une idée moins généreuse, qui appartient à l'histoire qui dit, ce sont les nations, ce sont les États qui s'occupent chacun des biens qu'ils sont acceptés de proposer sur la Liste du patrimoine mondial."

According to some, the Convention has not created sufficient accountabilities for States Parties. Francioni criticizes the Convention for failing to establish clear obligations:

> It is a Convention that is based on establishing a system of great advantages for the States ... international cooperation, visibility, exchange of information and giving great prestige to certain areas. But of course it is very weak from the point of view of the obligations ... I think more could be done without amending the Convention by simply having Operational Guidelines that would be more stringent, making the countries more responsible. But you need very strong leadership on the part of both UNESCO and the members of the Committee. If the Committee then turns out to be a cosy group of States that are very much concerned in preserving, in sheltering themselves from more penetrating scrutiny, then it would be difficult.[54]

Phillips underscores the importance of embedding in the public and political conscience the idea that World Heritage Sites "are there for posterity and for the whole world and we manage them with that responsibility and not exclusively for our own national, indeed even local interest. And again that comes back to how countries interpret the Convention within their own bounds."[55]

Fostering a Global Conversation

An outstanding achievement of the Convention is its catalytic role in raising public awareness of heritage issues. In evaluating the Convention, several interviewees noted the world-wide influence of the idea of World Heritage. As Robertson Vernhes puts it, success lies in raising public awareness: "That there are sites of outstanding universal value that the whole of humanity should somehow help to safeguard, I think that has been one of the most positive and powerful messages."[56] For Milne, it is "the first and only successful evolution of the national park idea. That was a benchmark. So in concept I think it is an immense success as reflected by the number of States Parties."[57] Pressouyre remarks that such success was not predictable in the beginning: "The Convention, which in the 1980s did not seem destined to this future, is now known the entire world over and is, I would say, an extremely powerful tool for raising awareness in all countries."[58] Dumesnil

---

54   Canada Research Chair, interview Francioni.

55   Canada Research Chair on Built Heritage, Université de Montréal, audio interview of Adrian Phillips by Christina Cameron, London, 24 January 2008.

56   Canada Research Chair, interview Robertson Vernhes.

57   Canada Research Chair, interview Milne.

58   Canada Research Chair, interview Pressouyre. "La Convention qui, dans les années 80 ne paraissait pas appelée à cet avenir est maintenant connue dans le monde entier et c'est, je dirais un instrument de sensibilisation extrêmement puissant dans tout les pays."

underscores its contribution to making heritage known, not only globally but sometimes even within a country: "It is certainly important to raise consciousness of heritage at an international scale but also at the national level and I think that the Convention played a large role in this."[59]

Some emphasize the positive impact of World Heritage in encouraging dialogue and a better understanding of the rich diversity of the planet's cultures and natural resources. For Stovel, the Convention has been "an extraordinarily positive thing" because it motivates people in different countries "to talk to each other about their common things in a positive way."[60] He contrasts it with other international UNESCO cultural conventions which use the language of control to curtail negative behaviour, an approach that fails to win hearts and minds:

> What World Heritage does, because it focuses on the celebration of what has outstanding universal value, it gives a positive spin to what we have in common, it gives a positive spin to every group, every community, every nation, every region who is involved with a particular property going on the list, because they can say "we are sharing this with you. And it's yours. As well, you know, it's mine, but it's yours."[61]

He goes on to argues that the Convention has been an "international bridge" that fosters cooperation and a culture of peace:

> It's about stories that are so important that it doesn't matter whether it's inside one country or inside another country, it's a story; and we are all part of that story. So we can all celebrate together, and we can all take responsibility together for ensuring that that story survives or the expression, the forms of cultural expression that carry that story survive. This is very positive. This is all about peace on the planet. I'm speaking very abstractly but this is the way you go towards peace. This is why it's been possible frequently inside the Convention for those whose political leaders may be at war with each other to actually do a lot of things together.[62]

Heritage Theory and Practice

Some praise the Convention for its influence on constructing a theoretical framework for heritage. There is no doubt that the World Heritage system provides a platform where participants in the field of cultural and natural heritage

---

59   Canada Research Chair, interview Dumesnil. "C'est très important de faire prendre certes, à l'échelle internationale, conscience de la valeur d'un patrimoine mais également à l'échelle nationale et je crois que la Convention a joué un grand rôle à ce sujet."

60   Canada Research Chair, interview Stovel, April 2011.

61   Canada Research Chair on Built Heritage, Université de Montréal, audio interview of Herb Stovel by Christina Cameron, Ottawa, 3 February 2011.

62   Canada Research Chair, interview Stovel, April 2011.

can debate and deepen their understanding of the field. The exchange of ideas enriches those who participate in this endeavour and contributes to expanding heritage theories on a global scale.

François Leblanc characterizes the evolution of the concept of cultural heritage since the 1970s as moving from "elitist concepts about heritage to concepts that are much closer to the real humanitarian culture that touches things that are closer to what we are as human beings." He believes that "the notion of heritage is evolving not only in expert groups but with people working with heritage issues, who are interested in all sorts of questions that are much broader now than they were 30 years ago."[63] Mayor sums it up by pointing to the importance of "this human dimension of heritage which is added to the purely cultural, architectonic, historical dimensions of heritage."[64]

Cultural landscapes and the re-definition of authenticity are obvious examples of the influence of the Convention on global conservation practice. Luxen tells how his experience at Nara enlarged his understanding of different interpretations of heritage from one region to another. "This is a great enrichment because one discovers all of a sudden that there are ... values that other civilisations pay more attention to than we do and that it is important that we too open our eyes to these dimensions."[65] Beschaouch particularly values the intergenerational dialogue: "It is a transmission from experienced people to younger people. I know of no similar system in the world ... It does not exist in any other system in the world because if one leaves, one leaves for good ... People who love heritage do not leave."[66] Munjeri agrees: "I also can say that the Convention is a lot in terms of building,

---

63   Canada Research Chair on Built Heritage, Université de Montréal, audio interview of François Leblanc by Christina Cameron, Ottawa, 7 April 2009. "On est passé, depuis les années soixante-dix, à des concepts élitistes en matière de patrimoine, à des concepts qui sont beaucoup plus près de la véritable culture humanitaire qui touchent des choses qui sont beaucoup plus proches de ce qu'on est comme être humain. ... La notion de patrimoine est en train d'évoluer, non seulement au sein du groupe d'experts mais l'ensemble des gens qui tournent autour de la question du patrimoine mondial, s'intéresse à toutes sortes de questions qui sont beaucoup plus vastes maintenant qu'elles ne l'étaient il y a trente ans."

64   Canada Research Chair on Built Heritage, Université de Montréal, audio interview of Federico Mayor by Christina Cameron, Madrid, 18 June 2009."Je crois que cette dimension humaine du patrimoine s'est ajoutée à la dimension purement culturelle, architectonique, historique, etc. du patrimoine de l'héritage finalement culturel que nous avions jusqu'à ce moment."

65   Canada Research Chair, interview Luxen. "À Nara, c'est un des moments dans lequel on s'est rendu compte que les définitions du patrimoine pouvaient être différentes dans une région ou l'autre et ça c'est un grand enrichissement parce que on découvre tout d'un coup qu'il y a des choses, il y a des valeurs auxquelles d'autres civilisations sont plus attentives que nous et c'est important que nous aussi nous ouvrions aussi les yeux à ces dimensions."

66   Canada Research Chair on Built Heritage, Université de Montréal, audio interview of Azedine Beschaouch by Christina Cameron and Mechtild Rössler, Brasilia, 28 July 2010. "C'est un passage de relais entre les gens expérimentés et des gens plus jeunes. Je ne connais

bringing together a family spirit ... it's a forum in which we can all be at one in understanding the general principles of the Convention. That is the ideal."[67]

This exchange of ideas makes a significant contribution to scientific thought. Australian Professor Laurajane Smith uses the term "heritage discourse" to describe this critical analysis, noting that over time what she calls the authorized heritage discourse based on traditional European approaches has been challenged by alternative views.[68] Stovel sees this policy work as a "very, very positive output". He refers by way of example to discussions on monitoring, cultural landscapes and especially authenticity:

> The authenticity discussion opened up the possibility for the world as a whole, the conservation field as a whole to say "you must judge conservation decision-making in its cultural context." So beginning with that simple little word authenticity, the ripples in the water expanded to bring in this much larger idea, which is still with us. It was there before Nara, in the Nara document, but it was articulated in Nara, and it's never going to go away.[69]

Luxen also gives credit to these world-wide discussions for broadening the definition of heritage beyond an initial European one: "This dialogue also fostered an evolution of the concept of 'monuments and sites' towards something much more global which incorporates the whole of human activities with cultural landscapes, urban landscapes."[70] McNeely echoes several pioneers when he judges that one of the greatest successes is the concept of linking nature and culture, arguing that "conserving biodiversity is culture."[71] Luxen marvels at the rich connection between culture and natural heritage: "I find that the creators of this Convention were really innovators because it was this dimension that was not really sufficiently brought to light and this appears to me to be a really major point in the manner in which heritage is conceived today." He says it is the North Americans, Scandinavians, Japanese and Australians who helped Europeans appreciate the importance of nature.[72] Cleere echoes this sentiment: "One of the

---

pas au monde un système pareil ... ça n'existe dans aucun système au monde parce que si on quitte, on quitte pour toujours. ... Les gens qui aiment le patrimoine ne quittent pas."

67 Canada Research Chair, interview Munjeri.

68 Laurajane Smith, *The Uses of Heritage* (Abingdon, 2006), pp. 28–30.

69 Canada Research Chair, interview Stovel, April 2011.

70 Canada Research Chair, interview Luxen. "Ça a permis d'ailleurs d'évoluer du concept ' monuments et sites' vers quelque chose de beaucoup plus global qui englobe l'ensemble des activités humaines avec les paysages culturels, les paysages urbains."

71 Canada Research Chair on Built Heritage, Université de Montréal, audio interview of Jeff McNeely by Mechtild Rössler, Gland, 17 September 2010.

72 Canada Research Chair, interview Luxen. "J'ai découvert la richesse de l'articulation entre le patrimoine culture et le patrimoine nature. Je trouve que les initiateurs de cette Convention étaient vraiment des précurseurs, parce que c'est une dimension qui n'était pas suffisamment mise en évidence et que ça me parait vraiment un point majeur

things I did appreciate most in my eleven or twelve years was to see the way the thing got outside that Western European concept of culture. The rice terraces! That was quite a break-through. It really was."[73]

In terms of heritage practice, the Convention has undoubtedly stimulated the development of models and methodologies for conservation and can claim many successes. For some countries it marked the beginning of conservation activities. However, some worry about the unintended consequences of such a singular focus on World Heritage Sites. They argue that there is a system of heritage conservation "at two speeds," observing that implementation of the Convention may inadvertently result in neglect for the rest of a country's sites. This was not the intention of the Convention which addresses general conservation needs in article 5. Indeed the simultaneous adoption of the much ignored 1972 UNESCO Recommendation concerning the Protection, at National Level, of the Cultural and Natural Heritage was meant to address precisely this issue.

Azedine Beschaouch speaks to the vulnerability of important sites that have not been listed internationally. Although the Convention in article 12 makes it clear that not all qualifying sites will be listed internationally, "developing countries put all their limited resources to prestigious World Heritage Sites. Therefore we have heritage at two speeds."[74] With reference to African properties, Luxen concurs: "I think that Djenné, Timbuktu and Bandiagara are better conserved because they are on the World Heritage List ... But unfortunately, the fact of the matter is that all the resources go to these sites and at the same time others are pillaged."[75] Phillips characterizes the issue this way: "The danger is that they become rather special places treated to a much higher standard than anywhere else on a world scale ... [that] all the attention might go onto the World Heritage Sites and to the detriment of others."[76]

On the other hand, Thorsell celebrates the "significant positive results for conservation" that World Heritage has achieved. He refers to his report on the Convention's effectiveness in conserving natural sites, recalling that "35 sites were essentially saved." He applauds the successful mobilization in the 1990s of additional financial support, noting that the World Heritage Fund is itself modest but

---

dans la manière dont on doit concevoir aujourd'hui le patrimoine Ce sont les américains du nord, les scandinaves, les japonais, les australiens qui leur ont fait ouvrir les yeux sur l'importance de la nature."

73   Canada Research Chair on Built Heritage, Université de Montréal, audio interview of Henry Cleere by Christina Cameron, London, 24 January 2008.

74   Canada Research Chair, interview Beschaouch . "Dans les pays en développement, comme le patrimoine mondial est prestigieux tous les moyens qui ne sont pas énormes sont mis pour le patrimoine mondial. Et donc, nous avons un patrimoine à deux vitesses."

75   Canada Research Chair, interview Luxen. "Je crois que Djenné, Tombouctou et Bandiagara sont mieux conservés parce qu'ils sont sur la Liste du patrimoine mondial. ... Mais malheureusement, les faits c'est que toutes les ressources vont sur ces sites là et que pendant ce temps là, il y a des sites qui se font piller."

76   Canada Research Chair, interview Phillips.

Figure 6.3   **Three Directors-General of UNESCO who oversaw implementation of the World Heritage Convention from left to right: Koichiro Matsurra, Japan (1999–2010), Amadou-Mahtar M'Bow, Senegal (1974–1987) and Federico Mayor, Spain (1987–1999). Predecessors on screen, from left to right: Jaime Torres Bodet, Mexico (1948–1952), Julian Huxley, United Kingdom (1946–1948), René Maheu, France (1961–1974), Luther Evans, United States of America (1953–1958) and Vittorino Veronese, Italy (1958–1960)**

the system is attractive to donors. In his report he observes that "substantial amounts have been provided through the mechanism of the Global Environment Facility and the United Nations Foundation (UNF)" and concludes that World Heritage "is obviously an attractive target for donor funds which reflects confidence that other institutions have that the Convention's governance principles are sound."[77]

---

77 Canada Research Chair on Built Heritage, Université de Montréal, audio interview of Jim Thorsell by Christina Cameron, Banff, 11 August 2010; Thorsell, *World Heritage Convention*, p. 21.

Concluding Remarks

At the end of the twentieth century, the World Heritage Convention ranked as UNESCO's flagship program. The President of the 1999 General Assembly of States Parties called it "the most visible activity of UNESCO."[78] Rooted in the idealism and enthusiasm of the 1960s, the Convention adapted itself to changing circumstances as implementation proceeded. Although the text of the Convention remains constant, its application has changed with the evolving understanding of heritage conservation theory and practice.

Key policies and site-specific decisions over the years have shaped and developed the initial concept. The understanding of outstanding universal value has evolved; a formal monitoring and reporting system contributes to better conservation. Specific cases have forced deeper consideration about sites of conscience, sites associated with persons, the interpretation of the concept of creativity and the limitations to powers and authorities. Implementation has produced mixed results although most would argue that, at the turn of the century, its strengths exceeded its shortcomings.

In the name of the World Heritage Convention, innovations in heritage theory and conservation practice surpassed the expectations of the founders. Standards were set; properties were protected; heritage issues became a matter of public discourse. Several pioneers use the word "passionate" in their interviews to describe their feelings for the Convention. M'Bow attributes the success of the Convention to its openness, "this universal vision, this desire to inscribe the heritage belonging to different parts of the world."[79] Mayor concurs: "When we speak of heritage, we should always add, because there are so many magnificent works, there are places that are wonderful from the viewpoint of beauty, but we should also speak of culture, of history, of the creative effort, sometimes, of courage."[80] For Guo, the real message of the Convention lies in its ability to achieve harmony between human beings and the environment. "We have to keep the masterpieces made by our ancestors. And when we have no worry about the food, the house, about life,

---

78   UNESCO, Summary record of the twelfth General Assembly of States Parties to the Convention concerning the protection of the world cultural and natural heritage, Paris, 28-29 October 1999, Paris, 8 November 1999, WHC-99/conf-206/7, para. 10. Retrieved from http://whc.unesco.org/archive/1999/whc-99-conf206-7e.pdf

79   Canada Research Chair on Built Heritage, Université de Montréal, audio interview of Amadou-Mahtar M'Bow by Mechtild Rössler and Petra Van Den Born, Paris, 22 October 2009. "Ce que j'appellerais le succès, c'est l'ouverture, l'ouverture, l'universalité dans la conception du patrimoine mondial, cette vision universelle, cette volonté d'inscrire des patrimoines appartenant à différentes parties du monde."

80   Canada Research Chair, interview Mayor. "C'est pour cela qu'on a dit: les valeurs doivent être exceptionnelles et ça, c'est une ... à mon avis, une chose que nous devons, quand nous parlons du patrimoine, nous devons toujours ajouter parce que, il y a tellement des œuvres magnifiques, il y a des endroits qui sont formidables du point de vue de la beauté, mais il faut parler aussi de la culture, de l'histoire, de l'effort de créativité, parfois de courage."

we need to improve our quality of the life for our generation, especially for the next generation. So this is the real meaning."[81]

At the millennium, the challenges encountered during the first decades led to the reform agenda at the Cairns 2000 meeting of the World Heritage Committee. This account of the origins and early implementation of the World Heritage Convention provides the springboard for examining the ongoing evolution of the Convention in the twenty-first century.

---

81 Canada Research Chair, interview Guo.

Appendix
Vignettes of Interviewees

Carmen Añón Feliu

Figure A.1   Carmen Añón Feliu in Madrid, 2009

Trained in Spain as a landscape architect, Carmen Añón Feliu became a professor of the history of gardens and restorer of historic gardens. She directed a Masters course on landscape architecture at the Agricultural Engineering School at the Politécnica University of Madrid and a graduate course on the restoration of cultural landscapes and historic gardens for the Technical High School of Architecture of Madrid. As a practitioner, she directed conservation activities at the Spanish royal historic gardens for twenty five years. A long-standing member of ICOMOS, she has been Vice-President of the ICOMOS National Spanish Committee as well as President of the international ICOMOS Advisory Committee from 1992 to 1997. She served as President of the International Scientific Committee on Cultural Landscapes ICOMOS-IFLA and signed the ICOMOS Historic Gardens Charter known as the Florence Charter in 1981. Her involvement with World Heritage began informally with her participation in discussions of cultural nominations at ICOMOS headquarters in the 1980s and formally in 1993 when she attended Committee meetings as an executive board member of ICOMOS until 1999. From 2001, she joined the State Party delegation as a technical advisor. She attended several expert meetings including the Nara conference on authenticity in 1994. Teacher, consultant and scholar, she has written many books and articles including *Jardins en Espagne* published in collaboration with her daughters in 1999. In 2008 she received the ICOMOS Piero Gazzola Prize.

Nationality

Spanish

Dates of Participation in World Heritage

1980s–2009

Role

Carmen Añón Feliu reviewed nominations for ICOMOS in the 1980s. She attended Committee meetings in the 1993 to 1999 period on behalf of ICOMOS. In the twenty-first century she attended as a member of the State Party delegation. She also coordinated and carried out ICOMOS evaluation missions for cultural landscape nominations.

Azedine Beschaouch

Figure A.2   Azedine Beschaouch in Brasilia, 2010

After graduate studies in archaeology, Roman history and ancient languages at the École normale supérieure in Paris, Azedine Beschaouch worked as an inspector of antiquities before being appointed Director of Tunisia's National Institute of Archaeology and Art from 1975 to 1990. He began his long involvement with World Heritage as the State Party representative from 1980 to 1993. During that period he served as rapporteur in 1981 and chaired the Committee twice in 1989 and 1991. In addition, he prepared an evaluation of the implementation of the Convention in 1991 as part of the strategic review on the twentieth anniversary and participated in the 1994 Nara conference. In recognition of his international expertise in heritage and archaeology, he began in 1993 to work as the representative of the Assistant Director-General for Culture on delicate and sensitive issues at cultural sites in Bosnia-Herzegovina, Palestine and Cambodia. From 2005 to 2011 he served as special advisor to the Director-General of ICCROM. He continues to participate as the permanent secretary for the international coordinating committee for the safeguarding of the monuments at Angkor. Among his many publications on archaeological subjects, he has particularly focused on the archaeological site at Carthage and the monuments at Angkor. Among his awards, Azedine Beschaouch has been named an Officer in the French Legion of Honour and Commander of the Order of the Tunisian Republic.

Nationality

Tunisian

Dates of participation in World Heritage

1980 to the present

Role

From 1980 to 1993, Azedine Beschaouch represented the State Party and served as rapporteur in 1981 and Chairperson in 1989 and 1991. He subsequently worked for UNESCO on special projects related to World Heritage Sites, in particular Angkor, Cambodia.

Gérard Bolla

Figure A.3   Gérard Bolla in Paris, 2007

Gérard Bolla studied law and economy in Switzerland before coming to UNESCO in 1955. After working in the copyright law and human resources sectors, he became chief of staff to then Director-General René Maheu between 1969 and 1971. His general awareness of the proposals for an international conservation convention changed dramatically when he was appointed Deputy Assistant Director-General for Culture and Communication in November 1971. He immediately was sent to Washington to negotiate with the Stockholm summit team and others to ensure that the World Heritage Convention would be housed at UNESCO. With his colleague Michel Batisse, he oversaw the negotiations for the Convention text and its final adoption at the General Conference in 1972. Active in the early stages, Gérard Bolla attended the first General Assembly of States Parties in Nairobi in 1976 and most World Heritage statutory meetings until 1981, retiring from UNESCO in 1984. On the twentieth anniversary of the Convention in 1992, he attended the Washington meeting to prepare the strategic orientations and was feted at the 1992 Committee session in Santa Fe. With Batisse, he recounted his experience in *The Invention of World Heritage* in 2005.

Nationality

Swiss

Dates of Participation in World Heritage

1971–1981

Role

As a UNESCO staff member, Gérard Bolla was active in the negotiations leading up the adoption of the World Heritage Convention. Thereafter, he was directly involved in providing secretariat support for cultural heritage until 1981.

Mounir Bouchenaki

After studies in history at the University of Alger, Mounir Bouchenaki completed a doctorate in archaeology and ancient history at the University of Aix-en-Provence. From 1969 to 1981, he worked for the Algerian government as a conservation archaeologist and later as the Director of Antiquities, Museums and Historic Monuments in the Ministry of Culture and Information. He joined UNESCO's Cultural Heritage Division in 1982, working on international safeguarding campaigns for cultural sites. His involvement with World Heritage began early with the preparation of World Heritage nominations for six Algerian sites. When he became Director of UNESCO's Cultural Heritage Division in 1992, he attended World Heritage statutory meetings on a regular basis, continuing his operational

Figure A.4   Mounir Bouchenaki in Paris, 2009

oversight of sensitive files like Jerusalem. He served briefly as acting Director of the World Heritage Centre from 1999 to 2000 before assuming the post of Assistant Director-General for Culture. On his retirement, he was elected Director-General of ICCROM, thereby continuing his involvement with World Heritage from an advisory body perspective. Among other honours, he received the 2000 ICCROM Award in recognition of his contribution to cultural heritage.

Nationality

Algerian

Dates of Participation in World Heritage

1976 to the present

Role

Mounir Bouchenaki prepared Algerian cultural nominations until he joined UNESCO in 1982 when he oversaw conservation issues at specific sites. From 2005 to 2011 he represented ICCROM at World Heritage Committee sessions. Now an advisor to UNESCO and ICCROM, he contributes to sensitive files such as Jerusalem and Angkor.

Hans Caspary

Hans Caspary graduated in 1961 from studies in history, archaeology and art history at the University of Munich. After internships in Munich and Florence, he worked in Mainz from 1966 to 2001 as a conservator of historic monuments for the

Figure A.5   Hans Caspary in Mainz, 2011

Rhineland-Palatinate state in the Federal Republic of Germany. He participated at World Heritage Committee meetings from 1983 to 2001 as a representative of the State Party with expertise in cultural heritage. Until the unification of Germany in 1990, he was the only representative. In that year, in a highly symbolic gesture, a second representative from the former German Democratic Republic joined him at the fourteenth Committee session in Banff to nominate together the Palaces and Parks of Potsdam and Berlin. As a long-time member of ICOMOS with World Heritage experience, he has undertaken several evaluation and monitoring missions for cultural properties in Eastern Europe.

Nationality

German

Dates of participation in World Heritage

1983–2001

Role

Hans Caspary represented the Federal Republic of Germany from 1983 to 1990, then the unified State Party of Germany until 2001. He also worked occasionally for ICOMOS.

Lucien Chabason

Lucien Chabason studied sociology, public law and political science at the University of Paris before graduating from the École Nationale d'Administration. Following two

Figure A.6   Lucien Chabason in Paris, 2012

decades of public service in senior posts in the French Ministry of the Environment, he joined the United Nations Environment Program (UNEP) in 1994 to coordinate the Mediterranean Action Plan in Athens. He currently serves as a senior advisor to the Institute for Sustainable Development and International Relations in Paris. He participated in World Heritage Committee meetings as a member of the French delegation from 1979 to 1986 when he was responsible for site conservation and landscape planning in the French Ministry of the Environment. He was a member of the 1979 working group that developed inscription criteria for natural sites. His concern about the gap between monumental cultural heritage and pristine wilderness led him to present his ideas on rural landscapes to the 1984 Committee session in Buenos Aires. The ensuing discussion led to eight years of debate on what eventually became known as cultural landscapes. Because of his enduring interest in landscapes, he was invited by UNESCO to participate in the 1992 expert meeting at La Petite Pierre (France) which prepared proposals for new guidance on definitions and categories of cultural landscapes.

Nationality

French

Dates of Participation in World Heritage

1979–1986; 1992

Role

Lucien Chabason was a member of the French delegation to the World Heritage Committee from 1979 to 1986. He served as rapporteur for the 1984 World Heritage session. In 1992 he participated in the expert meeting on cultural landscapes at la Petite Pierre, France.

Henry Cleere

Figure A.7   Henry Cleere in London, 2008

After completing a Masters in English, Henry Cleere worked for over 20 years in the iron and steel industry before he resumed his studies in archaeology, obtaining a Ph.D. in 1980 from University College London for his research on the iron industry of Roman Britain. He was Director of the Council for British Archaeology from 1974 to 1991. Although an active member of the ICOMOS executive committee in the 1980s, his involvement with World Heritage only began in 1992 when he became ICOMOS World Heritage coordinator. For the next eleven years, he brought his professional knowledge and experience to improve the evaluation process for cultural properties, adding a greater focus on conservation and management considerations as well as introducing site visits. During that period, he presented about 350 site evaluations to the World Heritage Committee and participated in expert meetings including the Nara meeting on authenticity. Since that time he has been a consultant on the management aspects of World Heritage and a professor in archaeological heritage management at University College London. Of particular note among his publications are *Approaches to the Archaeological Heritage: a comparative study of world cultural resource management systems* in 1984 and *Archaeological Heritage Management in the Modern World* in 1989. First Secretary General of the European Association of Archaeologists, he was awarded its European archaeological heritage award in 2002.

Nationality

British

Dates of Participation in World Heritage

1992–2003 (full time) and from 2003 to the present (occasionally)

Role

As World Heritage coordinator for ICOMOS from 1992 to 2003, Henry Cleere undertook the technical evaluation of cultural site nominations and presented the ICOMOS position to the World Heritage Committee. He also contributed to state of conservation reports for cultural heritage sites.

Jim Collinson

Figure A.8   Jim Collinson in Windsor, 2010

Following studies in agricultural economics at the University of Manitoba, Jim Collinson graduated with a Masters degree in resource economics from the University of Michigan. During his long public service career in Canada, he was appointed in 1986 as Assistant Deputy Minister for the Canadian Parks Service, a position that included responsibility for the World Heritage Convention. At his first Committee meeting that year, he was elected Chairperson in the absence of a delegate who could not attend at the last minute and was re-elected for a second term in 1987. Among the sensitive files in those sessions were Australia's Wet Tropics of Queensland and China's Panda Reserves. Three initiatives stand as a legacy to his chairmanship. First, he proposed a working group to streamline the Committee's agenda and to explore the possibility of organizing an international cultural tentative list through the mechanism of a global study. He personally prepared a brief document scoping the task. Secondly he met with the World Bank in an effort to bring more resources into the World Heritage system for site projects. Thirdly, he initiated a formal request to the Director-General of UNESCO for more resources for the secretariat, a request that eventually resulted in the approval of new posts in the late 1990s.

Nationality

Canadian

Dates of Participation in World Heritage

1986–1990

Role

Jim Collinson headed the State Party delegation to World Heritage from 1986 to 1990. He was elected Chairperson of the tenth session in 1986 and the eleventh session in 1987.

Bernd von Droste

Figure A.9   Bernd von Droste

Bernd von Droste studied forestry at the University of Gottingen and the University of Munich, where he completed his Ph.D. in forest ecology in 1969. After joining UNESCO in 1973, he later became Director of the Division of Ecological Sciences and Secretary of the intergovernmental programme on Man and the Biosphere (MAB) from 1983 to 1991. He began work with the World

Heritage Convention in 1976, responsible for the natural heritage part of the secretariat for the World Heritage Convention. In 1992 Bernd von Droste became the first Director of UNESCO World Heritage Centre. He has been a member of IUCN's Commission for Parks and Protected Areas, now the World Commission for Protected Areas for more than 20 years. Following his retirement in 1999, he served as an advisor to UNESCO's Assistant Director-General for Culture as well as to the European Commission as an independent expert for project evaluation. In 2002 he was appointed Honorary Professor by the European University Viadrina/ Frankfurt (Oder) where he continues to teach as a professor at the postgraduate European heritage course. Teacher, consultant and author, Bernd von Droste has produced many books and articles on World Heritage issues.

Nationality

German

Dates of participation in World Heritage

1976–1999 (full time) and 1999 to the present (occasionally)

Role

Responsible for the natural part of the World Heritage Convention in the Division of Ecological Sciences, Bernd von Droste was alternating secretary from 1977 to 1991. From 1992 to 1999 he directed the amalgamated secretariat as the first Director of the World Heritage Centre.

Catherine Dumesnil

Figure A.10   Catherine Dumesnil in Paris, 2009

In her role of technical advisor on culture for the French National Commission for UNESCO, Catherine Dumesnil was assigned to work with World Heritage in 1996. When she began, there was no one at the French permanent delegation to UNESCO who had responsibility for following the World Heritage Convention. She therefore took on a more diplomatic role as a member of the State Party delegation. While other French experts carried out the technical and professional duties to prepare World Heritage files, she was responsible for ensuring that all the procedures were followed. In addition, she served as the focal point for advice on the World Heritage Convention, its statutory bodies and its rules of procedure. As such Catherine Dumesnil developed an in-depth knowledge of the procedural aspects of the World Heritage Convention and UNESCO's cultural programmes. She attended most Committee meetings from 1996 until 2009. She currently represents the European Union at UNESCO and thus remains involved in cultural heritage matters.

Nationality

French

Dates of Participation in World Heritage

1996–2009

Role

Catherine Dumesnil oversaw the procedural aspects of the World Heritage system for the State Party of France.

Regina Durighello

Figure A.11   Regina Durighello in Paris, 2009

Regina Durighello began working at the ICOMOS secretariat in 1990 as the assistant coordinator for the World Heritage Convention, eventually becoming Director of the World Heritage programme for this advisory body in 2002. Her responsibilities initially focused on the World Heritage inscriptions process, including the coordination of site visits and the finalization of texts for presentation to the Committee. During the 1990s she supported the work of a sequence of ICOMOS coordinators beginning with Léon Pressouyre, then Herb Stovel and Roland Silva followed by Henry Cleere. An occasional presenter of ICOMOS evaluations, she represented the organization at many expert meetings on conservation doctrine, World Heritage values and monitoring methodologies. She published an article in 2002 on the operational aspects of monitoring the state of conservation of World Heritage properties. In the twenty-first century, Regina Durighello has taken on additional responsibilities for coordinating state of conservation reporting.

Nationality

Italian/French

Dates of Participation in World Heritage

1990 to the present

Role

On behalf of ICOMOS, Regina Durighello coordinates the evaluation process for World Heritage nominations and state of conservation reporting for cultural properties.

Hal Eidsvik

Hal Eidsvik studied forestry and park management, graduating in 1964 with a Masters degree from the University of Michigan. He worked at Parks Canada for over 30 years as a park planner, then responsible for National Parks planning from 1968 to 1975 before becoming director of policy with responsibility for coordinating Parks Canada's international activities. Since 1968, he has been actively involved with IUCN's Commission on National Parks and Protected Areas, later called the World Commission on Protected Areas, as a member, as an executive officer from 1977 to 1980 and as its voluntary Chairperson from 1983 to 1990. As part of an exchange programme, he worked in Switzerland with IUCN from 1977 to 1980, during which time he participated in evaluations of natural World Heritage nominations and began drafting the 1982 global inventory of natural sites entitled *The World's Greatest Natural Areas: an indicative inventory of natural sites of World Heritage quality*. Following his retirement from Parks

Figure A.12   Hal Eidsvik in Ottawa, 2009

Canada he was seconded to the World Heritage Centre from 1993 to 1996. He was awarded the 1990 Fred M. Packard International Parks Merit Award by the Commission on National Parks and Protected Areas and the 1995 gold medal from the Royal Canadian Geographic Society.

Nationality

Canadian

Dates of Participation in World Heritage

1977–1996

Role

Hal Eidsvik worked first for IUCN as an evaluator of natural World Heritage nominations from 1977 to 1980. He participated in World Heritage Committee sessions first for IUCN, then as a member of the secretariat staff from 1993 to 1996.

Sir Bernard Feilden

After graduating as an architect in 1949 from the Architectural Association in London, Sir Bernard Feilden had a large architectural practice in Norwich. He came to architectural conservation in mid-career in 1963 when he assumed responsibility for Norwich Cathedral. A fervent advocate of education and training, for decades he gave regular lectures on architectural conservation at the Institute of Advanced Architectural Studies at York University and at ICCROM. Although

Figure A.13 Sir Bernard Feilden

he never attended a World Heritage Committee meeting, his involvement stemmed from his time as ICCROM Director-General from 1977 to 1981 when Anne Raidl introduced him to the nascent World Heritage Convention. He undertook many missions for UNESCO to World Heritage Sites. Consultant, teacher and author, he produced pioneering works in the field including *Conservation of Historic Buildings* in 1982, a standard text on conservation principles and practice, and *Guidelines for the Management of World Cultural Heritage Sites*, first written for the World Heritage Committee in 1983 and published in 1993 with co-author Jukka Jokilehto. In 1986, he received the Aga Khan award for architecture for the conservation of the dome of Al Aqsa mosque in Jerusalem. A pioneer in conservation architecture, he received the 1993 ICOMOS Piero Gazzola Prize and the 1995 ICCROM Award for his contribution to the field of conservation, protection and restoration of cultural heritage.

Nationality

British

Dates of Participation in World Heritage

1977–1993

Role

Sir Bernard Feilden oversaw ICCROM's contribution to the start-up phase of World Heritage and wrote the 1983 report on management guidelines for cultural properties.

Francesco Francioni

Figure A.14   Francesco Francioni in Rome, 2010

Francesco Francioni studied law at the University of Florence and Harvard University, awarded a Master of Laws degree from Harvard   in 1968. Law professor at the University of Sienna from 1980 to 2003, he held the Jean Monnet Chair in International Law there from 1999 to 2003, followed by his appointment as Professor of International Law at the European University Institute in Florence. His involvement with World Heritage began in 1992 when he was engaged by the governments of Italy and France to provide legal advice on the proposed autonomy of the newly created World Heritage Centre. He subsequently attended World Heritage Committee sessions between 1993 and 1998, serving as Chairperson of the twenty-first session in Naples in 1997. During his tenure, he made an official visit to Kakadu National Park in Australia to assess the potential impacts from the proposed Jabiluka mine. Teacher and prolific author on international cultural law, he initiated and edited an important book *The 1972 World Heritage Convention: a Commentary* published in 2008.

Nationality

Italian

Dates of Participation in World Heritage

1992–1998

Role

Francesco Francioni provided legal advice on the administrative structure of the World Heritage Centre in 1992. He attended World Heritage Committee meetings from 1993 to 1998 as a member of the State Party delegation, chairing the 1997 session in Naples.

Guo Zhan

Figure A.15 Guo Zhan in Brasilia, 2010

After studies in archaeology at Peking University, Guo Zhan graduated in 1982 from the Chinese Academy of Social Sciences with a Masters degree in the history of Mongolia and the Yuan dynasty. Starting in 1976, he was employed for over three decades by the State Administration of Cultural Heritage in various positions related to cultural heritage protection and management, culminating in his appointment as a commissioner of the department. He became the focal point for World Heritage when China ratified the Convention in 1985. In this role, he introduced basic documents like the Venice Charter and the World Heritage Convention to Chinese professionals, helping to integrate those principles into Chinese cultural resource management practices. He also coordinated the preparation of China's nominations to the World Heritage List and established communications networks with UNESCO, ICOMOS and ICCROM. In his professional capacity, Guo Zhan carried out international missions for ICOMOS and attended expert meetings including the important 1994 Nara conference on authenticity. A founding member of ICOMOS China and active in ICOMOS international, he was elected as a vice-president in 2005. Teacher and author, he has published several books on World Heritage in China and heritage management practices.

Nationality

Chinese

Dates of Participation in World Heritage

1985 to the present

Role

Focal point for World Heritage in China, Guo Zhan coordinated Chinese cultural nominations and fostered international communications for World Heritage. Since 1994, he has attended most World Heritage Committee meetings as an advisor to the State Party delegation. In recent years, he has attended as a technical advisor to ICOMOS.

Natarajan Ishwaran

Figure A.16   Natarajan Ishwaran in Paris, 2009

Natarajan Ishwaran studied zoology and animal ecology at the University of Peradeniya, Sri Lanka, before obtaining a Ph.D. in wildlife biology and management from Michigan State University. He joined UNESCO in 1986, working as a specialist in environmental science and natural heritage first in Paris from 1986 to 1993, then at the UNESCO office in Jakarta, Indonesia from 1993 to 1996. Right from the start, he worked on the implementation of the natural resources aspect of the World Heritage Convention, returning from Jakarta to eventually lead the natural heritage section of the World Heritage Centre until 2004 when he became the Director of the Ecological and Earth Sciences Division and Secretary of the Man and the Biosphere programme until his retirement in 2012. In his work, Natarajan Ishwaran promoted the use of UNESCO designated sites such as World Heritage Sites, biosphere reserves and geoparks as globally significant areas for learning sustainable development practices. He is co-author of scientific technical articles and books on subjects dealing with ecology, conservation and protected area management.

Nationality

Sri Lankan

Dates of Participation in World Heritage

1986–1993; 1996–2004

Role

In close collaboration with IUCN, Natarajan Ishwaran provided secretariat support for the natural heritage properties, verifying the completeness of nominations and providing substantive input to state of conservation reports.

Nobuo Ito

Figure A.17   Nobuo Ito in Kyoto, 2012

Nobuo Ito studied architecture at Tokyo University, obtaining a doctoral degree in 1960 from the Department of Engineering for his thesis on Japanese mediaeval architecture. He enjoyed a 40-year career in Japan's public service, working on heritage legislation and many conservation projects. Beginning at the Tokyo National Museum, Ito then directed the Architecture Division in the Agency for Cultural Affairs in 1971, becoming Chief Inspector in 1977, then Director-General of Tokyo National Research Institute of Cultural Properties from 1978 till 1987. On his retirement, he was awarded the title of Researcher Emeritus. He subsequently moved to the academic field, becoming Professor at Kobe Design University from 1989 to 1995 and is now Professor Emeritus. He represented Japan at the April 1972 intergovernmental experts meeting in Paris to prepare the final draft of the World Heritage Convention. He also participated in the 1994 Nara meeting on authenticity and the 1996 session of the World Heritage Committee in Merida, the year that Hiroshima was inscribed. He was a council member at ICCROM from 1978 to 1989 and has belonged to ICOMOS from 1975 to the present, serving as Vice-President from 1993 to 1996. In 2011 he was awarded the ICOMOS Piero Gazzola Prize.

Nationality

Japanese

Dates of Participation in World Heritage

1972, 1994-1996

Role

Nobuo Ito attended the April 1972 expert meeting on behalf of Japan. He participated in the Nara meeting on authenticity in 1994 and was a member of the Japanese delegation to the World Heritage Committee in 1996.

Jukka Jokilehto

Figure A.18   Jukka Jokilehto in Rome, 2010

A graduate in architecture and town planning from Helsinki Polytechnic in 1966, he was a practicing architect and urban planner in the 1960s. He began at ICCROM in 1972 as a teacher and coordinator for the architectural and urban conservation programme. Following specialized studies at the Institute of Advanced Architectural Studies in England and Rome, he obtained the degree of Doctor in Philosophy at York University in 1986. He was appointed Assistant

Director-General of ICCROM from 1995 to 1998. He attended Committee sessions on an intermittent basis in the 1980s, all meetings from 1988 to 1997 on behalf of ICCROM, then from 2000 to 2006 as an ICOMOS evaluator and presenter of nominations. An expert on authenticity, the concept of outstanding universal value and cultural site management, he regularly participated in World Heritage expert meetings including one on management guidelines in 1983, the Nara meeting on authenticity in 1994 and the Zimbabwe meeting in 2000. A prolific lecturer and writer, Jukka Jokilehto co-authored the ground-breaking *Guidelines for the Management of World Cultural Heritage Sites* in 1993 and compiled the 2008 ICOMOS volume *The World Heritage List: What is OUV? Defining the Outstanding Universal Value of Cultural World Heritage Properties.* He was the recipient of the 2000 ICCROM Award in recognition of special merit in the field of conservation, protection and restoration. In 2009 Nicolas Stanley Price and Joe King published *Conserving the Authentic: Essays in honour of Jukka Jokilehto.*

Nationality

Finnish

Dates of Participation in World Heritage

1982–2006

Role

From 1982 to 1997 Jukka Jokilehto provided training and conservation advice on behalf of ICCROM and from 2000 to 2006 he presented nomination evaluations for ICOMOS at World Heritage Committee meetings.

François Leblanc

Educated as an architect at the University of Montreal, François Leblanc then undertook specialized studies in historic building conservation in 1973 at the Institute of Advanced Architectural Studies at York University, England. He had a distinguished career both at home and abroad. In Canada, he held leadership positions in conservation architecture and engineering with Parks Canada from 1971 to 1979, then served as Vice-President of Heritage Canada from 1983 to 1992 with responsibility for heritage properties and several heritage programmes, and finally occupied the post of Chief Architect at the National Capital Commission from 1992 to 2001. On the international front, he served as Director of the ICOMOS secretariat in Paris from 1979 to 1983 and later as head of field projects for the Getty Conservation Institute from 2001 to 2007. For over 20 years, he held executive

Figure A.19 François Leblanc in Ottawa, 2009

positions in professional organizations like ICOMOS Canada, the Association for Preservation Technology and international ICOMOS. His involvement in World Heritage began at the early stages of implementation when he was sent by the government of Canada to work in the ICOMOS secretariat in Paris from 1979 to 1983. A highlight during this period was the inscription of Jerusalem. His many publications range from philosophical musings about heritage to technical conservation reports emerging from his hundreds of field projects.

Nationality

Canadian

Dates of Participation in World Heritage

1979–1983

Role

In his role as Director of the ICOMOS secretariat, François Leblanc was responsible for coordinating the evaluation of World Heritage cultural nominations, attending statutory meetings and managing the office.

Francisco Lopez Morales

Francisco Lopez Morales holds a degree in architecture from the Autonomous University of Mexico and was granted a doctorate in urban studies from the University of Grenoble II in 1980. Following a period of urban restoration projects in Montreal and Paris, he became Secretary General for ICOMOS Mexico from

Figure A.20   Francisco Lopez Morales in Brasilia, 2010

1991 to 1997. On the international stage, he became a member of ICOMOS in 1991, serving as General Secretary of the 15th ICOMOS International General Assembly in Xi'an, China in 2005 and Vice-President from 2010 to 2011. His involvement with World Heritage began in 1995 when he became an advisor to the incoming Chairperson, Teresa Franco, who presided over the 1996 session in Merida. He later served as rapporteur of the twenty-fifth session of the World Heritage Committee in 2001. Since that time, Francisco Lopez Morales has directed the World Heritage unit at the National Institute for Anthropology and History in Mexico. A professor and author of a number of works on cultural landscapes, vernacular architecture and the monuments of Mexico, he coordinated a study in 2003 on representativity on the World Heritage List in Latin America, Canada and the United States. In 1987, he received the Juan Pablos national award for his book *Arquitectura vernácula en México.*

Nationality

Mexican

Dates of Participation in World Heritage

1995 to the present

Role

Francisco Lopez Morales provided cultural heritage expertise to the Mexican delegation at World Heritage sessions from 1996 to the present.

Jean-Louis Luxen

Figure A.21   Jean-Louis Luxen in Leuven, 2009

After graduating with a doctorate in law from the Catholic University of Leuven in 1963, Jean-Louis Luxen studied economics at Stanford University from 1964 to 1965. He had a distinguished career at the Belgian Ministry of Cultural Affairs, first as chief of staff to various ministers from 1972 to 1981, then as Secretary General in charge of the cultural portfolio from 1981 to 2006. During that period he represented his government in various international organizations and conferences, including an appointment as President of the Cultural Heritage Committee of the Council of Europe from 1989 to 1993. His involvement with World Heritage stems from his election as Secretary General of ICOMOS for three terms from 1993 to 2002. In this capacity, he oversaw the advisory body's participation at World Heritage sessions and participated in various expert meetings including the pivotal Nara meeting on authenticity in 1994, and meetings on the intangible dimensions of World Heritage at Zimbabwe in 2000 and Harare in 2003. Professor, lecturer and author of many papers on cultural heritage and intercultural dialogue, Jean-Louis Luxen has recently explored issues relating to heritage economics, conservation funding and legal frameworks to prevent illicit traffic of cultural properties. Since 2002, he is the senior legal expert for the Euromed Heritage programme.

Nationality

Belgian

Dates of Participation in World Heritage

1993–2002

Role

As ICOMOS Secretary General, Jean-Louis Luxen participated in World Heritage sessions from 1993 to 2002 and various expert meetings.

Koichiro Matsuura

Figure A.22   Koïchiro Matsuura in Paris, 2009

Koïchiro Matsuura of Japan was educated at the Law Faculty of the University of Tokyo and at the Faculty of Economics of Haverford College in Pennsylvania. He began his diplomatic career in 1959, holding positions including Director-General of the Economic Co-operation Bureau of Japan's Ministry of Foreign Affairs in 1988, Director-General of the North American Affairs Bureau, Ministry of Foreign Affairs in 1990 and Deputy Minister for Foreign Affairs. During this period he was involved in Japan's ratification of the World Heritage Convention in 1992. He served as Ambassador of Japan to France from 1994 to 1999. He was appointed by UNESCO's General Conference on November 12 1999 to serve a six-year term as Director-General of UNESCO; he completed a second term in 2009. Elected as Chairperson of the World Heritage Committee in 1998 for a one-year term, he presided over the twenty-second session in Kyoto in 1998. In 2008 he published a book in Japanese that has been translated as *World Heritage: a Personal Reflection by the Director-General of UNESCO*.

Nationality

Japanese

Dates of Participation in World Heritage

1998–2009

Role

Koichiro Matsuura was Chairperson of the World Heritage Committee from 1998 to 1999. As Director-General of UNESCO from 1999 to 2009, he participated in the opening sessions of most General Assemblies of States Parties and World Heritage Committee meetings during his mandate.

Federico Mayor

Figure A.23   Federico Mayor in Madrid, 2009

Federico Mayor completed a Ph.D. in pharmacy at the Complutense University in Madrid in 1958. After professorships in biochemistry at universities in Granada and Madrid, he served as Director of the Severo Ochoa Molecular Biology Centre in Madrid from 1973 to 1978. His political career began on the national front when he became Under-Secretary of the Spanish Ministry of Education and Science in 1974, then a member of Parliament and Chairman of the Parliamentary Commission for Education and Science in 1977 and 1978, an advisor to the Prime Minister and finally Minister of Education and Science in 1981 and 1982. In 1987 he was elected a Member of the European Parliament. After being Deputy Director-General of UNESCO from 1978 to 1981, he returned to the organization in 1983 and 1984 as special advisor to Director-General M'Bow, whom he succeeded for two terms from 1987 to 1999. A scientist, author and poet, he then returned to Spain to create the Foundation for a Culture of Peace.

Nationality

Spanish

Dates of Participation in World Heritage

1978–1999

Role

Federico Mayor oversaw UNESCO's secretariat support for the World Heritage Convention, deciding in 1992 to create the World Heritage Centre. As Director-General he attended two Committee sessions in Santa Fe (1992) and Merida (1996) during which he promoted his vision of a culture of peace.

Amadou-Mahtar M'Bow

Figure A.24   Amadou-Mahtar M'Bow in Paris, 2009

After completing his higher education in geography at the Sorbonne in Paris, Amadou-Mahtar M'Bow taught history and geography in Senegal, where he directed basic education from 1952 to 1957. Appointed to the position of Minister of Education and Culture during his country's transitional period of internal autonomy in 1957 and 1958, he resigned in order to engage in the struggle for independence. After this had been achieved, he became a member of the National Assembly of Senegal, serving as Minister of Education from 1966 to 1968, then Minister of Cultural and Youth Affairs from 1968 to 1970. Elected in 1966 to UNESCO's Executive Board as a representative from Senegal, he joined the organization in 1970 as Assistant Director-General for Education. Appointed Director-General in 1974, he held this position until 1987.

Nationality

Senegal

Dates of participation in World Heritage

1974–1987

Role

As Director-General of UNESCO, Amadou-Mahtar M'Bow oversaw the secretariat support for the World Heritage Convention. He participated in the start-up phase, receiving country ratifications that allowed the Convention to come into force and witnessing the inscription of the first sites to the World Heritage List.

Jeff McNeely

Figure A.25   Jeff McNeely

Following studies in anthropology at the University of California at Los Angeles, Jeff McNeely worked in Asia for different organizations, in particular as the representative for the World Wildlife Fund-IUCN, thereby establishing IUCN's first country programme. He joined IUCN headquarters in 1980 as executive officer of the Commission for National Parks and Protected Areas, where he was responsible for relations with UNESCO including World Heritage. From 1981 to 1983 he produced evaluations of natural heritage nominations and completed the 1982 global inventory of natural sites entitled *The World's Greatest Natural Areas: an indicative inventory of natural sites of World Heritage quality*. He led the IUCN secretariat to the third World Parks Congress in Bali in 1982 and was Secretary General of the fourth World Congress on National Parks and Protected

Areas in Caracas in 1992. He later became the Director of IUCN's biodiversity programme and Deputy Director-General of the organization. He continued to work with IUCN as chief scientist until his retirement in 2009 and as senior science advisor until March 2012.

Nationality

American

Dates of participation in World Heritage

1980–1983 (and occasionally until 2012)

Role

Jeff McNeely represented the IUCN secretariat at World Heritage Committee meetings from 1981 to 1983 and carried out evaluations of natural World Heritage Sites. Since then, he has occasionally participated in evaluations and monitoring activities at World Heritage properties.

Rob Milne

Figure A.26 Rob Milne in Paris, 2009

After completing a Masters degree in ecology from North Carolina State University in 1963, Rob Milne worked for the United States National Park Service, first as an ecologist and park naturalist, then for two decades as the senior official at the Office of International Affairs responsible for international policy and programmes, including implementation of the World Heritage Convention. His leadership of hundreds of training and professional development activities in all continents helped to build a world-wide network of park professionals. In partnership with Parks Canada and the University of Michigan, he organized in the 1980s the annual Parks International Seminar that brought together site managers from all continents to learn about management of parks, sites and protected areas. He chaired the 1993 Bureau of the World Heritage Committee. After his retirement in 1996, Rob Milne was seconded from the American government to the World Heritage Centre where he provided policy advice to the Director and subsequently undertook occasional monitoring missions to threatened sites on behalf of UNESCO. In recognition of his outstanding commitment to conservation, Rob Milne received IUCN's Fred M. Packard International Parks Merit Award in 1984 and UNESCO's Dubrovnik Gold Medal for World Heritage in 1997.

Nationality

American

Dates of participation in World Heritage

1977–1997 (and occasionally until 2007)

Role

Rob Milne attended almost every World Heritage Committee meeting between 1977 and 1995. As a result of the change from the Bush to the Clinton administrations, he took over the Committee chairmanship mid-year following the resignation of the Chairperson of the 1992 session in Santa Fe, Jennifer Salisbury. In 1996-1997 he was principal advisor to the Director of the World Heritage Centre.

Dawson Munjeri

Dawson Munjeri was educated in history and an African language at the University of Zimbabwe followed by studies in information systems at the University of Wales. He recently completed doctoral studies on international heritage protection frameworks. He joined ICOMOS in the 1980s, serving a term as Vice-President from 1999 to 2002. His first exposure to World Heritage came when he was site manager at Great Zimbabwe which became a World Heritage Site in 1986. His in-depth involvement began in 1995 when, as Deputy Executive Director of National Museums and Monuments of Zimbabwe, he hosted the first global strategy meeting for African cultural heritage and the World Heritage Convention in Harare. He subsequently participated in further expert meetings on the global strategy and on revision to the Operational Guidelines. As a member of the World Heritage Committee from 1997 to 2003, Dawson Munjeri led the country's delegation and served as rapporteur to

Figure A.27   Dawson Munjeri in Brasilia, 2010

the 2000 session in Cairns. In 2003 he hosted the ICOMOS General Assembly and Scientific Symposium at Victoria Falls which studied the conservation of intangible values in monuments and sites. His publications and lectures have focused on oral history, tangible and intangible heritage, authenticity, cultural landscapes and legal frameworks for heritage protection. Dawson Munjeri has served as Deputy Permanent Delegate of Zimbabwe to UNESCO from 2002 to the present.

Nationality

Zimbabwean

Dates of participation in World Heritage

1986 to the present

Role

Dawson Munjeri has been site manager at Great Zimbabwe World Heritage Site, head of Zimbabwe's delegation to World Heritage Committee meetings, a participant in expert meetings and rapporteur to the 2000 session.

Adrian Phillips

Adrian Phillips was educated in geography at Oxford University and in planning at the University of London. His career spans the field of environmental conservation both within the United Kingdom and abroad. Nationally, he headed the Countryside Commission for over 10 years between 1981 and 1992. Internationally, his active involvement with World Heritage came from 20 years of work with IUCN. He was

Figure A.28   Adrian Phillips in London, 2008

elected Chair of IUCN's World Commission on Protected Areas from 1994 to 2000 and then Vice-Chair with special responsibility for World Heritage from 2000 to 2004. As a geographer he did not limit himself to natural heritage considerations, initiating discussions with ICOMOS on cultural landscapes. He has written numerous articles on conservation and landscape themes and edited a series on best practices for the management of protected areas. His recent participation with the National Trust provides a new perspective on World Heritage issues within the United Kingdom. In 1998 he was awarded Commander of the Most Excellent Order of the British Empire (CBE) for services to the countryside and the environment.

Nationality

British

Dates of participation in World Heritage

1994–2007

Role

Adrian Phillips participated regularly in IUCN evaluations of World Heritage nominations and attended several World Heritage Committee sessions as a member of IUCN's delegation.

Harald Plachter

Harald Plachter studied biology, zoology and chemistry, graduating with a Ph.D. in 1978 at the University of Erlangen, Germany. From 1978 to 1990 he headed

Figure A.29 Harald Plachter in Mainz, 2011

the Division for Species and Biotope Protection at the Bavarian State Agency for Environment Protection, Munich. During the same period, he lectured in nature conservation at the University of Erlangen and the University of Ulm. After completing post-doctoral studies (Habilitation) in 1987 at the University of Ulm, he was appointed in 1990 as full professor and Chair for Nature Conservation at the Faculty for Biology of Marburg University, Germany. His involvement with World Heritage began in 1992 when he joined the German delegation as its natural heritage expert. In addition to attending statutory meetings until 2000, he contributed to expert meetings, in particular to the important Vanoise meeting in 1996 on natural heritage principles and criteria. He also organized training seminars at the German Agency for Nature Protection and its institution on Vilm Island. Active in professional nature conservation organizations, Harald Plachter worked closely with IUCN and the World Commission on Protected Areas, becoming Vice-Chair for World Heritage for a short time until 2005. Among his more than 170 publications are thoughtful articles on World Heritage cultural landscapes. In recognition for his service, he received the Bruno H. Schubert Award for Nature Conservation in 1988.

Nationality

German

Dates of Participation in World Heritage

1992–2004 and occasionally until 2010

Role

Harald Plachter attended Committee meetings as the natural heritage expert in the German delegation from 1992 until the turn of the century. He was Vice-Chair for World Heritage for the World Commission on Protected Areas from 2004 to 2005 and attended the Kazan meeting on outstanding universal value in that capacity. He carried out evaluations and monitoring missions, mainly to natural heritage sites in Eastern Europe, until 2010.

Léon Pressouyre

Figure A.30   Léon Pressouyre in Paris, 2008

Based in Paris, Léon Pressouyre was a Professor of mediaeval art at the Université de Paris 1 – Sorbonne, eventually becoming Vice-President of this institution. His expertise was developed through research on the mediaeval cloister fragments at Châlons-sur-Marne. His involvement with World Heritage began in 1980 when, on behalf of ICOMOS, he was asked to retrospectively document the cultural World Heritage Sites that had previously been inscribed in 1978 and 1979. For the next decade, he presented the cultural nominations at World Heritage Committee sessions on behalf of ICOMOS. In addition to his university career and work with World Heritage, Léon Pressouyre acted as a special advisor to the Director-General of UNESCO, chairing a UNESCO expert committee for monitoring the situation of monuments and cultural heritage in Bosnia and Herzegovina (1996-2001), and presiding over the scientific committee for the reconstruction of the Mostar Bridge (2002-2004). He was the author of many publications on the history of mediaeval monuments as well as a ground-breaking work *The World Heritage Convention, Twenty Years Later* published by UNESCO in French in 1993 and in English in 1996.

Nationality

French

Dates of Participation in World Heritage

1980–1997

Role

Léon Pressouyre served as ICOMOS coordinator for the World Heritage Convention from 1980 to 1990; subsequently he was part of the French delegation to the World Heritage Committee from 1990 to 1997.

Anne Raidl

Figure A.31   Anne Raidl in Vienna, 2008

Austrian by nationality and based at UNESCO headquarters in Paris, Anne Raidl was trained as a lawyer. She worked at the Cultural Heritage Division of UNESCO for her entire career. In addition to supporting the World Heritage Convention from its inception, her duties included responsibility for supporting two other UNESCO instruments, the Convention for the Protection of Cultural Property in the Event of Armed Conflict (1954) and the Convention on the Means of Prohibiting and Preventing the Illicit Import, Export and Transfer of Ownership of Cultural Property (1970). In addition, she collaborated with member states for drafting new UNESCO recommendations including notably the Recommendation concerning the Safeguarding and Contemporary Role of Historic Areas (Nairobi, 1976).

Nationality

Austrian

Dates of Participation in World Heritage

1972–1992

Role

As a staff member of UNESCO's Cultural Heritage Division, Anne Raidl was alternating secretary from 1977 to 1991 and responsible for the cultural part of the secretariat for the World Heritage Convention. She was appointed Director of the Cultural Heritage Division in 1981, a position that she held until her retirement in 1992.

John Reynolds

Figure A.32   John Reynolds in Springfield, 2011

Born when his father was a ranger at Yellowstone National Park, John Reynolds studied landscape architecture, obtaining a Master's degree from the State University College of Forestry at Syracuse University in the state of New York. In 1966, he began a lifetime career in the United States National Park Service as a landscape architect, moving through a variety of planning and management positions to reach the executive level. In addition to leading the establishment of two new parks and the expansion of a third, he initiated a sustainable design approach in the organization. His involvement with World Heritage began when he was Deputy Director of the National Park Service with oversight responsibility for national and international issues. He is the recipient of the Department of the Interior's highest honour, the Distinguished Service Award.

Nationality

American

Dates of Participation in World Heritage

1994–1999

Role

John Reynolds represented the State Party at World Heritage Committee and Bureau meetings. He provided technical advice for the development of American positions on issues under consideration by the Committee.

Jane Robertson Vernhes

Figure A.33   Jane Robertson Vernhes in Paris, 2009

After completing her doctoral studies in applied ecology at the University of Paris, Jane Robertson Vernhes joined the Ecological Sciences Division of UNESCO, first as a consultant, then as a staff member. She began by working on the early implementation of the World Heritage Convention and took on other duties in support of the Man and the Biosphere (MAB) programme. Working directly for Bernd von Droste, she oversaw the verification of nominations for World Heritage, organized statutory meetings and administered the technical aspects for international assistance from the World Heritage Fund. When the World Heritage Centre was created in 1992, she remained in the Ecological Sciences Division and became the specialist in charge of the coordination of the World Network of Biosphere Reserves. From this position, she promoted the interaction of biosphere reserves and natural World Heritage Sites as elements in an integrated system to protect natural resources.

Nationality

British

Dates of Participation in World Heritage

1979–1991

Role

Jane Robertson Vernhes provided secretariat support to the World Heritage system on behalf of the Ecological Sciences Division of UNESCO.

Roland Silva

Figure A.34   Roland Silva in Victoria, 2011

In the 1950s, Roland Silva studied architecture and archaeology at the Architectural Association School of Architecture, the Royal Archaeological Institute and the University of London. He obtained a doctorate in the same field at the University of Leiden in the Netherlands. He introduced a scientific approach for archaeological activities as head of the Department of Archaeology for the Sri Lankan government. He subsequently created a new robust organization to fund cultural activities, known as the Central Cultural Fund. In addition to preparing the first three World Heritage nominations for his country in the early 1980s, he also created ICOMOS Sri Lanka at the encouragement of Michel Parent. First elected at the ICOMOS General Assembly, Roland Silva had three consecutive mandates as President of ICOMOS from 1990 to 1999. During his tenure and with his encouragement, ICOMOS scientific committees grew from 11 to 24 groups. In 1999 he was awarded the Piero Gazzola Prize, established in 1979 in memory of ICOMOS' founding President and the highest distinction awarded by ICOMOS.

Nationality

Sri Lankan

Dates of Participation in World Heritage

1990–1999

Role

As President of ICOMOS he participated in Committee meetings during his tenure as well as expert meetings including the Nara conference on authenticity in 1994 and the first African global strategy session in Harare in 1995.

Herb Stovel

Figure A.35   Herb Stovel in Rome, 2002

Trained as an architect at Montreal's McGill University, Herb Stovel became a global leader in heritage conservation and management. His career included academic appointments, stints in two international organizations and consultancies. From 1990 to 1993 he was a professor in the graduate heritage conservation programme at the University of Montreal and later at Carleton University from 2006 until his death in 2012. In service to many conservation organizations, Stovel was President of the Association for Preservation Technology from 1989 to 1991, Secretary General of ICOMOS from 1990 to 1993, President of ICOMOS Canada from 1993 to 1997 and Director of the Heritage Settlements Unit at ICCROM from 1998 until 2004. Introduced to World Heritage in 1982 by fellow Canadian François Leblanc, he began his official involvement at the fourteenth session of the World Heritage Committee in Banff in 1990. He was one of the key drafters of the Nara Document on Authenticity in 1994, a text that clarified the concept of authenticity and promoted the concept of cultural diversity in conservation. At ICCROM he coordinated the organization's advice to the World Heritage Committee, including the preparation of the first global training strategy for cultural properties. A highlight among his many publications is *Risk Preparedness: A Management Manual for World Cultural Heritage* published in 1998. He was the recipient of the 2011 ICCROM Award in recognition of special merit in the field of conservation, protection and restoration of cultural heritage.

Nationality

Canadian

Dates of Participation in World Heritage

1990–2012

Role

During his term as Secretary General of ICOMOS, Herb Stovel coordinated the process for evaluating nominations of cultural properties, notably hiring Henry Cleere in 1992 as ICOMOS World Heritage coordinator. For ICCROM, he provided the advisory body's technical advice on training and management issues.

Jim Thorsell

Based in Banff, Alberta, Jim Thorsell completed an interdisciplinary doctoral programme in resource sciences at the University of British Columbia in 1971 with emphasis on planning and park management. Beginning as a park ranger in Banff National Park in 1962, Thorsell worked as a researcher, planner, trainer and project manager in Canada before spending the period from 1979 to 1999 working on international conservation projects. He has carried out field missions to over 600 protected areas in 90 countries. Beginning in 1983, he prepared and presented on behalf of IUCN more than 150 evaluations of natural site nominations for UNESCO's World Heritage Committee, introducing site visits as a necessary component for assessing properties. An early pioneer in sustainable development and natural resource conservation, he is an authority on international peace parks, particularly in

Figure A.36   Jim Thorsell in Banff, 2010

mountain areas. Jim Thorsell has authored or co-authored more than 300 publications on subjects ranging from the management of tropical forest habitats to World Heritage Sites, Biosphere Reserves and the World Conservation Strategy, including *World Heritage Twenty Years Later* in 1992 and *World Heritage Convention: effectiveness 1992–2002 and lessons for governance* in 2003. In recognition of his impressive contribution to nature conservation, Jim Thorsell was awarded the IUCN Park International Merit Award and Canada's 2007 Harkin Conservation Medal as well as an honorary doctorate in 2009 from the University of Alberta.

Nationality

Canadian

Dates of Participation in World Heritage

1983–1998 (full time) and 1998 to present (occasionally)

Role

Jim Thorsell represented IUCN at World Heritage Committee meetings under various job titles beginning as executive officer responsible for World Heritage, then as head of the natural heritage programme and finally as senior advisor for natural heritage.

Abdelaziz Touri

Based in Rabat, Abdelaziz Touri completed a doctoral programme in Islamic art and archaeology at the Université Paris IV-Sorbonne in 1987. A long-time public

Figure A.37   Abdelaziz Touri in Paris, 2011

servant, he worked for the Moroccan Department of Cultural Affairs, serving as Director of Cultural Heritage from 1987 to 2000 and then as Secretary General of the Department of Cultural Affairs from 2000 to 2007. He also carried out professional activities in research and archaeological excavation, notably serving as Assistant Director of the National Institute for archaeological sciences and heritage (l'Institut National des Sciences de l'Archéologie et du Patrimoine) from 1986 to 1988. He has published extensively on subjects relating to World Heritage Sites in Morocco, archaeology, heritage and intercultural dialogue in the Mediterranean basin. He participated in the special expert meeting on the concept of outstanding universal value in Kazan, Russian Federation in 2005.

Nationality

Moroccan

Dates of Participation in World Heritage

1992–2005

Role

Abdelaziz Touri led the Moroccan delegation for World Heritage and served as Chairperson of the twenty-third session of the World Heritage Committee in 1999.

Russell Train

A graduate of Princeton and educated as a lawyer at Columbia University, Russell Train worked as a legal advisor to Congressional committees and as a judge of

Figure A.38   Russell Train in Springfield, 2008

the United States Tax Court. His lifelong interest in wildlife and environmental conservation led to several leadership appointments. He became President of the Conservation Foundation from 1965 to 1969, Under-Secretary of the Department of the Interior from 1969 to 1970, and Administrator of the Environmental Protection Agency (EPA) under Presidents Nixon and Ford from 1973 to 1977. He had a long-lasting involvement with the World Wildlife Fund as a founding member, President and Chairman from 1978 to 1990 and Chairman of the National Council from 1994 to 2001. With regard to World Heritage, he participated in the 1965 White House Conference on International Cooperation where the idea of a World Heritage Trust took form. He promoted the concept in international circles and eventually got an opportunity, when he was appointed first Chairman of the Council on Environmental Quality (CEQ) in the Executive Office of the President from 1970 to 1973, to insert a commitment to a World Heritage initiative into the 1971 Presidential message.

Nationality

American

Dates of Participation in World Heritage

1965–1972; 1992; 2002

Role

Russell Train promoted the concept of a World Heritage Trust in the years leading up to the creation of the World Heritage Convention. He participated in meetings to celebrate the 20th and 30th anniversaries of the World Heritage Convention in Santa Fe and Venice respectively.

Licia Vlad Borrelli

Figure A.39   Licia Vlad Borrelli in Rome, 2010

Licia Vlad Borrelli studied archaeology at the University of Florence and the Archaeological School of Rome and Athens from 1946 to 1950. At the Istituto Centrale per Restauro, she was responsible for the archaeological sector from 1950 to 1974 before joining the Department of Cultural Affairs as an archaeological inspector, a position she held until 1991. Her involvement with World Heritage included regular attendance at statutory meetings and leadership of the Committee as Chairperson in 1983 and rapporteur in 1990. She noted the importance of hosting the 1983 session in Florence, at that time a World Heritage Site, because it demonstrated to the Committee the challenges of conserving heritage values in a living city. She published often on subjects related to classical archaeology and technical conservation including *Il restauro archeologico: Storia e materiali* in 2003. In recognition of her stellar career in heritage conservation, Licia Vlad Borrelli was awarded a gold medal of merit for culture by the government of Italy.

Nationality

Italian

Dates of Participation in World Heritage

1982–1999

Role

Licia Vlad-Borrelli participated as a cultural heritage expert in the State Party delegation to World Heritage. She chaired the seventh session of the World Heritage Committee in Florence in 1983 and was rapporteur of the fourteenth session in Banff in 1990. She also attended the 1992 Washington expert meeting to prepare the strategic orientations.

Raymond Wanner

Figure A.40   Raymond Wanner in Springfield, 2011

Ray Wanner has a Ph.D. in educational history and comparative education from the University of Pennsylvania. Dr Wanner entered diplomatic service, working for the State Department of the United States of America for three decades from 1972 to 2002. During that time he served at the United States permanent delegation to UNESCO in Paris until the country's withdrawal from that organization in 1984. He then became the UNESCO desk officer in Washington until his retirement from federal public service. He subsequently became senior advisor to the United Nations Foundation (UNF), working closely with the World Heritage Centre in an innovative way to pursue one of the UNF's major priorities, namely the preservation of biodiversity. Throughout his career, he retained his interest in international educational systems, producing scholarly publications and engaging actively in the work of the International Institute for Educational Planning. He has been honoured with the Secretary of State's Career Achievement Award and the UNESCO Human Rights Medal for his commitment to international intellectual cooperation.

Nationality

American

Dates of Participation in World Heritage

1982–2010

Role

As part of the American mission to UNESCO, Ray Wanner supported American World Heritage delegations from 1982 to 1984. He then served in Washington as the UNESCO desk officer at the State Department from 1984 to 2002. Following his retirement from government, he attended World Heritage sessions on behalf of the United Nations Foundation from 2002 to 2010.

Bibliography

Manuscript Sources

IUCN, Gland, Resolutions of the first to the nineteenth General Assembly of IUCN, 1948–1994.

John F. Kennedy Presidential Library and Museum, Washington, Samuel E. Belk personal papers, International Cooperation Year committee reports, box 10, 1965–1966.

Nixon, Richard, Special message to the Congress proposing the 1971 environmental program, 8 February 1971, art. IV.

Parks Canada, Gatineau, World Heritage archives, file 1972–1978.

UNESCO, Paris, Records of the third session to the seventeenth session of the General Conference of UNESCO, 1948–1972.

UNESCO Archives, Paris, World Heritage files, 1968–2000.

Université de Montréal, Canada Research Chair on Built Heritage, interviews with World Heritage pioneers, 2006–2012.

University of Maryland, Hornbake Library, Papers of Ernest A. Connally, 1967–1997.

United States National Parks Service, Washington, World Heritage archives, file 1973–1975.

Printed Primary Sources

Adams, Alexander B., *First World Conference on National Parks* (Washington: US Department of the Interior, 1962).

Batisse, Michel, "The struggle to save our world heritage," *Environment*, 34/10 (1992): 12–32.

Eidsvik, Harold K., "Guest Comment: Plitvice National Parks, World Heritage Site and the Wars in the Former Yugoslavia," *Environmental Conservation*, 20 (1993): 293.

Feilden, Bernard and Jukka Jokilehto, *Management Guidelines for World Cultural Heritage Sites* (Rome: ICCROM, 1993).

Van Hooff, Herman, "The monitoring and reporting of the state of properties inscribed on the World Heritage List," *ICOMOS Canada Bulletin*, 4/3 (1995): 12–14.

ICOMOS Australia, *The Australia ICOMOS guidelines for the conservation of places of cultural significance* (Burra Charter) (n.p.: Australia ICOMOS, 1979).

IUCN, *Proceedings of the ninth General Assembly at Lucerne 25 June to 2 July 1966* (Morges: IUCN, 1967).

IUCN Commission on National Parks and Protected Areas, *The World's Greatest Natural Areas: an indicative inventory of natural sites of World Heritage quality* (Gland: IUCN, 1982).

IUCN International Commission on National Parks, *United Nations List of National Parks and Equivalent Reserves: Part Two and Addenda to Part One* (Morges: IUCN, 1962).

Larsen, Knut Einar (ed.), *Nara Conference on authenticity in relation to the World Heritage Convention: Proceedings, Nara, Japan, 1–6 November 1994* (Paris: UNESCO World Heritage Centre/Japan Agency for Cultural Affairs, 1995).

Linstrum, Derek, "An alternative approach? An interview with Anne Raidl," *Momentum,* special issue (1984): 50–55.

Mishra, Hemanta and N. Ishwaran, "Summary and conclusions of the workshop on the World Heritage Convention held during the IV World Congress on national parks and protected areas Caracas, Venezuela, February, 1992," *World Heritage: twenty years later. Based on papers presented at the World Heritage and other workshops held during the IVth World Congress on National Parks and Protected Areas, Caracas, Venezuela, February 1992* (Gland and Cambridge: IUCN, 1992).

Parent, Michel, "La problématique du Patrimoine Mondial Culturel," *Momentum,* special issue (1984): 33–49.

Purdum, Todd S., "Clinton unveils plan to halt gold mine near Yellowstone," *The New York Times,* 13 August 1996.

Saouma-Ferero, Galia (ed.), *L'Authenticité et l'intégrité dans un contexte africain: réunion d'experts, Grand Zimbabwe, 26–9 mai 2000/Authenticity and integrity in an Africain context: expert meeting, Great Zimbabwe, 26–9 May 2000* (Paris: UNESCO, 2001).

Slatyer, Ralph O., "The Origin and Evolution of the World Heritage Convention," *Ambio,* 12/3–4 (1983): 138–40.

Stovel, Herb, *Safeguarding historic urban ensembles in a time of change: a Management Guide* (Hull: Parks Canada, 1991).

Stovel, Herb, "Monitoring world cultural heritage sites," *ICOMOS Canada Bulletin,* 4/3 (1995): 15–20.

Stovel, Herb, *Risk Preparedness: a Management Manual for World Cultural Heritage* (Rome: ICCROM, 1998).

Thorsell, Jim, "From Strength to Strength: World Heritage in its 20th Year," *World Heritage twenty years later. Based on papers presented at the World Heritage and other workshops held during the IVth World Congress on National Parks and Protected Areas, Caracas, Venezuela, February 1992* (Gland and Cambridge: IUCN, 1992): 19-26.

Thorsell, Jim, "Nature's Hall of Fame: IUCN and the World Heritage Convention," *Parks,* 7/2 (1997): 7.

Thorsell, Jim and Jacqueline Sawyer (eds), *World Heritage twenty years later. Based on papers presented at the World Heritage and other workshops held during the IVth World Congress on National Parks and Protected Areas, Caracas, Venezuela, February 1992* (Gland and Cambridge: IUCN, 1992).

Thorsell, Jim and Jim Paine, "An IUCN/WCMC perspective on safeguarding the integrity of cultural heritage properties," *ICOMOS Canada Bulletin*, 4/3 (1995): 21–3.

The Athens Charter for the Restoration of Historic Monuments (Athens: First International Congress of Architects and Technicians of Historic Monuments, 1931).

The monument for the man: records of the II International Congress of Restoration (Venice: ICOMOS, 1964).

UNESCO, IUCN and ICME, *Technical Workshop: World Heritage and Mining, Technical Workshop, 21–23 September 2000 Gland, Switzerland* (Paris: UNESCO, 2001).

Wells, Roderick T., *Earth's Geological History: a contextual framework for assessment of World Heritage fossil site nominations* (Gland: IUCN, 1996).

Secondary Sources

Association des anciens fonctionnaires de l'UNESCO (ed.), *L'UNESCO racontée par ses anciens* (Paris: UNESCO, 2006).

Batisse, Michel and Gérard Bolla, *The Invention of World Heritage* (Paris: Association of Former UNESCO Staff Members, 2005).

Cameron, Christina, "From Warsaw to Mostar: the World Heritage Committee and Authenticity," *Bulletin of the Association for Preservation Technology*, 39/ 2–3 (2008): 19–24.

Cameron, Christina, "The Evolution of the Concept of Outstanding Universal Value," *Conserving the Authentic: Essays in Honour of Jukka Jokilehto*, Nicholas Stanley-Price and Joseph King (eds) (Rome: ICCROM, 2009): 127–36.

Cameron, Christina and Mechtild Rössler, "Voices of the Pioneers: UNESCO's World Heritage Convention 1972–2000," *Journal of Cultural Heritage Management and Sustainable Development*, 1/1 (2011): 42–54.

Cameron, Christina and Mechtild Rössler, "The shift towards conservation: early history of the 1972 World Heritage Convention and global heritage conservation," *Understanding Heritage. Perspectives in Heritage Studies*, Marie-Theres Albert, Roland Bernecker and Britta Rudolff (eds) (Berlin/ Boston: De Gruyter, 2013), 69-76.

Von Droste, Bernd, "A Gift from the Past to the Future. Natural and cultural world heritage," in UNESCO, *Sixty Years of Science at UNESCO 1945–2005* (Paris: UNESCO, 2006): 389–400.

Von Droste, Bernd et al., *Cultural Landscapes of Universal Value: Components of a Global Strategy* (New York: Gustav Fischer Verlag Jena, 1995).

Von Droste, Bernd et al. (eds), *Linking Nature and Culture: Report of the Global Strategy Natural and Cultural Heritage Expert Meeting 25 to 29 March 1998, Amsterdam, Netherlands* (The Hague: UNESCO/Ministry of Foreign Affairs, 1998).

Francioni, Francesco, "Thirty Years on: is the World Heritage Convention ready for the 21ˢᵗ century?,"*The Italian Yearbook of International Law*, 12 (2002): 1–38.

Francioni, Francesco and Federico Lenzerini (eds), *The 1972 World Heritage Convention: a Commentary* (Oxford: Oxford University Press, 2008).

Gillespie, Alexander, *Protected Areas and International Environmental Law* (Leiden/Boston: Martinus Nijhoff, 2008).

Hazen, Helen Diane, *The Role of the World Heritage Convention in Protecting Natural Areas* (St. Paul: University of Minnesota unpublished doctoral thesis, 2006).

Holdgate, Martin, *The Green Web: A Union for World Conservation* (London: Earthscan, 1999).

Van Hooff, Herman, "Monitoring and Reporting in the Context of the World Heritage Convention and its Application in Latin America and the Caribbean," in *Monitoring World Heritage, World Heritage 2002, Shared Legacy, Common Responsibility, Associated Workshops 11–12 November 2002, Vicenza, Italy* (Paris: UNESCO World Heritage Centre, 2004): 32–8.

IUCN, *50 Years of Working for Protected Areas: a brief history of IUCN World Commission on Protected Areas* (Gland: IUCN, 2010).

Jerome, Pamela, (ed.), "Special edition on authenticity," *Bulletin of the Association for Preservation Technology*, 39/ 2–3 (2008): 1–71.

Jokilehto, Jukka, "Considerations on authenticity and integrity in World Heritage context," *City and Time*, 2/1 (2006): 1–16.

Jokilehto, Jukka, "World Heritage: defining the outstanding universal value," *City and Time*, 2/2 (2006): 1–10.

Jokilehto, Jukka, et al., *The World Heritage List: What is OUV? Defining the Outstanding Universal Value of Cultural World Heritage Properties* (Berlin: Hendrik Bässler Verlag, 2008).

Jokilehto, Jukka, *ICCROM and the Conservation of Cultural Heritage: A History of the Organization's First 50 Years, 1959–2009* (Rome: ICCROM, 2011).

Keune, Russell V., "An interview with Hiroshi Daifuku," *CRM: The Journal of Heritage Stewardship*, 8/1 and 2 (2011): 31–45.

Labadi, Sophia, "A review of the Global Strategy for a Balanced, Representative and Credible World Heritage List 1994–2004," *Conservation and Management of Archaeological Sites*, 7/2 (2005): 89–102.

Lemaire, Raymond, "Report of the President of ICOMOS Piero Gazzola 1965–1975: a tribute to Piero Gazzola," *Thirty Years of ICOMOS* (Paris: ICOMOS, 1995).

McCormick, John, *Reclaiming Paradise: the Global Environmental Movement* (Bloomington: Indiana University Press, 1991).

McCormick, John, *The Global Environmental Movement* (Chichester: Wiley, 1995).

Maurel, Chloé, *Histoire de l'UNESCO – les trente premières années, 1945–1974* (Paris: l'Harmattan, 2012).

Mitchell, Nora J., "Considering the Authenticity of Cultural Landscapes," *Bulletin of the Association for Preservation Technology*, 39/ 2–3 (2008): 25–32.

Musitelli, Jean, "Opinion: World Heritage, between Universalism and Globalization," *International Journal of Cultural Property*, 11/2 (2002): 323–36.

Pederson, Arthur, *Managing Tourism at World Heritage Sites: a Practical Manual for World Heritage Site Managers* (Paris: UNESCO World Heritage Centre, 2002).

Poisson, Olivier, "Le patrimoine mondial, trente ans," *Monumental, Revue scientifique et technique des monuments historiques*, 1 (2008): 8–12.

Pressouyre, Léon, *The World Heritage Convention, Twenty Years Later* (Paris: UNESCO, 1996).

Reeves, Richard, *President Nixon alone in the White House* (New York: Simon & Schuster, 2001).

Rosabal, Pedro and Mechtild Rössler, "A model of teamwork: El Vizcaino," *World Conservation*, 2 (2001): 21.

Rössler, Mechtild, "World Heritage Cultural Landscapes: A UNESCO Flagship Programme 1992–2006," *Landscape Research*, 31/4 (2006): 333–53.

Rössler, Mechtild, "Applying Authenticity to Cultural Landscapes," *Bulletin of the Association for Preservation Technology*, 39/ 2–3 (2008): 47–52.

Rössler, Mechtild, "La liste du patrimoine mondial en péril; un outil de la Convention de 1972," *Monumental, Revue scientifique et technique des monuments historiques*, 1 (2008): 33.

Shen, Juan, *Patrimoine et audiovisuel, l'enjeu de la médiation scientifique dans la communication des connaissances* (Paris: Université Panthéon-Assas Paris II doctoral thesis, 2010).

Smith, Laurajane, *The Uses of Heritage* (Abingdon: Routledge, 2006).

Stanley-Price, Nicholas and Joe King (eds), *Conserving the Authentic: Essays in honour of Jukka Jokilehto* (Rome: ICCROM, 2009).

Stott, Peter H., "The World Heritage Convention and the National Park Service, 1962–1972," *The George Wright Forum*, 28/3 (2011): 279–90.

Stott, Peter H., "The World Heritage Convention and the National Park Service: The First Two Decades, 1972–1992," *The George Wright Forum*, 29/1 (2012): 148–175.

Stovel, Herb, "An advisory body view of the development of monitoring for World Cultural Heritage," in *Monitoring World Heritage, World Heritage 2002, Shared Legacy, Common Responsibility, Associated Workshops 11–12 November 2002, Vicenza, Italy* (Paris: UNESCO World Heritage Centre, 2002): 17–21.

Stovel, Herb, "Effective use of authenticity and integrity as World Heritage qualifying conditions," *City & Time*, 2/3 (2007): 21–36.

Stovel, Herb, "Origins and influence of the Nara Document on Authenticity," *Bulletin of the Association for Preservation Technology*, 39/ 2–3 (2008): 9–18.

Thorsell, Jim, *World Heritage Convention: effectiveness 1992–2002 and lessons for governance* (Gland: IUCN, 2003).

Titchen, Sarah, *On the construction of outstanding universal value: UNESCO's World Heritage Convention (Convention concerning the Protection of the World Cultural and Natural Heritage, 1972) and the identification and assessment of cultural places for inclusion in the World Heritage List* (Canberra: Australian National University unpublished doctoral thesis, 1995).

Train, Russell E., "World Heritage: A Vision for the Future," *World Heritage 2002: Shared Legacy, Common Responsibility* (Paris: UNESCO World Heritage Centre, 2003).

Turtinen, Jan, *Globalising Heritage: On UNESCO and the Transnational Construction of a World Heritage* (Stockholm: Stockholm Center for Organizational Research, 2000).

UNESCO, *Investing in World Heritage: past achievements, future ambitions. A guide to International Assistance* (Paris: UNESCO World Heritage Centre, 2002).

UNESCO, *Linking Universal and Local Values: Managing a Sustainable Future for World Heritage* (Paris: UNESCO World Heritage Centre, 2004).

UNESCO, *60 ans d'histoire de l'UNESCO. Actes du colloque international, Paris, 16–18 novembre 2005* (Paris: UNESCO, 2007).

"UNESCO's world heritage sites: a danger list in danger," *The Economist*, 26 August 2010.

World Heritage 2002: Shared Legacy, Common Responsibility (Paris: UNESCO, 2003).

Index

References to illustrations are in **bold**

Abu Simbel and Philae (Nubian
    monuments), Egypt 12, 104
advisory bodies 174–201
Africa, cultural heritage 91–2
Air and Ténéré Nature Reserves, Niger 70
ALECSO 217
Algeria, Tipasa 113
American Draft Convention 17–20, 28, 124
Angkor ruins, Cambodia 3, **138**, 139–41
    on Danger List 140, 141
Arch Foundation 218
Archaeologists, European Association of 254
Athens Charter for the Restoration of
    Historic Monuments 1
Auschwitz-Birkenau, Poland 36, 167, 174,
    223, 225
Australia 133–4
    Great Barrier Reef 60
    Kakadu National Park 145, 147, 172,
        174, 185, 218, 223, 229–31, **229**
    Wet Tropics of Queensland 170
    Willandra Lakes Region 119
authenticity
    attributes 90
    concept evolution 101
    cultural, dependency of 88
        landscapes 68, 240, 241
        sites 85–90
        value 97–8
    definition 39, 85, 89
    Nara conference
        discussions 86–90
        participants **86**
    Nara document 87
    Warsaw, reconstructed market square
        40–42, **40**
    World Heritage List 39–42

Barthélemy, Jean 191
Batisse, Michel 4, 7, 8, 10, 11, 17, 20, 23,
    24, 49, 62, 196, 250
    photograph **9**
    proposals for rural landscapes 65, **65**
Benin, Royal Palaces of Abomey 112
Bennett, Peter 25, 163
Beschaouch, Azedine 50, 80, 88, 140, 240,
    242
    awards 249
    photographs **66**, **129**, **248**
    role 249
    vignette 248–9
Biodiversity, Convention on 234
Bolla, Gérard 21, 22, 23, 24, 25, 26
    photographs **48**, **249**
    role 250
    vignette 250–51
Borobudur, Indonesia 104
Bouchenaki, Mounir 205, 208, 213–14
    ICCROM Award 251
    photograph **251**
    role 251
    vignette 250–51
Brasilia, Brazil 77
Brichet, Robert 14, 16, 17
Budowski, Gerardo 175

Cambodia, Angkor ruins 3, **138**, 139–41
Cameron, Christina, photographs **66**, **129**
Canaima National Park, Venezuela **180**
canals, heritage 95
Caspary, Hans 227, 232
    photograph **252**
    role 252
    vignette 251–2
Causeway Coast, Northern Ireland 60

Chabason, Lucien 60–61
    photograph **253**
    role 253–4
    vignette 252–3
Chamberlain, James, photograph **66**
Chastel, André 188
China, Sichuan Giant Panda Sanctuaries
        121
civil society, and World Heritage
        Committee 216–19
Cleere, Henry 55, 82, 98–9, 140, 192–3,
        193–4, 208–9, 224, 241–2
    archaeological heritage award 254
    photographs **88, 192, 199, 254**
    publications
        *Approaches to the Archaeological
            Heritage* 254
        *Archaeological Heritage
            Management in the Modern
            World* 254
    role 255
    vignette 254–5
climate change xv
Clinton, President 146
Collinson, Jim 53, 56, 66, 69, 121, 166,
        217, 232
    photograph **255**
    role 256
    vignette 255–6
Connally, Ernest 188
Coolidge, Harold J. 5, 8
Côte d'Ivoire/Guinea
    Mount Nimba Strict Nature Reserve
        111, 137, 145
    Tai National Park 111
Croatia
    Dubrovnik, Old City 136, 223, 227,
        227–8, **227**
    Plitvice Lakes National Park 136
cultural heritage
    Africa 91–2
    concept 240
    intangible 123
    qualified personnel 152
cultural heritage sites 12, 120
    global inventory 194
    and natural heritage sites, balance
        58–71

cultural landscapes 66, 122, 222
    authenticity 68, 240, 241
    definition problems 69–70
    global survey 68–9
    types 67–8
    see also rural landscapes

Daifuku, Hiroshi 13, 14, 185
Dalibard, Jacques 126
Daoulatli, Abdelaziz 191
De Angelis d'Ossat, Guglielmo 14
de Guichen, Gael 198
Democratic Republic of the Congo
        (formerly Zaire)
    Garamba National Park 111, 115, 137
    Kahuzi-Biega National Park 115, **117**
    Okapi Wildlife Reserve 115
    Salonga National Park 115
    Virunga National Park 115
    World Heritage in Danger List, sites 116
Di Castri, Francesco, photograph **9**
Diaz-Berrio, Salvador, photograph **66**
Djoudj National Bird Sanctuary, Senegal
        111
DoCoMoMo 96
Doñana National Park, Spain 145, 148
Droste, Bernd von 27, 49, 51–2, 55, 63, 88,
        124–5, 128–9, 137, 146, 163, 167,
        169, 172, 201, 204, 208, 209, 213,
        224–5
    criticism of ICOMOS 194
    photographs **129, 202, 216, 256**
    role 257
    vignette 256–7
    World Heritage Centre
        director 207
        vision for 205
Dubrovnik, Old City, Croatia 136, 223,
        227, 227–8
    photograph **227**
Dumesnil, Catherine 222, 238–9
    photograph **257**
    role 258
    vignette 258
Durighello, Regina 56–7
    photograph **258**
    role 259
    vignette 259

Durmitor National Park, Montenegro 111

East Rennell, Solomon Islands 93, 123
ECOSOC 3
Ecuador, Galapagos Islands **138**, 139,
        141–4, 185
Egypt, Abu Simbel and Philae (Nubian
        monuments) 12, 104
Eidsvik, Hal 50–51, 108, 136, 144, 146,
        147, 164, 178, 184, 234
    IUCN International Parks Merit Award
        260
    photograph **260**
    role 260
    Royal Canadian Geographic Society,
        gold medal 260
    vignette 259–60
El Vizcaino Whale Sanctuary, Mexico 115,
        116–17, **118**, 218
Environmental Diplomacy Institute 218
Erder, Cevat 191
Ethiopia, Simien National Park 111, 151
European Landscape Convention 94
Evans, Luther, photograph **243**

FAO 2, 177
Feilden, Bernard 55, 120–21
    Aga Khan award 261
    ICCROM Award 261
    ICOMOS Piero Gazzola Prize 261
    photograph **261**
    publications
        Conservation of Historic Buildings
            261
        *Guidelines for the Management of
            World Cultural Heritage Sites*
            261
    role 261
    vignette 260–61
Feliu, Carmen Añón 70, 233
    ICOMOS Piero Gazzola Prize 247
    Jardins en Espagne 247
    photographs **199**, **247**
    role 248
    vignette 247–8
Firouz, Eskandar 175
Fisher, Joseph 4–5, 5–6, 6–7
    photograph **6**

Florence Charter (1981) 247
Fondation patrimoine historique
        international 218
Ford, President 289
France-UNESCO Cooperation Agreement
        153
Francioni, Francesco 56, 147, 165, 181,
        196, 211, 228, 230–31, 236, 238
    photograph **262**
    role 262–3
    *The 1972 World Heritage Convention:
        a Commentary* 262
    vignette 262
Franco, Maria-Theresa 143, 269
Friends of the Earth 218

Galapagos Islands, Ecuador **138**, 139,
        141–4, 185
Garamba National Park, Democratic
        Republic of the Congo 111, 115, 137
Gaudi, Antonio 225
Gazaneo, Jorge, photograph **202**
Gazzola, Piero 16, 186, 187
    photograph **186**
    Prize 247, 261, 266, 285
Giant's Causeway and Causeway Coast,
        Northern Ireland 60
Global Environment Facility 234, 243
Great Barrier Reef, Australia 60
Greece
    Meteora 60, 181, **182**
    Mystras 106
Guo Zhan 56, 89, 209, 235, 244–5
    photograph **263**
    role 264
    vignette 263–4

Hales, David 33
heritage
    canals 95
    conservation 103–7
    and development 236
    and mass tourism 235–6
    routes 95
    theory/practice, and World Heritage
        Convention 239–43, 244
    see also cultural heritage; natural
        heritage

High Tech Visual Promotion Centre 218
Hillig, Jürgen, photograph **202**
Hiroshima Peace Memorial, Japan 224,
     225
Holdgate, Martin 3, 5, 175
Hooff, Herman van 131, 135
Huxley, Julian 175
     photograph **243**

IBE 211, 212
ICCROM 29, 43
     Award 261, 267, 286
     establishment 11, 197
     ITUC program 200
     publications
          *Management Guidelines for World
             Cultural Heritage Sites* 199
          Risk Preparedness 199
     training activities 199–200
     World Heritage, input 198–201
Iceland, Thingvellir, parliament site 3
Ichkeul National Park, Tunisia 106–9, **107**,
     109
     on Danger List 108
ICME 148, 149, 218
ICOM 23, 128, 186, 217
ICOMOS 29, 33, 35–6, 39, 43, 52
     architectural structures, study 195
     difficulties 196–7
     establishment 13, 185–7
     evaluations 190–91, 193, 194–5
     field missions 193
     Historic Gardens Charter (1981) 247
     on landscapes 62
     scientific committees 193
     von Droste's criticism of 194
     World Heritage
          input 187–97
          management handbook 195
          monitoring system 195–6
     see also Venice Charter
IIEP 211, 212, 291
IFLA 62, 217
IFPC 217
ILO 2
India 133
     Manas Wildlife Sanctuary 137
Indonesia, Borobudur 104

International Centre for Conservation 197
International Charter for the Conservation
     and Restoration of Monuments and
     Sites (Venice Charter) *see* Venice
     Charter
International Committee on Monuments,
     Artistic and Historical Sites and
     Archaeological Excavations (1951)
     11
International Cooperation Year (1965) 4
International Federation of Shingon
     Buddhism 218
International Fund for Animal Welfare 218
International Museums Office 1
International Organization for the
     Protection of the Arts 217
Ishwaran, Natarajan 57, 184, 234
     photograph **264**
     role 265
     vignette 264–5
Island of Gorée, slave site, Senegal 36,
     223, 225
Italy, Venice, threats to 104, 113
Ito, Nobuo 123
     ICOMOS Piero Gazzola prize 266
     photograph **265**
     role 266
     vignette 265–6
ITUC program, ICCROM 200
IUCN (formerly IUPN) 20, 29–30, 37–8,
     43, 178–85
     Convention on Conservation of the
          World Heritage 20
     establishment 175–8
     International Parks Merit Award 260,
          276, 287
     National Parks Conference (1962) 3–4
     and natural heritage sites 50–51, 181
     proposed convention 10–11
     research 183
     wilderness bias 59–60
     World Heritage, input 176–7
     on World Heritage Committee 178, 179
     World Heritage Sites, monitoring
          proposals 109–10
     The World's Greatest Natural Areas 51,
          59, 78, 179, 181
     see also WCPA

IUPN 2, 175
 foundation 104
 see also IUCN
IWGC 20

Japan, Hiroshima Peace Memorial 224,
 225
Japanese Trust Fund for the Preservation of
 the World Cultural Heritage 153
Jerusalem
 Old City and Walls 136, 167, **168**, 170,
 174
 sovereignty issue 168–9
Jokilehto, Jukka 55, 84, 89, 98, 121, 197,
 198, 222–3
 ICCROM Award 267
 photographs **199**, **266**
 publications
 Guidelines for the Management of
 World Cultural Heritage Sites
 267
 The World Heritage List 267
 role 267
 vignette 266–7
Jones, Barry 134

Kahuzi-Biega National Park, Democratic
 Republic of the Congo 115, **117**
Kakadu National Park, Australia 145, 147,
 172, 174, 185, 218, 223, 229–31,
 229
Khawaykie, Elisabeth, photograph **216**
Komatsu, Taro, photograph **216**

La Petite Pierre group 68, 70
Lake District National Park, UK **64**, 66, 67
landscapes
 Andean 94
 terrace 93
 see also cultural landscapes; rural
 landscapes
Las Médulas, Spain 226
l'Association pour la sauvegarde de la
 Casbah d'Alger 218
League of Arab States 15
Leblanc, François 52, 126, 163–4, 169,
 187, 188, 240
 photograph **268**

role 268
 vignette 267–8
Lemaire, Raymond 16, 50, 86, 89, 186,
 187, 188, 191
 photograph **87**
les Amis du patrimoine du Maroc 218
Levi-Strauss, Laurent 129
Loomis, Henry 17
Lopes, Henri 227
Lucas, Bing 67, 185
Luxen, Jean-Louis 57, 88–9, 172, 173, 194,
 233, 235–6, 240, 241
 photographs **199**, **270**
 role 271
 vignette 270

McCormick, John 2
McNeely, Jeff 51, 60, 119, 146, 153, 178,
 179, 180, 241
 photograph **274**
 role 275
 vignette 274–5
Maheu, René 12, 13, 21, 23
 photographs **22**, **243**
Makagiansar, Makaminan, photograph **202**
Man and the Biosphere (MAB),
 programme 9, 10, 256, 283
Manas Wildlife Sanctuary, India 137
Matsuura, Koïchiro xiii, 55, 87, 93, 99,
 118, 123, 173, 214, 218–19
 photographs **243**, **271**
 role 272
 vignette 271
 World Heritage 271
Matteucci, Mario 16, 17
Maurel, Chloé 8, 12
Mauretania 158
Mayor, Federico 139, 204, 205–6, 207,
 214, 240, 244
 photographs **243**, **272**
 role 273
 vignette 272
M'Bow, Amadou-Mahtar 77, 173, 197,
 244
 photographs **243**, **273**
 role 274
 vignette 273
Meteora, Greece 60, 181, **182**

Mexico, El Vizcaino Whale Sanctuary 115,
116–17, **118**, 218
military conflict, and World Heritage in
Danger List 136
Millon, Henry 188
Milne, Rob 49, 60, 164, 174, 184, 194,
215, 233–4, 238
Dubrovnik Gold Medal for World
Heritage 276
IUCN International Parks Merit Award
276
photograph **275**
role 276
vignette 275–6
mining
and World Heritage 145–50
World Heritage, workshop 149
Monastery of the Hieronymites, Portugal 113
Montenegro
Durmitor National Park 111
Natural and Culturo-Historical Region
of Kotor 135–6
on Danger List 136
Morales, Lopez 100
Arquitectura vernácula en México 269
photograph **269**
role 269
vignette 268–9
Mount Nimba Strict Nature Reserve, Côte
d'Ivoire/Guinea 111, 137
on Danger List 145
Munjeri, Dawson 91–2, 101, 196–7, 237,
240–41
photograph **277**
role 277
vignette 276–7
Mutal, Sylvio 131
Mystras, Greece 106

Nara conference
authenticity, discussions 86–90
participants **86**
national governments *see* States Parties
"natural", redefinition of 97
Natural and Culturo-Historical Region of
Kotor, Montenegro 135–6, 152
natural heritage 2, 3, 12
UNESCO 7

natural heritage sites 50, 51, 181, 185, 200
and cultural heritage sites, balance
58–71
Natural Resources Defence Council 218
New Zealand, Tongariro National Park **69**,
123
Ngorongoro Conservation Area, Tanzania
111, 152, 185
on Danger List 110
Nicholls, Frank 10, 175
Niger 97
Air and Ténéré Nature Reserves, Niger
70
W National Park 170, **171**, 174
Nixon, President 17, 18, 19, 289
Nordic World Heritage Office 92, 94, 153,
210, 216
Northern Ireland, Giant's Causeway and
Causeway Coast 60
Nubian monuments 12, 104

Okapi Wildlife Reserve, Democratic
Republic of the Congo 115
Olyff, Michel 58
OMMSA 217
OWHC 218

Pacific Island peoples 93
Palladio, Antonio, buildings 225
Palmyra, Syria 105–6
Parent, Michel 23, 24, 33, 34, 36, 41, 203,
225
photograph **34**
President, ICOMOS 75
Pavlic, Breda, photograph **216**
Philippot, Paul 197, 198
Phillips, Adrian 51, 67, 209, 238, 242
CBE award 278
photograph **278**
role 278
vignette 277–8
Plachter, Harald 236
Award for Nature Conservation 279
photograph **279**
role 280
vignette 278–9
Plitvice Lakes National Park, Croatia
136

Poland
  Auschwitz-Birkenau 167, 174, 223, 225
  Wieliczka Salt Mine 111
  see also Warsaw
Portugal
  Monastery of the Hieronymites 113
  Tower of Bélem 113
Pressouyre, Léon 52, 54, 60, 67, 70, 86,
    101–2, 114, 163, 166, 168, 169–70,
    181, 201, 231–2, 236, 237, 238
  photographs **189, 280**
  resignation from World Heritage 191
  role 281
  vignette 280
  working methods 188–90
  The World Heritage Convention,
    Twenty Years Later 280
Price, Nicholas Stanley, and Joe King,
    Conserving the Authentic: Essays
    in honour of Jukka Jokilehto 267
Pro Esteros Mexico 218

Queensland, Wet Tropics 170, 174

Raidl, Anne 48–9, 78, 80, 112, 120, 174,
    198, 199, 201, 202, 204
  photographs **48, 202, 281**
  role 282
  vignette 281
Ramsar Convention 109
Reynolds, John 224
  Distinguished Service Award 283
  photograph **282**
  role 283
  vignette 282–3
Rhône-Poulenc Foundation 216
Rivière, Georges-Henri xiii, 186
Robben Island prison, South Africa 224, 225
Robertson Vernhes, Jane 49, 61, 65–6, 120,
    163, 166, 170, 215, 233, 237, 238
  photograph **283**
  role 284
  vignette 283–4
Royal Palaces of Abomey, Benin 112
rural landscapes
  Batisse's proposal **65**, 66
    criticism of 66–7
  ICOMOS on 62

identification of 62–4
  World Heritage List 59–71
  see also cultural landscapes
Rwenzori Mountains National Park,
    Uganda 137

Sadat, Anwar, President 60
Salans, Carl 22, 24
Salonga National Park, Democratic
    Republic of the Congo 115
Scott, Peter 4
Second International Congress of
    Architects and Specialists of
    Historic Monuments (1964) 13
Senegal
  Djoudj National Bird Sanctuary 111
  Island of Gorée, slave site 36, 223, 225
Sheppard, David 185
Sichuan Giant Panda Sanctuaries, China
    121
Silva, Roland 86, 191, 193
  ICOMOS Piero Gazzola Prize 285
  photograph **284**
  role 285
  vignette 284–5
Simien National Park, Ethiopia 111, 151
sites of conscience 223–5, 244
Slatyer, Ralph 58, 165, 190, 230
Smith, Laurajane 241
Solomon Islands, East Rennell 93, 123
Sorlin, François 16
South Africa, Robben Island prison 224,
    225
Spain
  Doñana National Park 145, 148
  Las Médulas 226
Stantcheva, Magdalina, photograph **66**
States Parties 155–74, 221
  General Assembly of 155–62
  meaning 155
Steltzer, Helmut 191
Stockholm Conference on the Human
    Environment (1972) 3, 9, 10, 217
Stott, Peter 4, 10
Stovel, Herb 57, 79, 81–2, 84–5, 86, 91,
    99–100, 101, 126, 131–2, 166, 185,
    191–2, 193–4, 196, 200, 201–2,
    208, 209, 239, 241

ICCROM Award 286
  photographs **88, 285**
  *Risk Preparedness: A Management
    Manual for World Cultural
    Heritage* 286
  role 286
  vignette 285–6
Strong, Maurice 22, 175
Sturgill, R.G. 25
sustainable development 9
Sylla, Seydina, photograph **66**
Syria, Palmyra 105–6

Tai National Park, Cote d'Ivoire/Guinea 111
Talbot, Lee 8, 10, 11
Tanzania, Ngorongoro Conservation Area
    111, 152, 185
  on Danger List 110
Taralon, Jean 188
Taylor, Leslie, photograph **66**
Thingvellir, parliament site, Iceland 3
Thomas, Jean 25
Thorsell, Jim 57–8, 110, 165, 170, 178,
    179–81, 209, 242–3
  in Canaima National Park, Venezuela
    180
  Harkin Conservation Medal 287
  influence 184
  IUCN International Parks Merit Award
    287
  photographs **180, 287**
  publications
    World Heritage Convention 287
    World Heritage Twenty Years Later
      287
  role 287
  vignette 286–7
TICCIH 96
Tipasa, Algeria 113
Titchen, Sarah xiv, 1
  photograph **216**
Tongariro National Park, New Zealand **69**,
    123
Torres Bodet, Jaime, photograph **243**
Touri, Abdelaziz 70, 173, 196, 224
  photograph **288**
  role 288
  vignette 287–8

tourism, mass xv, 221
  drawbacks of 235
  and World Heritage Sites 234–6
Tower of Bélem, Portugal 113
towns, historic 76, 77, 187, 195
Train, Russell 4, 5, 7, 8, 17, 18, 19, 121–2
  photographs **18, 289**
  role 289
  vignette 288–9
Tunisia, Ichkeul National Park 106–9, **107,**
    109
  on Danger List 108

Uganda, Rwenzori Mountains National
    Park 137
UIA, World Heritage Committee,
    attendance at meetings 217
UK, Lake District National Park **64**, 66, 67
UN, *List of Protected Areas and Equivalent
    Reserves* 3, 10, 30, 51
UNDP 128, 131, 217
UNEP 128, 130, 235, 253
UNESCO xiii, 2, 201–16
  Cultural Heritage Division 112, 201,
    205, 208
  cultural heritage initiatives 12–14
  Cultural Sector 4, 11, 13, 17, 21
  draft convention 14–17
  experts meeting (1968) 8–9
  natural heritage 7
  and natural sciences 7–8
  World Heritage Committee, secretariat
    201–4
UNESCO Conventions
  International Protection of Monuments,
    Groups of Buildings and Sites of
    Universal Value (1971) 20
  Protection of Cultural Property in the
    Event of Armed Conflict (1954) 11,
    186, 281
  Safeguarding of the Intangible Cultural
    Heritage (2003) 123
  see also World Heritage Convention
    (1972)
UNESCO Recommendations
  International Principles applicable to
    Archaeological Excavations (1956)
    12

Preservation of Cultural Property
endangered by Public or Private
Works (1968) 13–14
Protection at National Level of the
Cultural and National Heritage
(1972) 26, 242
Safeguarding of the Beauty and
Character of Landscapes and Sites
(1962) 12–13
UNF 115–16
UNHCR 217
UNICEF 217
UNPROFOR 136
UNSCCUR 2, 103
US
Convention on the Establishment of a
World Heritage Trust 20
draft convention on the environment
17–20, 28, 124
Yellowstone National Park 19, 145–7,
177
on Danger List 146

Vanoise
meeting 96, 182
report 97
Venezuela, Canaima National Park **180**
Venice, Italy, threats to 104, 113
Venice Charter 13, 16, 41, 86, 89, 90, 104,
187
Veronese, Vittorino, photograph **243**
Virunga National Park, Democratic
Republic of the Congo 115
Vlad Borrelli, Licia 55, 61–2, 234–5
Gold Medal of Merit 290
*Il restauro archeologico: Storia e
materiali* 290
photographs **66**, **290**
role 290–91
vignette 290
Vliet, Margaret van, photograph **48**

W National Park, Niger 170, **171**, 174
Wanner, Ray 99, 193, 224
Career Achievement Award 291
photograph **291**
role 292
UNESCO Human Rights Medal 291

vignette 291
Warsaw, reconstructed market square **40**
authenticity of 40–42
WCPA 51, 148–9, 177
Wet Tropics, Queensland, Australia 170,
174
WHO 2
Wichiencharoen, Adul 133, 146
photograph **66**
Wieliczka Salt Mine, Poland 111
Willandra Lakes Region, Australia 119
World Bank 234, 256
World Conservation Monitoring Centre
130
Protected Areas Data Unit 184
World Heritage
decentralization 94
and mining 145–50
workshop 149
overlapping international agreements,
reconciliation 21–4
research on xiv
symbol 58
World Heritage Centre 91, 99, 114, 115,
130, 144, 145, 146
advisory bodies, attempts to absorb
208–9
decentralization issues 210–12
establishment 205–6
functions 207
logo 212–13
negative issues 207–9
resources 207–8
staff photograph **206**
von Droste's vision for 205
see also Nordic World Heritage Office
World Heritage Committee 27–8, 44, 52,
162–74, 204–16
advisory bodies 174–201
and civil society 216–19
collegiality 163
expertise, lack of 164–5
functions 162
ICOM attendance 217
IFLA attendance 217
IUCN in 178
membership 71–2, 156–62
qualifications 163

Operational Guidelines 30–31, 42–3,
    68, 73, 76, 84, 93, 95, 96, 97
    advocacy rules 165, 170–72
    management planning 119, 120, 223
    monitoring 160–61, 223
    revision 99, 113
    politicization of 165–74
    St Petersburg session xv
    secretariat 201–4, 215
    UIA attendance 217
World Heritage Convention (1972)
    achievements 238–9, 244–5
    adoption xiii, xiv, 26
    funding 152–3
    and heritage theory/practice 239–43,
        244
    individuals, recognition of 225–6
    interviews with pioneers xiv
    limitations 226–31
    origins 1–2, 14–17
    politicization 236–8
    preamble 104
    purpose 104–5
    ratification 26, 155
    scope 27
    success 221
    and the zeitgeist 1
World Heritage in Danger List 106–7, 111,
        112, 133, 135–45
    Angkor ruins 140, 141
    deletions from 150–51
    Democratic Republic of the Congo
        sites 116
    example 135
    and military conflict 136
    Mount Nimba 145
    sparing use of 137
    statistics **116**
    Yellowstone National Park 146
World Heritage Fund 25, 92, 96, 151, 213
    lack of resources 233–4
World Heritage List xiii
    balance of sites **46**, 47
        cultural and natural heritage 58–71
        regional distribution 71–102
    criteria for inclusion
        authenticity 39–42
        cultural heritage 35–7

    integrity 42–3
    management and conservation
        43–4
    natural heritage 37–8
    selection 31–2, 33–8
cultural sites 45–6
    authenticity 85–90
    global strategy 82–5
    global study 75–82
    increased representation 90–102
delisting, procedure 150–51
Europe, dominance of 100
growth 45, **46**
landscapes *see* cultural landscapes;
    rural landscapes
"outstanding universal value" 32–3
regional distribution of sites 47, **47**
selection issues 73–4, 231–2
size 48–58
    concerns about 53–5
    and credibility 56
    proposals to limit 55–6, 57–8
    vision for 28–32
World Heritage Oral Archives project xiii
World Heritage Sites 3
    examples 6
    management
        planning 118–22
        traditional 122–3
    and mass tourism 234–6
    monitoring 109–35, 183
        definition 130
        formalization of 115
        IUCN proposals 109–10
        reactive 109–18
        systematic, and reporting 123–35
        types of 130–31
    threats to 135–51
World Heritage Trust, proposal 3, 4–5, 17,
        19–20, 151
World Heritage in Young Hands kit 215–16
    launch **216**
World Parks Congress, Fourth 68
WWF 217

Yellowstone National Park, US 19, 145–7,
        177
    on Danger List 146

Yugoslavia
    Balkan Wars 227
    see also Croatia; Montenegro

Zachwatowicz, Jan 14, 16
Zaire *see* Democratic Republic of the
    Congo